2nd Edition

English Bus H

Notable Independents

British Bus Publishing

Body codes used in the Bus Handbook series:

Type:

A	Articulated vehicle
B	Bus, either single-deck or double-deck
BC	Interurban - high-back seated buses.
C	Coach
M	Minibus with design capacity of sixteen seats or less
N	Low-floor bus (*Niederflur*), either single-deck or double-deck
O	Open-top bus (CO = convertible; PO = partial open-top)
S	High-capacity school transport 3+2 seated vehicles.

On trams, ST are single-deck trams and DT are double-deck trams.

Seating capacity is then shown. For double-decks the upper deck capacity is followed by the lower deck.

Door position:

C	Centre entrance/exit
D	Dual doorway.
F	Front entrance/exit
R	Rear entrance/exit (no distinction between doored and open)
T	Three or more access points

Equipment:

L	Lift for wheelchair	TV	Training vehicle.
M	Mail compartment	RV	Used as tow bus or engineer's vehicle.
T	Toilet	w	Vehicle is withdrawn and awaiting disposal.

e.g. - B32/28F is a double-deck bus with thirty-two seats upstairs, twenty-eight down and a front entrance/exit. N43D is a low-floor bus with two or more doorways.

Re-registrations:

Where a vehicle has gained new index marks, the details are listed at the end of each fleet showing the current mark, followed in sequence by those previously carried starting with the original mark. Marks carried more than once are not always repeated.

Regional books in the series:

The English Bus Handbook: Smaller Groups
The English Bus handbook: Notable Independents
The English Bus Handbook: Coaches
The Ireland & Islands Bus Handbook
The Welsh Bus Handbook
The Scottish Bus Handbook
The London Bus Handbook

Annual books are produced for the major groups:

The Stagecoach Bus Handbook
The Go-Ahead Bus Handbook
The First Bus Handbook
The Arriva Bus Handbook
The National Express Coach Handbook (bi-annual)
Most editions for earlier years are available direct from the publisher.

Associated series:

The Hong Kong Bus Handbook
The Malta Bus Handbook
The Leyland Lynx Handbook
The Mailvan Handbook
The Postbus Handbook
The Overall Advertisement Bus Handbook - Volume 1
The Toy & Model Bus Handbook - Volume 1 - Early Diecasts
The Fire Brigade Handbook (fleet list of each local authority fire brigade)
The Police Range Rover Handbook

Some earlier editions of these books are still available. Please contact the publisher on 01952 255669.

Contents

English Bus Handbook: Notable Independents

This second edition of the English Bus Handbook: Notable Independents is part of a series that details the fleets of certain bus and express coach operators throughout Britain. A list of current editions is shown on page 2. The operators included here are more of those who provide tendered and commercial services, primarily in England. The Smaller Groups book comprises those operators that so far have not featured in their own dedicated volume. There are two other parts to the English Bus Handbook set; this book and a third volume that caters for the coach and touring fleets. Information and suggestions that will help us to develop these titles would be welcome.

Quality photographs for inclusion in the series are welcome, for which a fee is payable. Unfortunately the publishers cannot accept responsibility for any loss and they require that you show your name on each picture or slide. High-resolution digital images of six megapixels or higher are also welcome on CD or DVD.

To keep the fleet information up-to-date, the publishers recommend the Ian Allan magazine *Buses* published monthly, or for more detailed information, the PSV Circle monthly news sheets.

The writer and publisher would be glad to hear from readers should any information be available which corrects or enhances that given in this publication.

Series Editor: Bill Potter.
Principal Editors for *The English Majors: Notable Independents* : David Donati and Bill Potter.

Acknowledgments:
We are grateful to the operating companies, Bob Downham, Tom Johnson, Peter Marley and Stuart Martin for their assistance in the compilation of this book.

The cover photographs are by Richard Godfrey.

ISBN 9781904875635

Published by British Bus Publishing Ltd, 16 St Margaret's Drive, Telford, TF1 3PH

Telephone: 01952 255669 - *www.britishbuspublishing.co.uk*

HALTON TRANSPORT

Halton Borough Transport Ltd, Moor Lane, Widnes, Halton, WA8 7AF

1-8				ADL Dart 4 11.1m		MCV Evolution		N40F*	2007-08	*5-8 are N41F	
1	AE57FAJ	3	AE57FAM	5	AJ58PZH			7	AJ58PZL		
2	AE57FAK	4	AE57FAO	6	AJ58PZK			8	AJ58PZM		

11-14				Dennis Dart SLF 10.8m		Marshall Capital		N39F	1998-99		
11	T760LFM	12	V993LLG	13	V994LLG			14	V995LLG		

15-19				Dennis Dart SLF 11.3m		Marshall Capital		N43F	2000	*18/9 are N44F	
15	W471VMA	17	W985XFM	18	W986XFM			19	X965ULG		
16	W987XFM										

20	X966ULG	Dennis Dart SLF 8.8m	Marshall Capital	N28F	2000
21	X967ULG	Dennis Dart SLF 11.3m	Marshall Capital	N41F	2000
22	X968ULG	Dennis Dart SLF 11.3m	Marshall Capital	N41F	2000

23-33				Dennis Dart SLF 11.3m		Marshall Capital		N42F	2001-02		
23	Y203PFM	26	Y204PFM	29	DG02WXV			32	DF02EHY		
24	Y207PFM	27	DG02WXU	30	DA02PUX			33	DF02EKC		
25	Y202PFM	28	DG02WXT	31	DA02PUY						

34	DE52URZ	Dennis Dart SLF 11.3m	East Lancs Myllennium	N43F	2002
35	DE52USB	Dennis Dart SLF 11.3m	East Lancs Myllennium	N43F	2002
36	DE52USC	Dennis Dart SLF 11.3m	East Lancs Myllennium	N43F	2002

Halton's fleet which was once dominated by the Leyland National and Lynx has been transformed to one where the Dart holds sway. The latest choice of bodywork is MCV and 1, AE57FAJ, is seen in Liverpool. *Alan Blagburn*

East Lancs Myllennium bodywork was fitted to the Darts delivered in the early years of this decade. Seen arriving in Liverpool is 41, PG03YYW. *Mark Doggett*

37-43

TransBus Dart 11.3m — East Lancs Myllennium — N43F — 2003

37	DK03NTD	39	DK03TNN	41	PG03YYW	43	PG03YYZ
38	DK03NTE	40	DK03TNL	42	PG03YYX		

44-50

TransBus Dart 11.3m — East Lancs Myllennium — N43F — 2004

44	DK04MKG	46	DK04MKJ	48	DK54JPJ	50	DK54JPO
45	DK04MKF	47	DK04MKE	49	DK54JPU		

51-54

ADL Dart 11.3m — East Lancs Myllennium — N43F — 2005

51	PN05SYG	52	PN05SYH	53	PN05SYJ	54	PN05SYO

57	K853MTJ	Leyland Lynx LX2R11C15Z4R	Leyland Lynx 2	B51F	1992	
60	J923MKC	Leyland Lynx LX2R11C15Z4R	Leyland Lynx 2	B51F	1992	
61	J924MKC	Leyland Lynx LX2R11C15Z4R	Leyland Lynx 2	B51F	1992	
62	J926MKC	Leyland Lynx LX2R11C15Z4R	Leyland Lynx 2	B51F	1992	
63	J927MKC	Leyland Lynx LX2R11C15Z4R	Leyland Lynx 2	B51F	1992	
64	K852MTJ	Leyland Lynx LX2R11C15Z4R	Leyland Lynx 2	B51F	1992	
65	N421ENM	Dennis Dart 9.8m	Marshall C37	B40F	1995	Uno, Hatfield, 2008
66	N422ENM	Dennis Dart 9.8m	Marshall C37	B40F	1995	Uno, Hatfield, 2008
67	N423ENM	Dennis Dart 9.8m	Marshall C37	B40F	1995	Uno, Hatfield, 2008

Web: www.haltontransport.co.uk

IPSWICH BUSES

Ipswich Buses Ltd, 7 Constantine Road, Ipswich, IP1 2DL

9	MRT9P	Leyland Atlantean AN68/1R	Roe	O43/26D	1976	
	PO51UMK	Dennis Trident	East Lancs Lolyne	N46/20D	2001	Docklands Minibuses, 2008
34	MUH287X	Leyland Olympian ONLXB/1R	Eastern Coach Works	B45/32F	1982	Galloway, Mendlesham, 2007
35	D201FYM	Leyland Olympian ONLXB/1R	Eastern Coach Works	B42/26D	1986	Arriva London, 2003
37	D150FYM	Leyland Olympian ONLXB/1R	Eastern Coach Works	B42/26D	1986	Arriva London, 2003
38	H705PTW	Leyland Olympian ON2R50C13Z4	Alexander RH	B47/33F	1990	Dublin Bus, 2003
39	H775PTW	Leyland Olympian ON2R50C13Z4	Alexander RH	B47/33F	1990	Dublin Bus, 2003
40	M640EPV	Volvo Olympian	East Lancs	B49/31D	1995	
41	M41EPV	Volvo Olympian	East Lancs	B49/31D	1995	
42	M42EPV	Volvo Olympian	East Lancs	B49/31D	1995	
44	C722NNN	Leyland Olympian ONLXB/1RV	Eastern Coach Works	B45/30F	1985	Trent, 1998
45	C100CHM	Leyland Olympian ONLXB/1RH	Eastern Coach Works	B46/22D	1986	London Traveller, Neasden, 1999
46	C101CHM	Leyland Olympian ONLXB/1RH	Eastern Coach Works	B46/22D	1986	London Traveller, Neasden, 1999
47	D138FYM	Leyland Olympian ONLXB/1RH	Eastern Coach Works	B46/22D	1986	London Traveller, Neasden, 1999
48	C97CHM	Leyland Olympian ONLXB/1RH	Eastern Coach Works	B46/22D	1986	Stagecoach London, 1999
50	P442SWX	DAF DB250	Optare Spectra	B43/26F	1997	Optare demonstrator, 1998
51	X151LBJ	DAF DB250	Optare Spectra	N47/25F	2000	
52	X152LBJ	DAF DB250	Optare Spectra	N47/25F	2000	
53	X153LBJ	DAF DB250	Optare Spectra	N47/25F	2000	
54	PF04WML	VDL Bus DB250	East Lancs Lowlander	N43/29F	2004	

55-59		DAF DB250	East Lancs Lowlander	N43/29F	2002-03	
55	PN52XBO	57	PN52XBJ	58	PJ53OLA	59	PJ53OLB
56	PN52XBK					

60	PJ54YZT	Scania N94UD OmniDekka	East Lancs	N47/33F	2005	
61	PJ54YZU	Scania N94UD OmniDekka	East Lancs	N47/33F	2005	
62	PJ54YZV	Scania N94UD OmniDekka	East Lancs	N47/33F	2005	

70-75		Scania N230UB OmniCity	East Lancs Esteem	N36F	2007	
70	YN56NVB	72	YN56NVD	74	YN56NVF	75	YN56NVG
71	YN56NVC	73	YN56NVE			

84-88		TransBus Dart 11.1m	East Lancs Myllennium	N41F	2003	
84	PJ53OLC	86	PJ53OLG	87	PJ53OLH	88	PJ53OLK
85	PJ53OLE					

89-99		Dennis Dart SLF 11.1m	East Lancs Spryte	N41F	2000	
89	X89LBJ	92	X92LBJ	95	X195LBJ	98	X98LBJ
90	X901LBJ	93	X93LBJ	96	X96LBJ	99	X199LBJ
91	X91LBJ	94	X94LBJ	97	X97LBJ	

100	PN52XBM	DAF DB250	East Lancs Lowlander	N43/29F	2002	

130-134		Dennis Dart SLF	East Lancs Spryte	N42F*	1997-98	*133/4 are N39F
130	P130PPV	132	P132PPV	133	R133FBJ	134	R134FBJ
131	P131PPV					

135	T135KPV	Dennis Dart SLF	East Lancs Spryte	N36F	1999	
136	T136KPV	Dennis Dart SLF	East Lancs Spryte	N36F	1999	
137	T137KPV	Dennis Dart SLF	East Lancs Spryte	N36F	1999	
160	J160LPV	Dennis Lance 11m	East Lancs EL2000	B53F	1992	
162	L162ADX	Dennis Lance 11m	East Lancs EL2000	B49F	1994	
170	YG52DGO	Optare Excel L1180	Optare	N42F	2002	
171	YG52DHF	Optare Excel L1180	Optare	N42F	2002	
172	YG52DGE	Optare Excel L1180	Optare	N42F	2002	
173	YG52DGF	Optare Excel L1180	Optare	N42F	2002	

188-196		Optare Excel L1150	Optare	N37F*	1997	*seating varies
188u	R197DDX	191u	P191SGV	192	P192SGV	196u	P196SGV
189u	R189DDX					

Ipswich Buses feature a name emblazoned within a banner on the front of each vehicle, a tradition it continues. Dutch-built DAF DB250 59, PJ53OLB, carries East Lancs Lowlander bodywork that is named 'Frank Ayton'.
Mark Doggett

231-236		Optare Solo M920		Optare		N33F	2001	
231	X231MBJ	233	X233MBJ	235	X235MBJ		236	X236MBJ
232	X232MBJ	234	X234MBJ					

237-243		Optare Solo M920		Optare		N33F	2002-04	
237	YG02FVZ	239	YN03NDU	241	YN04LWK		243	YN04LWJ
238	YN03NDK	240	YN03NDL	242	YN04XZA			

244	YN53ELW	Optare Solo M950	Optare			N33F	2004	Optare demonstrator, 2004
245	YJ05XNY	Optare Solo M950 DE Hybrid	Optare			N33F	2005	

Previous registrations:

D138FYM	D138FYM, WLT838		H775PTW	90D1028
H705PTW	90D1038		WOI607	B117LDX

Named vehicles: 9 *Eastern Belle*; 35 *Leigh Fellgett*; 38 *Western Sovereign*; 39 *Western Hussar*; 40 *Beatrice Maud*; 41 *May*; 42 *Vigilant*; 44 *Orion*; 45 *Royal Oak*; 46 *Southern Belle*; 47 *Ironsides*; 48 *Challenger*; 50 *Delight*; 51 *Atlantis*; 52 *Veronica*; 53 *Mi Amigo*; 54 *Sunbeam*; 58 *Pride of Ipswich*; 59 *Frank Ayton*; 60 *Olaf Tryggvasson*; 61 *Robert Ransome*; 62 *Simon A Graham*; 74 *Sir Bobby Robson*; 84 *Peter Bruff*; 87 *Lady Diana, Princess of Wales*; 88 *Cedric*; 89 *Memory*; 90 *HN (Jimmy) James*; 91 *Centaur*; 92 *Ceres*; 93 *Dannebrog*; 94 *Dauntless*; 95 *Edith May*; 96 *Evening Star*; 97 *Gladys*; 98 *Glen Way*; 99 *Kimberley*; 100 *Centennial*; 130 *HMS Grafton*; 131 *Will Everard*, 132 *Raybel*; 133 *Coronation*; 134 *Caroline 1964 to 2004*; 135 *Team Ipswich*; 136 *Ena*; 137 *Sir Alf Ramsey*; 160 *Barbara Jean*; 162 *Doris*; 170 *King John*; 171 *Leonard Squirrell*; 188 *Lady Helen*; 189 *Gentle Giant*; 191 *Gusford Primary School*, 192 *Handford Hall Primary School & Nursery*; 193 *Orwell Junior Traveller*, 194 *Saint Matthew's School*, 196 *St Mark's RC Primary School*, 231 *Christchurch Park*; 232 *Holywells Park*; 233 *Chantry Park*; 234 *Alexandra Park*; 235 *Gippeswyck Park*; 236 *Bourne Park*; 237 *Jubilee*; 238 *Morning Star*; 239 *Minerva*; 240 *Leading Light*; 241 *Teutonic*; 242 *Conqueror*; 243 *Spirit of Endeavour*; 244 *Britannia*.

Web: www.ipswichbuses.co.uk

METRO COASTLINES

Blackpool Transport Services Ltd, Rigby Road, Blackpool, FY1 5DD

101-133 DAF SB220LC550 Optare Delta BC46F 1990-93 131-3 are BC48F

101	G101NBV	108	G108NBV	115	H115YHG	124	J124GRN
102	G102NBV	109	H109YHG	116	H116YHG	125	J125GRN
103	G103NBV	110	H110YHG	118	H118CHG	126	J126GRN
104	G104NBV	112	H112YHG	119	H119CHG	127	K127UFV
105	G105NBV	113	H113YHG	122	H122CHG	132	H2FBT
106	G106NBV	114	H114YHG	123	J123GRN	133	H3FBT
107	G107NBV						

210-218 Optare Excel L1070 Optare N40F 1999

210	T210HCW	213	T213HCW	215	T215HCW	217	T217HCW
211	T211HCW	214	T214HCW	216	T216HCW	218	T218HCW
212	T212HCW						

219	T880RBR	Optare Excel L1150	Optare	N41F	1999	Go-Ahead Northern, 2004
220	T881RBR	Optare Excel L1150	Optare	N41F	1999	Go-Ahead Northern, 2004
221	T884RBR	Optare Excel L1150	Optare	N41F	1999	Go-Ahead Northern, 2004

222-226 Optare Excel L1150 Optare N39F 1999 Reading Buses, 2008

222	T932EAN	224	T934EAN	225	T935EAN	226	T936EAN
223	T933EAN						

240-246 Optare Solo M950 Optare N33F On order for route 2

240		242		244		246	
241		243		245			

247	YJ08PFK	Optare Solo M950	Optare	N33F	2008
248	YJ08PFA	Optare Solo M950	Optare	N33F	2008

249-260 Optare Solo M950 Optare N33F 2004

249	YN53ZWK	252	YN53ZWP	255	YN53ZWU	258	YN53ZWX
250	YN53ZWL	253	YN53ZWR	256	YN53ZWV	259	YN53ZWY
251	YN53ZWM	254	YN53ZWT	257	YN53ZWW	260	YN53ZWZ

Metro Coastlines is the name currently used by Blackpool Transport which provides some services into the Wyre and Fylde districts as well as Blackpool. It also operates the tram service from Starr Gate to Fleetwood. Optare Delta 115, H115YHG carries the livery for route 11. *Alan Blagburn*

The Optare Solo has been adopted for many of the services that operate inland. Eight of the type carry the colours of *Lifestyle Line* 16, including 286, YJ07EJA, seen passing Woolworths. *Tony Wilson*

261-276

		Optare Solo M850		Optare			N28F	1999	
261	V261HEC	265	V265HEC	269	V269HEC		274	V274HEC
262	V262HEC	266	V266HEC	271	V271HEC		275	V275HEC
263	V263HEC	267	V267HEC	272	V272HEC		276	V276HEC
264	V264HEC	268	V268HEC	273	V273HEC			

277-284

		Optare Solo M850		Optare			N29F	2002-03	
277	PL03BPZ	279	YG02FVR	281	YG02FVT		283	PN03UGG
278	YG02FVP	280	YG02FVS	282	YG02FVU		284	PN03UGH

285	YN04XYZ	Optare Solo M850		Optare		N27F	2004	Optare demonstrator, 2006

286-297

		Optare Solo M950		Optare			N33F	2007-08	
286	YJ07EJA	289	YJ07EJE	292	YJ07EJL		295	YJ08PFE
287	YJ07EJC	290	YJ07EJG	293	YJ07EJN		296	YJ08PFF
288	YJ07EJD	291	YJ07EJK	294	YJ08PFD		297	YJ08PFG

301-309

		Dennis Trident		East Lancs Myllennium Lolyne N47/30F			2002	
301	PJ02PYD	304	PJ02PYH	306	PJ02PYO		308	PJ02PYS
302	PJ02PYF	305	PJ02PYL	307	PJ02PYP		309	PJ02PYT
303	PJ02PYG							

310-318

		Dennis Trident		East Lancs Myllennium Lolyne N47/30F			2003	
310	PJ03TFF	313	PJ03TFU	315	PJ03TFX		317	PJ03TFZ
311	PJ03TFK	314	PJ03TFV	316	PJ03TFY		318	PJ03TGE
312	PJ03TFN							

319-327

		TransBus Trident		East Lancs Myllennium Lolyne N47/30F			2004	
319	PN04XDE	323	PN04XDG	325	PN04XDJ		327	PN04XDL
322	PN04XDF	324	PN04XDH	326	PN04XDK			

Five of the longer-length Tridents that were new to Blackburn Buses are now in the Metro fleet. These are used on route 1, which runs along the promenade from Fleetwood to Pontins close to Starr Gate, and thus competes with the trams. Number 334, PN52XKG, is seen at Fleetwood Ferry. *Mark Lyons*

328-332

		ADL Trident		East Lancs Myllennium Lolyne N51/30F		2006		
328	PF06EZL		330	PF06EZN	331	PF06EZO	332	PF06EZP
329	PF06EZM							

333-337

		ADL Trident		East Lancs Myllennium Lolyne N53/37F		2002		Lancashire United, Blackburn, '07
333	PN52XKF		335	PN52XKH	336	PN52XKJ	337	PN52XKK
334	PN52XKG							

338	PO51UMG	ADL Trident	East Lancs Myllennium Lolyne N46/20D	2001	Go-Ahead London, 2008
339	PO51UMJ	ADL Trident	East Lancs Myllennium Lolyne N46/20D	2001	Go-Ahead London, 2008
365	UWW5X	Leyland Olympian ONLXB/1R	Roe	B47/29F	1982	West Yorkshire PTE, 1986
366	UWW11X	Leyland Olympian ONLXB/1R	Roe	B47/29F	1982	West Yorkshire PTE, 1986
367	UWW15X	Leyland Olympian ONLXB/1R	Roe	B47/29F	1982	West Yorkshire PTE, 1986

368-373

		Leyland Olympian ONCL10/1RZ		East Lancs		B45/31F	1989	
368	F368AFR		370	F370AFR	372	F372AFR	373	F373AFR
369	F369AFR		371	F371AFR				

374-379

		Volvo Olympian		Northern Counties Palatine 2		B43/29F	1994	
374	M374SCK		376	M376SCK	378	M378SCK	379	M379SCK
375	M375SCK		377	M377SCK				

401-410

		Leyland Olympian ONLXB/1R		Eastern Coach Works		B43/32F*	1983-84	Trent, '96 *407-10 are B45/30F
401	XAU701Y		404	XAU704Y	407	A707DAU	409	A709DAU
402	XAU702Y		405	XAU705Y	408	A708DAU	410	A710DAU
403	XAU703Y		406	XAU706Y				

411	D367JJD	Leyland Olympian ONLXB/1RH	Eastern Coach Works	B42/30F	1987	Lancashire United, 2007

412-418

		Leyland Olympian ONLXB/1R	Eastern Coach Works	B51/32D	1983-84	Lancashire United, 2007
412	OFS684Y	414	A704YFS	416	A720YFS	418	B740GSC
413	A703YFS	415	A719YFS	417	A721YFS	

Blackpool's tram service has seen a decline over the last decade, with around half the fleet currently mothballed, and the future of many of these cars is being discussed. However, some movement between the reserve fleet and the operations fleet has taken place. Rarely seen outside high summer are the boats, represented by 600, which was operating for the Fylde Tramway Society when pictured. *Mark Doggett*

501-518

				Optare MetroRider MR37		Optare		B25F	1996-98	
501w	P501UFR		506w	S506LHG		511	S511LHG		515	S515LHG
502w	P502UFR		507w	S507LHG		512	S512LHG			
503w	P503UFR		508w	S508LHG		513	S513LHG		517	S517LHG
504w	P504UFR		509w	S509LHG		514	S514LHG		518	S518LHG
505w	S505LHG		510w	S510LHG						

590	N590GRN	Optare MetroRider MR37	Optare	B25F	1995	
593	N593LFV	Optare MetroRider MR37	Optare	B25F	1995	
813	D213FYM	Leyland Olympian ONLXB/1RH	Eastern Coach Works	PO42/26D	1987	Ensign Bus, Purfleet, 2007
849	D149FYM	Leyland Olympian ONLXB/1RH	Eastern Coach Works	PO42/26D	1987	Ensign Bus, Purfleet, 2007
857	D257FYM	Leyland Olympian ONLXB/1RH	Eastern Coach Works	PO42/26D	1987	Ensign Bus, Purfleet, 2007
858	D258FYM	Leyland Olympian ONLXB/1RH	Eastern Coach Works	PO42/26D	1987	Ensign Bus, Purfleet, 2007

Special event vehicles:

364	B364UBV	Leyland Atlantean AN68D/2R	East Lancs	BC49/29F	1984	
507	HFR507E	Leyland Titan PD3A/1	MCCW	B41/30R	1967	preservation, 2008

Ancillary vehicles:

873	D173FYM	Leyland Olympian ONLXB/1RH	Eastern Coach Works	PO42/28D	1987	Ensign Bus, Purfleet, 2007
969	K129UFV	DAF SB220	Optare Delta	TV	1993	
970	K130UFV	DAF SB220	Optare Delta	TV	1993	

Previous registrations;

D173FYM	D173FYM, VLT173		PN52XKG	PN52XKG, 428EXA
D244FYM	D244FYM, VLT244		PN52XKH	PN52XKH, 213CNU
D367JJD	D263FUL, VLT9		PN52XKJ	PN52XKJ, LUF549
PN52XKF	PN52XKF, BCB340		PN52XKK	PN52XKK, PCB24

Trams:

8		BET M4d		Blackpool Corporation (1974)	ST48D	1935	
147		Preston McGuire		Hurst Nelson	ST46/32D	1924	Vintage car

600-607

		English Electric M4d		English Electric		OST56C	1934-35	602 is OST52C
600		602		604		605		607u

611		LETS A66		LETS		N--C	1997	
619u		English Electric M4d		Bolton Group (1987)		T58	1973	

621-637

EMB M4d — Brush — ST48C — 1937 — 626 is ST46C, 633 is a Trawler

621u	623	626	630	631	632u	633u	634u	636u	637u
622	625u	627u						

641-647

Primrose/Brush M4d — East Lancs — ST52D — 1984-88 — 641/2/6 is ST53D

641	642	643	644	645w	646	647

648		Maley & Taunton/Brush M4d	East Lancs	ST52D	1985
660		Maley & Taunton M4d	Roberts	ST56C	1953

671-676

English Electric, 1935 M4s — English Electric (refurbished 1958-62) — ST53C — 1935

671	672	673	674	675	676u

678	English Electric, 1935 M4d	English Electric	ST48C	1961
680	English Electric, 1935 M4d	English Electric	ST48C	1960

681-687

Maley & Taunton M4d — MCW — ST61C — 1960

681	682	683	684	685	686u	687u

700-726

English Electric M4d — English Electric — DT54/40C — 1934-35 — 706 is ODT54/40C, seating varies

700	703	707	710	713	716u	718	720	722	724
701	704u	708u	711	715u	717u	719	721	723	726
702	706	709	712					

732u	Dick Kerr M4s	Blackpool Corporation	48-seat	1960	*Moon Rocket*
733u	English Electric M4d	Blackpool Corporation	35-seat	1962	*Wild West Loco*
734u	Dick Kerr B4s	Blackpool Corporation	60-seat	1962	*Wild West coach*
735u	English Electric M4d	Blackpool Corporation	57/42	1963	*Hovercraft*
736	Dick Kerr M4d	Blackpool Corporation	71-seat	1965	*Frigate*
761	English Electric (1934) M4d	Blackpool/East Lancs	DT56/44F	1979	
762	English Electric (1934) M4d	Blackpool/East Lancs	DT56/34D	1982	

Web: www.blackpooltransport.co.uk

Many trams are in overall advert colours but those trams in the new scheme carry route T branding as shown on double-deck car 710. *Mark Doggett*

NETWORK WARRINGTON

Warrington Borough Transport Ltd, Wilderspool Causeway, Warrington, WA4 6PT

1-18 Dennis Dart SLF 10.8m Marshall Capital N40F 1999

1	T201AFM	6	T206AFM	11	V211JLG	15	V215JLG
2	T202AFM	7	T207AFM	12	V212JLG	16	V216JLG
3	T203AFM	8	T208AFM	13	V213JLG	17	V217JLG
4	T204AFM	9	V209JLG	14	V214JLG	18	V218JLG
5	T205AFM	10	V210JLG				

19-28 Dennis Dart SLF 10.8m Marshall Capital N41F 2001

19	Y619GFM	22	Y622GFM	25	Y631GFM	27	Y627GFM
20	Y629GFM	23	Y623GFM	26	Y626GFM	28	Y628GFM
21	Y621GFM	24	Y624GFM				

29 DE02URX Dennis Dart SLF 10.8m Marshall Capital N41F 2002

31-34 Dennis Dart SLF MCV Capital N40F 2003

31	DA52ZVK	32	DA52ZVL	33	DF52ABU	34	DF52AXG

35-40 DAF SB120 Wrightbus Merit N39F 2003

35	DG53FLH	37	DG53FJY	39	DG53FJV	40	DG53FJU
36	DG53FLJ	38	DG53FJX				

41-46 VDL Bus SB120 Wrightbus Merit N39F 2004

41	DE04YNG	43	DE04YNH	45	DE04YNC	46	DE04YND
42	DE04YNF	44	DE04YNB				

47-58 VDL Bus SB120 Wrightbus Merit N39F 2005-06

47	DK55HMH	50	DK55HMJ	53	DK55OMM	56	DK55OMR
48	DK55HMG	51	DK55HMO	54	DK55OMO	57	DK55OPL
49	DK55HMF	52	DK55HMU	55	DK55OMP	58	DK55OPM

59-65 VDL Bus SB120 Wrightbus Merit N39F 2006

59	DK56MLJ	61	DK56MLL	63	DK56MLV	65	DK56MLU
60	DK56MLX	62	DK56MLN	64	DK56MLO		

Network Warrington was voted the *'Operator of the Year'* in 2008 in the Bus and Coach awards. 45, DE04YNC, is a VDL Bus SB120 with Wrightbus Merit bodywork. The Merit was the model name for buses sold through Volvo.
Alan Blagburn

New to Blazefield's Keighley and District operation before passing to Harrogate, H513RWX is now 167 in the Network Warrington fleet. It is one of four Volvo Olympians to arrive from Blazefield in 2004, while four Alexander-bodied examples new to Dublin Bus also joined the fleet that year. *Alan Blagburn*

66-77 — VDL Bus SB120, Wrightbus Cadet, N39F, 2007

66	DK07EZJ	69	DK07EZG	72	DK07EZO	75	DK07EZL
67	DK07FWH	70	DK07EZR	73	DK07EZN	76	DK07EZH
68	DK07FWJ	71	DK07EZP	74	DK07EZM	77	DK07FWL

78-82 — VDL Bus SB120, Wrightbus Pulsar, N39F, 2008

78	YJ57BPK	80	YJ57DPU	81	YJ57BRX	82	YJ57BRZ
79	YJ57BPO						

112-123 — Optare MetroRider MR35, Optare, B25F, 1998-99

112	S112GUB	115	S115GUB	118	S118GUB	121	S121GUB
113u	S113GUB	116	S116GUB	119	S119GUB	122	T322ELG
114u	S114GUB	117	S117GUB	120	S120GUB	123	T323ELG

No.	Reg	Chassis	Body	Layout	Year	Origin
150	C100UBC	Dennis Dominator DDA1010	East Lancs	B46/33F	1986	Leicester, 1989
151	C101UBC	Dennis Dominator DDA1010	East Lancs	B46/33F	1986	Leicester, 1989
152	C102UBC	Dennis Dominator DDA1010	East Lancs	B46/33F	1986	Leicester, 1989
153	C103UBC	Dennis Dominator DDA1010	East Lancs	B46/33F	1986	Leicester, 1989

159-162 — Volvo Olympian, Alexander RL, B47/33F, 1991, Dublin Bus, 2004

159	H215PVW	160	H146PVW	161	H187PVW	162	H191PVW

No.	Reg	Chassis	Body	Layout	Year	Origin
163	H514RWX	Volvo Olympian	Northern Counties Palatine	B47/30F	1990	Harrogate & District, 2003
164	H515RWX	Volvo Olympian	Northern Counties Palatine	B47/30F	1990	Harrogate & District, 2003
165	K711ASC	Volvo Olympian	Alexander RH	B47/32F	1992	Huntingdon & District, 2003
166	K712ASC	Volvo Olympian	Alexander RH	B47/32F	1992	Huntingdon & District, 2003

167-170 — Volvo Olympian, Northen Counties Palatine, B47/30F, 1990, Harrogate & District, 2004

167	H513RWX	168	H516RWX	169	H517RWX	170	H519RWX

No.	Reg	Chassis	Body	Seating	Year	Notes
171-180		Volvo Olympian	Alexander RL	B47/31F	1996	Dublin Bus, 2008

171	N584JND	**174**	P324SWC	**177**	P384SWC	**179**	P368SWC
172	P314SWC	**175**	P302SWC	**178**	P411SWC	**180**	P462SWC
173	P268SWC	**176**	P337SWC				

No.	Reg	Chassis	Body	Seating	Year
181	F121XEM	Dennis Dominator DDA1018	East Lancs	B51/31F	1988
182	F122XEM	Dennis Dominator DDA1018	East Lancs	B51/31F	1988
193w	F103XEM	Dennis Dominator DDA1017	East Lancs	B51/37F	1988
194w	F104XEM	Dennis Dominator DDA1017	East Lancs	B51/37F	1988

195-199 — Dennis Dominator DDA1017 — East Lancs — B51/37F — 1989

195w	F95STB	**197**u	F97STB	**198**u	F98STB	**199**w	F99STB
196w	F96STB						

239-243 — Dennis Dart 9m — Marshall C36 — B35F — 1995

239w	M239YCM	**241**w	M241YCM	**242**w	M242YCM	**243**	M243YCM
240w	M240YCM						

No.	Reg	Chassis	Body	Seating	Year	Notes
244	M593HKH	Dennis Dart 9.8m	Plaxton Pointer	B40F	1994	Plaxton demonstrator, 1995
245	M284HRH	Dennis Dart 9.8m	Plaxton Pointer	B40F	1994	Plaxton demonstrator, 1995
246	M246YWM	Dennis Dart 9.8m	Plaxton Pointer	B40F	1995	
247	M247YWM	Dennis Dart 9.8m	Marshall C37	B39F	1995	
248	M248YWM	Dennis Dart 9.8m	Marshall C37	B39F	1995	

Special event vehicle:

No.	Reg	Chassis	Body	Seating	Year
148	BED729C	Leyland Titan PD2/40	East Lancs	B34/30F	1965

Ancillary vehicle:

No.	Reg	Chassis	Body	Seating	Year
30	DA52ZVM	Dennis Dart SLF	MCV C39 (2004)	TV	2003

Previous registrations:

H146PVW	91D1076	P312SWC	96D313
H187PVW	91D1099	P314SWC	96D307
H191PVW	91D1092	P324SWC	96D312
H215PVW	91D1079	P337SWC	96D324
H513RWX	H513RWX, H12SDW	P368SWC	96D321
N584JNO	96D296	P384SWC	96D314
P268SWC	96D311	P411SWC	96D319

Web: www.warringtonboroughtransport.co.uk; **On order**: 12 Volvo B7RLE with Wrightbus Eclipse Urban bodies.

Pictured on a rainy day in Warrington, 75, DK07EZL, is a VDL Bus SB120 with one of the last Wrightbus Cadet bodies. The 2008 intake features the new Pulsar model. *Dave Heath*

NOTTINGHAM CITY TRANSPORT

Nottingham City Transport - South Notts - Pathfinder

Nottingham City Transport Ltd, Lower Parliament Street, Nottingham, NG1 1GG

120-129		Optare Solo M920			Optare			N33F	2000	
120	TB	W599PTO	123	TB	W603PTO	126	TB	W606PTO	128	TB	W608PTO
121	TB	W601PTO	124	TB	W604PTO	127	TB	W607PTO	129	TB	W609PTO
122	TB	W602PTO	125	TB	W605PTO					

130-139		Optare Solo M920			Optare			N33F	2001	
130	TB	FP51AOH	133	TB	YJ51XST	136	TB	YJ51ZVW	138	TB	YJ51ZVY
131	TB	FP51AOJ	134	TB	YJ51XSU	137	TB	YJ51ZVX	139	TB	YJ51ZVZ
132	TB	YJ51XSR	135	TB	YJ51ZVV					

140	TB	FP02XMA	Optare Solo M920		Optare			N33F	2002	

201-207		Scania N94UB OmniTown			East Lancs			N33F	2004	
201	TB	YN04AMK	203	TB	YN04AMV	205	TB	YN04ANP	207	TB	YN04AMX
202	TB	YN04AMU	204	TB	YN04ANF	206	TB	YN04ANR			

208-222		Scania CN94UB OmniCity			Scania			N42F	2004	
208	GM	YN54AHA	212	GM	YN54AHF	216	GM	YN54AHP	220	GM	YN54AHU
209	GM	YN54AHC	213	GM	YN54AHG	217	GM	YN54AHL	221	GM	YN54NXL
210	GM	YN54AHD	214	GM	YN54AHJ	218	GM	YN54AHO	222	GM	YN54NXK
211	GM	YN54AHE	215	GM	YN54AHK	219	GM	YN54AHV			

Nottingham's choice of supplier in recent years has focused on Scania products. First of the 2004 intake of Scania OmniCity buses is 208, YN54AHA, seen here in Park & Ride colours. *Mark Lyons*

Several of the early Optare Solo buses supplied to Nottingham have now moved on to independents following the delivery of a batch of Versa buses in 2008. Seen lettered for the *blue line* is 258, V258DRB. *Alan Blagburn*

234-241

			Optare Solo M920		Optare		N33F	2001-02		
234	TB	FP02XMB	236	TB	Y236LRR	238	TB	Y238LRR	240	TB	X947HBC
235	TB	FP02XMC	237	TB	Y237LRR	239	TB	X239HBC	241	TB	X241HBC

242-251

			Optare Solo M920		Optare		N33F	2000		
242	TB	W242PAU	245	TB	W958PAU	248	TB	W248PAU	250	TB	W959PAU
243	TB	W243PAU	246	TB	W246PAU	249	TB	W249PAU	251	TB	W251PAU
244	TB	W244PAU	247	TB	W247PAU					

252-261

			Optare Solo M920		Optare		N33F	2000		
252	u	V252JRR	255	TB	V255JRR	258	TB	V258DRB	260	TB	V260DRB
253	TB	V253JRR	256	TB	V256JRR	259	TB	V259DRB	261	TB	V261DRB
254	TB	V254JRR	257	TB	V257DRB					

262-276

			Optare Solo M920		Optare		N33F	1999		
262	TB	V262DRC	266	TB	V266DRC	268	TB	V268DRC	276	u	V276DRC
263	TB	V263DRC	267	TB	V267DRC	271	TB	V271DRC			

301	NM	YN07LDX	Scania OmniLink K270 UB	Scania	N53F	2007
302	NM	YN07LDY	Scania OmniLink K270 UB	Scania	N53F	2007
303	NM	YN07LDZ	Scania OmniLink K270 UB	Scania	N53F	2007
304	NM	YN57FYV	Scania OmniLink K270 UB	Scania	N45F	2008
305	NM	YN57FYW	Scania OmniLink K270 UB	Scania	N45F	2008
306	NM	YN57FYX	Scania OmniLink K270 UB	Scania	N45F	2008

307-326

			Optare Versa V1100		Optare		N38F	2008		
307	TB	YK08EPA	312	TB	YK08EPJ	317	TB	YK08EPU	322	TB	YK08EPZ
308	TB	YK08EPC	313	TB	YK08EPL	318	TB	YK08EPV	323	TB	YK08ESF
309	TB	YK08EPD	314	TB	YK08EPN	319	TB	YK08ERJ	324	TB	YK08ESG
310	TB	YK08EPE	315	TB	YK08EPO	320	TB	YK08EPX	325	TB	YK08ESN
311	TB	YK08EPF	316	TB	YK08EPP	321	TB	YK08EPY	326	TB	YK08ESO

405-416 — Dennis Trident — East Lancs Lolyne — N51/35F — 1999 — *414-6 are N49/34F

405	NM	T405BNN	408	NM	T408BNN	411	NM	T411BNN	414 NM T414BNN
406	NM	T406BNN	409	NM	T409BNN	412	NM	T412BNN	415 NM T415BNN
407	NM	T407BNN	410	NM	T410BNN	413	NM	T413BNN	416 NM T416BNN

417-436 — Dennis Trident — East Lancs Lolyne — N53/34F — 1999 — *seating varies

417	NM	T417XVO	422	GM	T422XVO	427	NM	V427DRC	432 NM V432DRC
418	NM	T418XVO	423	NM	V423DRC	428	NM	V428DRC	433 NM V433DRC
419	NM	T419XVO	424	NM	V424DRC	429	NM	V429DRC	434 NM V434DRC
420	NM	T420XVO	425	NM	V425DRC	430	NM	V430DRC	435 NM V435DRC
421	GM	T421XVO	426	NM	V426DRC	431	NM	V431DRC	436 NM V436DRC

458-465 — Volvo Olympian — East Lancs Pyoneer — B49/35F — 1998-99

458	u	S458ATV	460	NM	S460ATV	462	NM	S462ATV	464 NM S464ATV
459		S459ATV	461	NM	S461ATV	463	NM	S463ATV	465 NM S465ATV

466	NM	R466RRA	Volvo Olympian	East Lancs Pyoneer	B49/39F	1998
499	NM	YN05GWX	Scania OmniCity CN94UB	Scania	N42F	2005
500	NM	YN05GWY	Scania OmniCity CN94UB	Scania	N42F	2005

501-513 — Scania OmniCity CN94UB — Scania — N42F — 2002

501	NM	FH51LTX	505	NM	FD05SDZ	508	NM	FG52WFT	511 NM FG52WFW
502	NM	FD02SEY	506	NM	FD05YDV	509	NM	FG52WFU	512 NM FG52WFX
503	NM	FD02SDX	507	NM	FD05YDW	510	NM	FG52WFV	513 NM FG52WFY
504	NM	FD02SDY							

515-518 — Scania OmniCity CN94UB — Scania — N41F — 2007

515	NM	YN07EYA	516	NM	YN07EYB	517	NM	YN07EYC	518 NM YN07EYD

528	GM	X94USC	Scania L94UB	Wrightbus Solar	N43F	2000	Scania demonstrator, 2001

529-537 — Scania OmniCity CN94UB — Scania — N42F — 2003

529	NM	YS03ZLN	532	NM	YS03ZLX	534	NM	YS03ZHL	536 NM YS03ZHN
530	NM	YS03ZLU	533	NM	YS03ZLY	535	NM	YS03ZHM	537 NM YS03ZHP
531	NM	YS03ZLV							

545	NM	Y957DRR	Optare Excel L1180	Optare	N43F	2001
546	w	Y546DTO	Optare Excel L1180	Optare	N43F	2001
547	w	Y547LRB	Optare Excel L1180	Optare	N43F	2001
548	w	Y548LRB	Optare Excel L1180	Optare	N43F	2001

549-562 — Optare Excel L1180 — Optare — N41F — 2001-2002

549	NM	FD51EYR	553	NM	FD51EYV	557	NM	FD02SFE	560 TBt FD02SFK
550	NM	FD51EYS	554	NM	FD51EYW	558	NM	FD02SFF	561 TBt FD02SFN
551	NM	FD51EYT	555	NM	FD51EYX	559	TBt	FD02SFJ	562 TBt FD02SFO
552	NM	FD51EYU	556	NM	FD51EYY				

647-665 — Dennis Trident — East Lancs Lolyne — N53/34F* — 2000 — *656/8 are N49/34F

647	NM	W647SNN	652	NM	W652SNN	657	NM	W657SNN	662 NM X662WCH
648	NM	W648SNN	653	NM	W653SNN	658	NM	W658SNN	663 NM X663WCH
649	NM	W649SNN	654	NM	W654SNN	659	NM	W659SNN	664 NM X664WCH
650	NM	W941SNR	655	NM	W942SNR	660	NM	W943SNR	665 NM X665WCH
651	NM	W651SNN	656	NM	W656SNN	661	NM	X661WCH	

666	NM	Y966DRC	Dennis Trident	East Lancs Lolyne	N57/37F	2001
667	NM	Y667DRC	Dennis Trident	East Lancs Lolyne	N57/37F	2001
668	NM	Y668DRC	Dennis Trident	East Lancs Lolyne	N57/37F	2001
669	NM	YN06TGY	Scania N94 UD OmniDekka	East Lancs	N51/36F	2006
670	NM	YN06TGZ	Scania N94 UD OmniDekka	East Lancs	N51/36F	2006
671	NM	YN07EYO	Scania N94 UD OmniDekka	East Lancs	N51/36F	2007
672	NM	YN07EYP	Scania N94 UD OmniDekka	East Lancs	N51/36F	2007

Dennis Trident 658, W658SNN carries *purple line* branding. This batch has Lolyne bodies from East Lancs, suppliers of bodywork to the fleet for many years. *Richard Godfrey*

673-698

Scania N94 UD OmniDekka | East Lancs | N51/39F | 2005

673	NM	YN05WGD	680	NM	YN05WFV	687	NM	YN05WFS	693	NM	YN05NGG
674	NM	YN05WFB	681	NM	YN05WFW	688	NM	YN05WFT	694	NM	YN05NGJ
675	NM	YN05WFC	682	NM	YN05WFP	689	NM	YN05WFK	695	NM	YN05NGU
676	NM	YN05WFE	683	NM	YN05WFX	690	NM	YN05NGE	696	NM	YN05NGV
677	NM	YN05WFF	684	NM	YN05WFO	691	NM	YN05NGZ	697	NM	YN05NGX
678	NM	YN05WFG	685	NM	YN05WFU	692	NM	YN05NGF	698	NM	YN05NGY
679	NM	YN05WGC	686	NM	YN05WFR						

701	GM	Y701LRB	Scania L94 UA	Wrightbus Solar Fusion	AN59D	2001
702	GM	FP51EXN	Scania L94 UA	Wrightbus Solar Fusion	AN59D	2001
703	GM	FP51EXO	Scania L94 UA	Wrightbus Solar Fusion	AN59D	2001
704	GM	FE02AKV	Scania L94 UA	Wrightbus Solar Fusion	AN59D	2002
705	GM	YP02AAN	Scania L94 UA	Wrightbus Solar Fusion	AN59D	2002
708	GM	YN05WFD	Scania N94 UD OmniDekka	East Lancs	N51/39F	2005
709	GM	YN05WGG	Scania N94 UD OmniDekka	East Lancs	N51/39F	2005

710-732

Scania N94 UD OmniDekka | East Lancs | N51/39F | 2003

710	GM	YN53CFX	716	GM	YN53CFG	722	NM	YN53CFE	728	NM	YN53CFL
711	GM	YN53CFV	717	GM	YN53CFZ	723	NM	YN53CFO	729	NM	YN53CEU
712	GM	YN53CFP	718	NM	YN53CFF	724	NM	YN53CFK	730	NM	YN53CEF
713	GM	YN53CFU	719	NM	YN53CFD	725	NM	YN53CEO	731	NM	YN53CEJ
714	GM	YN53CFA	720	NM	YN53CHF	726	NM	YN53CEK	732	NM	YN53CEV
715	GM	YN53CFY	721	NM	YN53CFJ	727	NM	YN53CEA			

733-747

Scania N94 UD OmniDekka | East Lancs | N51/39F | 2004

733	NM	YN04UJU	737	NM	YN04UJZ	741	NM	YN04UJR	745	NM	YN04UJH
734	NM	YN04UJV	738	NM	YN04UJT	742	NM	YN04UJP	746	NM	YN04UJJ
735	NM	YN04UJW	739	NM	YN04UJL	743	NM	YN04UJF	747	NM	YN04UJK
736	NM	YN04UJS	740	NM	YN04UJY	744	NM	YN04UJG			

The latest double-decks for Nottingham are Scania N270s with East Lancs bodies. The 2004 intake of the OmniDekka model is represented here by 739, YN04UJL, in *purple line* colours. *Richard Godfrey*

748-754

Scania N94 UD OmniDekka   East Lancs   N51/39F   2006

| 748 | NM | YN06TGF | 750 | NM | YN06TGK | 752 | NM | YN06TGU | 754 | NM | YN06TGX |
| 749 | NM | YN06TGJ | 751 | NM | YN06TGO | 753 | NM | YN06TGV | | | |

755-761

Scania N94 UD OmniDekka   East Lancs   N51/36F   2007

| 755 | NM | YN07EYR | 757 | NM | YN07EYU | 759 | NM | YN07EYX | 761 | NM | YN07EYZ |
| 756 | NM | YN07EYT | 758 | NM | YN07EYW | 760 | NM | YN07EYY | | | |

775-779

Scania CN94 UB OmniCity   Scania   N32F   2004

| 775 | GM | YN04GMU | 777 | GM | YN04GMX | 778 | GM | YN04GMY | 779 | GM | YN04GMZ |
| 776 | GM | YN04GMV | | | | | | | | | |

940-944

Scania N270 UD OmniDekka   East Lancs   N51/35F   2007

| 940 | NM | YN07EYH | 942 | NM | YN07EYK | 943 | NM | YN07EYL | 944 | NM | YN07EYM |
| 941 | NM | YN07EYJ | | | | | | | | | |

945-960

Scania N270 UD OmniDekka   East Lancs   N51/35F   2008

945	NM	YN08MSO	949	NM	YN08MSY	953	NM	YN08MLJ	957	NM	YN08MLU
946	NM	YN08MSU	950	NM	YN08MTE	954	NM	YN08MLK	958	NM	YN08MLV
947	NM	YN08MSV	951	NM	YN08MLE	955	NM	YN08MLL	959	NM	YN08MLX
948	NM	YN08MSX	952	NM	YN08MLF	956	NM	YN08MLO	960	NM	YN08MLY

Web: www.nottinghamcitytransport.co.uk
On order: 11 Scania N230 with Optare Omnidekka bodies

Nottingham Express Transit - Joint venture between Nottingham City Transport and Transdev

201-215

Bombardier Incentro   Bombardier (Derby)   N62D   2004

| 201 | 202 | 203 | 204 | 205 | 206 | 207 | 208 | 209 | 210 |
| 211 | 212 | 213 | 214 | 215 | | | | | |

PLYMOUTH CITYBUS

Plymouth Citybus Ltd, Milehouse, Plymouth, Devon, PL3 4AA

1-12				Dennis Dart SLF		Plaxton Pointer		N39F	1996
1	N101UTT	4	N104UTT	8	N108UTT			10	N110UTT
2	N102UTT	5	N105UTT	9	N109UTT			12	N112UTT
3	N103UTT	7	N107UTT						

13-27				Dennis Dart SLF		Plaxton Pointer 2		N39F	1998-99
13	R113OFJ	17	R117OFJ	21	R121OFJ			25	R125OFJ
14	R114OFJ	18	R118OFJ	22	R122OFJ			26	R126OFJ
15	R115OFJ	19	R119OFJ	23	R123OFJ			27	S127FTA
16	R116OFJ	20	R120OFJ	24	R124OFJ				

28-40				Dennis Dart SLF		Plaxton Pointer SPD		N43F	1999
28	T128EFJ	32	T132EFJ	35	T135EFJ			38	T138EFJ
29	T129EFJ	33	T133EFJ	36	T136EFJ			39	T139EFJ
30	T130EFJ	34	T134EFJ	37	T137EFJ			40	T140EFJ
31	T131EFJ								

41-48				Dennis Dart SLF		Plaxton Pointer SPD		N41F	2000-01
41	X141CDV	43	X143CFJ	45	Y645NYD			47	Y647NYD
42	X142CDV	44	Y644NYD	46	Y646NYD			48	Y648NYD

51w	M51HOD	Volvo B6-9.9M		Plaxton Pointer		B40F	1994
52w	M52HOD	Volvo B6-9.9M		Plaxton Pointer		B40F	1994
53w	M53HOD	Volvo B6-9.9M		Plaxton Pointer		B40F	1994

55-59				Dennis Dart SLF 11.3m		Plaxton Pointer SPD		N41F	2001
55	WA51ACO	57	WA51ACV	58	WA51ACX			59	WA51ACY
56	WA51ACU								

60-71				TransBus Dart 11.3m		TransBus Super Pointer		N41F	2002-03
60	WJ52GNY	63	WJ52GOC	66	WJ52GOK			69	WA03BHY
61	WJ52GNZ	64	WJ52GOE	67	WA03BHW			70	WA03BHZ
62	WJ52GOA	65	WJ52GOH	68	WA03BHX			71	WA03BJE

72-79				ADL Dart 10.7m		ADL Pointer		N37F	2004
72	WA54JVV	74	WA54JVX	76	WA54JVZ			78	WA54JWD
73	WA54JVW	75	WA54JVY	77	WA54JWC			79	WA54JWE

80-94				Mercedes-Benz Citaro O530		Mercedes-Benz		N42F	2005-07
80	WJ53HLG	84	WJ53HLN	88	WA56OZM			92	WA56OZS
81	WJ53HLH	85	WJ53HLO	89	WA56OZO			93	WA56OZT
82	WJ53HLK	86	WJ53HLP	90	WA56OZP			94	WA56OZU
83	WJ53HLM	87	WJ53HLR	91	WA56OZR				

102-110				Dennis Dart 9.8m		Plaxton Pointer		B40F	1992
102	K102SFJ	104w	K104SFJ	107w	K107SFJ			110w	K110SFJ
103w	K103SFJ	105w	K105SFJ	109w	K109SFJ				

112-126				Dennis Dart 9.8m		Plaxton Pointer		B40F	1993
112	L112YOD	116	L116YOD	120	L120YOD			124w	L124YOD
113	L113YOD	117	L117YOD	121	L121YOD			125w	L125YOD
114	L114YOD	118	L118YOD	122	L122YOD			126	L126YOD
115	L115YOD	119	L119YOD	123w	L123YOD				

Plymouth Citybus was one of the first of the former municipals to take the Enviro 200 model. Based on the Euro4 model of the Dart, 141, WA08LDU, illustrates the styling in Plymouth's colours. *Mark Bailey*

127-132

127-132		Dennis Dart 9.8m		Plaxton Pointer		B40F	1994	
127	M127HOD		129	M129HOD	131	M131HOD	132	M132HOD
128	M128HOD		130	M130HOD				

133	WA56HHO	ADL Dart 4 10.7m	ADL Enviro 200	N37F	2006	
134	WA56HHP	ADL Dart 4 10.7m	ADL Enviro 200	N37F	2006	
135	WA56HHN	ADL Dart 4 10.7m	ADL Enviro 200	N37F	2006	

136-146

136-146		ADL Dart 4 10.7m		ADL Enviro 200		N38F	2008	
136	WA08LDF		139	WA08LDL	142	WA08LDV	145	WA08LEF
137	WA08LDJ		140	WA08LDN	143	WA08LDX	146	WA08LEJ
138	WA08LDK		141	WA08LDU	144	WA08LDZ		

173	G643CHF	Volvo Citybus B10M-50	East Lancs	B49/39F	1989	Arriva Southern Counties, 2000
174	G640CHF	Volvo Citybus B10M-50	East Lancs	B49/39F	1989	Arriva Southern Counties, 2000
175	B175VDV	Volvo Citybus B10M-50	East Lancs	B42/35F	1984	
176	B176VDV	Volvo Citybus B10M-50	East Lancs	B42/35F	1984	
177	H177GTT	Volvo Citybus B10M-50	East Lancs	BC48/30F	1991	
178	H178GTT	Volvo Citybus B10M-50	East Lancs	BC48/30F	1991	

179-190

179-190		Volvo Citybus B10M-50		Alexander RV		B47/37F	1988-89	Trent Buses, 1999-2000
179	G612OTV		182	G621OTV	185	F602GVO	188	F605GVO
180	G614OTV		183	F600GVO	186	F603GVO	189	F606GVO
181	G615OTV		184	F601GVO	187	F604GVO	190	F607GVO

197	G623OTV	Volvo Citybus B10M-50	Alexander RV	B47/37F	1989	Chambers, Bures, 2000
201	X201CDV	Dennis Dart SLF	Plaxton Pointer MPD	N29F	2000	
202	X202CDV	Dennis Dart SLF	Plaxton Pointer MPD	N29F	2000	
203	X203CDV	Dennis Dart SLF	Plaxton Pointer MPD	N29F	2000	
204	X204CDV	Dennis Dart SLF	Plaxton Pointer MPD	N29F	2000	
205	WA03BJF	TransBus Dart 8.8m	TransBus Mini Pointer	N29F	2003	

The Plymouth Citybus double-deck fleet is comprised entirely of Volvo Citybuses with either Alexander or East Lancs bodywork. One of the latter, 173, G643CHF is seen on a private hire. The bus was new to the North Western fleet, then part of the Drawlane group. *Mark Bailey*

273-289

		Mercedes-Benz 709D	Plaxton Beaver	B25F	1993-95	
273w	M273HOD	277 N277PDV	281 N281PDV	286 N286PDV		
274w	M274HOD	278 N278PDV	282 N282PDV	287 N287PDV		
275	N275PDV	279 N279PDV	283 N283PDV	288 N288PDV		
276	N276PDV	280 N280PDV	284 N284PDV	289 N289PDV		

301	K301WTA	Volvo B10M-60	Plaxton Première ière 350	C51F	1993	
302	L302YOD	Volvo B10M-60	Plaxton Première 350	C51F	1993	
304	M304KOD	Volvo B10M-62	Plaxton Première 350	C49FT	1995	
305	M305KOD	Volvo B10M-62	Plaxton Première 350	C49FT	1995	
307	JSK261	Volvo B10M-62	Plaxton Première 350	C49FT	1996	
308	JSK262	Volvo B10M-62	Plaxton Première 350	C49FT	1997	
309	JSK263	Volvo B10M-62	Plaxton Première 350	C49FT	1998	
311	JSK264	Volvo B10M-62	Plaxton Première 350	C49FT	2000	
312	JSK265	Volvo B10M-62	Plaxton Première 350	C53F	2000	
313	Y313NYD	Volvo B10M-62	Plaxton Paragon	C49FT	2001	
314	Y314NYD	Volvo B10M-62	Plaxton Paragon	C49FT	2001	
315	WA03MGE	Volvo B12M	TransBus Paragon	C49FT	2003	
316	WA03MGJ	Volvo B12M	TransBus Paragon	C49FT	2003	
340	F973HGE	Volvo B10M-60	Plaxton Paramount 3500 III	C53F	1989	Fishwick, Leyland, 1992
346	F988HGE	Volvo B10M-60	Plaxton Paramount 3500 III	C53F	1990	Park's of Hamilton, 1990

Special event vehicle:

358	MCO658	Leyland Titan PD2/12	Metro-Cammell	O30/26R	1956	

Previous registrations:

F973HGE	F973HGE, JSK261	JSK264	W311SDV
JSK261	N307UTT	JSK265	W312SDV
JSK262	P308CTT	MCO658	MCO658, ADV935A
JSK263	R309STA		

Web: www.plymouthcitybus.co.uk

PRESTON BUS

Preston Bus Ltd, 221 Deepdale Road, Preston, PR1 6NY

1-8		Optare Solo M920 SR		Optare		N32F	2008	
1	PN08SVK	3	PN08SVO	5	PN08SVR		7	PN08SVT
2	PN08SVL	4	PN08SVP	6	PN08SVS		8	PN08SVU

29-39		Optare MetroRider MR17		Optare		B29F	1996-98	
29w	N429GBV	34w	R434NFR	36	R436NFR		38	R438RCW
30w	N430GBV	35	R435NFR	37w	R437NFR		39	R439RCW
32	R432NFR							

51-58		Optare Solo M850		Optare		N29F	2001-02	
51	PE51YHF	53	PE51YHH	55	PE51YHK		57	PE51YHM
52	PE51YHG	54	PE51YHJ	56	PE51YHL		58	PE51YHN

59-67		Optare Solo M850		Optare		N29F	2002	
59	PN52ZVH	62	PN52ZVL	64	PN52ZVO		66	PN52ZVR
60	PN52ZVJ	63	PN52ZVM	65	PN52ZVP		67	PN52ZVS
61	PN52ZVK							

68-76		Optare Solo M850		Optare		N29F	2006	
68	PO56RNY	71	PO56ROU	73	PO56RPV		75	PO56RPY
69	PO56RNZ	72	PO56RPU	74	PO56RPX		76	PO56RPZ
70	PO56ROH							

77-80		Optare Solo M880		Optare		N28F	2007	
77	PN07NTK	78	PN07NTL	79	PN57NFF		80	PN57NFG

The Preston operation is expected to pass to Stagecoach by the time this edition is published. Recent arrivals are a batch of Solo SRs which commenced a new numbering series. Seen in Park & Ride colours is 4, PN08SVP.
Dave Heath

81-91 — Optare Solo M920 — Optare — N32F — 2005-06

81	PE55WMD	84	PO56RRU	87	PO56RRY	90	PO56RSV
82	PE55WMF	85	PO56RRV	88	PO56RRZ	91	PO56RSX
83	PE55WMG	86	PO56RRX	89	PO56RSU		

92-97 — Optare Solo M950 — Optare — N32F — 2007

92	PN07NTJ	94	PN07NTO	96	PN07NTU	97	PN07NTV
93	PN07NTM	95	PN07NTT				

101	H101BFR	Leyland Olympian ON2R50C13Z4	Northern Counties Palatine	B47/30F	1991
102	H102BFR	Leyland Olympian ON2R50C13Z4	Northern Counties Palatine	B47/30F	1991
103	H103BFR	Leyland Olympian ON2R50C13Z4	Northern Counties Palatine	B47/30F	1991
104	H104BFR	Leyland Olympian ON2R50C13Z4	Northern Counties Palatine	B47/30F	1991
106	J976PRW	Leyland Olympian ON2R50C13Z4	Leyland	B47/31F	1991

107-114 — Leyland Olympian ON2R50C13Z4 — Leyland — B47/31F — 1992 — 107/14 are BC43/29F

107	J107KCW	109	J109KCW	112	J112KCW	114	J114KCW
108	J108KCW	110	J110KCW	113	J113KCW		

123-132 — Leyland Olympian ONCL10/2RZ — Alexander RL — B51/34F — 1988 — Lothian Buses, 2005

123	E323MSG	126	E326MSG	129	E329MSG	131	E331MSG
124	E324MSG	127	E327MSG	130	E330MSG	132	E332MSG
125	E325MSG	128	E328MSG				

133	A33MRN	Leyland Olympian ONTL11/2R	Eastern Coach Works	B47/27F	1984
134	G34OCK	Leyland Olympian ONCL10/1RZ	Leyland	BC43/29F	1990
135	G35OCK	Leyland Olympian ONCL10/1RZ	Leyland	BC43/29F	1990
136	G36OCK	Leyland Olympian ONCL10/1RZ	Leyland	B47/31F	1990
137	G37OCK	Leyland Olympian ONCL10/1RZ	Leyland	B47/31F	1990
151	PO56RSY	Scania N94UD OmniDekka	East Lancs	N47/33F	2007
152	PO56RSZ	Scania N94UD OmniDekka	East Lancs	N47/33F	2007

182-199 — Dennis Trident — East Lancs Pyoneer — N45/30F — 1999-2001

182	X182RRN	187	X187RRN	192	V192EBV	196	V196EBV
183	X183RRN	188	X188RRN	193	V193EBV	197	X197RRN
184	X184RRN	189	X189RRN	194	V194EBV	198	X198RRN
185	X185RRN	190	V190EBV	195	V195EBV	199	X199RRN
186	X186RRN	191	V191EBV				

200	PRN909	Scania N230 UB	East Lancs Esteem	N40F	2007

201-207 — Scania N94 UB — East Lancs Esteem — N40F — 2006

201	PL06RYO	203	PO56JDF	205	PO56JDK	207	PL56JDX
202	PL06RYP	204	PO56JDJ	206	PO56JDU		

208-211 — Scania N230 UB — East Lancs Esteem — N40F — 2007-08

208	PN57NFA	209	PN57NFC	210	PN57NFD	211	PN57NFE

212	F212YHG	Leyland Lynx LX112L10ZR1R	Leyland	B45F	1989
215	G215KRN	Leyland Lynx LX2R11C15Z4R	Leyland	B49F	1989
218	G218KRN	Leyland Lynx LX2R11C15Z4R	Leyland	B45F	1989
228	H28YBV	Leyland Lynx LX2R11C15Z4R	Leyland	BC45F	1990

Ancillary vehicle:

T128	K128UFV	DAF SB220	Optare Delta	TV	1993	Metro, Blackpool, 2007

Web: www.prestonbus.co.uk

READING BUSES

Reading Buses - Newbury Buses - Goldline

Reading Transport Ltd, Great Knollys Street, Reading, RG1 7HH

1-5			Scania CN270UB OmniCity		Scania		N41F	2006		
1	GL	YN56FBF	3	GL	YN56FBJ	4	GL	YN56FBK	5	GL	YN56FBL
2	GL	YN56FBG									

6-12			Scania CN230UB OmniCity		Scania		N41F	2006		
6	NB	YN56FAA	8	NB	YN56FAJ	10	NB	YN56FAM	12	NB	YN56FAU
7	NB	YN56FAF	9	NB	YN56FAK	11	NB	YN56FAO			

51	NB	RK08CYL	MAN NL273F		Wrightbus Meridian	N44F	2008
52	NB	RK08CYJ	MAN NL273F		Wrightbus Meridian	N44F	2008

101-108			Optare Solo M850		Optare		N30F	1999		
101	RG	S101LBL	103	RG	S103LBL	105w	RG	S105LBL	107	RG	S107LBL
102	RG	S102LBL	104w	RG	S104LBL	106	RG	S106LBL	108w	NB	V108DCF

109	NB	V109DCF	Optare Solo M920	Optare	N34F	2000
110	NB	V110DCF	Optare Solo M920	Optare	N34F	2000
111	NB	V946DCF	Optare Solo M920	Optare	N34F	2000
112	RG	V112DCF	Optare Solo M850	Optare	N30F	2000
113	RG	V113DCF	Optare Solo M850	Optare	N30F	2000
114	RG	V114DCF	Optare Solo M850	Optare	N30F	2000
115	RG	V115DCF	Optare Solo M850	Optare	N30F	2000
116	NB	W116SRX	Optare Solo M920	Optare	N34F	2000
117	NB	W117SRX	Optare Solo M920	Optare	N34F	2000
120	NB	Y594HPK	Optare Solo M920	Optare	N32F	2001
122	RG	YG02FVV	Optare Solo M850	Optare	N28F	2002
123	NB	YG02FVW	Optare Solo M850	Optare	N28F	2002
124	NB	YG02FVX	Optare Solo M850	Optare	N28F	2002
125	NB	YG02FVY	Optare Solo M920	Optare	N32F	2002
126	NB	Y595HPK	Optare Solo M920	Optare	N32F	2001

Recent years have seen a significant change to the fleet policy at Reading. Recent arrivals include twelve Scania OmniCity buses, five with the higher-powered engine for Goldline service. Number 7, YN56FAF, is one of the low-powered CN230 model. *Richard Godfrey*

Following on from several batches of OmniDekka buses, the latest East Lancs-bodied Scania vehicles have Olympus bodies. Lettered for the Woodley service is 849, YN57FXK. *Mark Lyons*

212	GL	YN04AHC	Scania K114EB4	Irizar Century 12.35	C49FT	2004	
213	GL	YN04AHD	Scania K114EB4	Irizar Century 12.35	C49FT	2004	
234	RGt	J786KHD	DAF SB2700	Van Hool Alizée	C51F	1992	Chesterfield, 1994

471-476
Dennis Trident — Plaxton President — N47/29F — 2001 — Pete's Travel, West Bromwich, '03

| 471 | RG | PO51WNF | 473 | RG | PO51WNJ | 475 | RG | PO51WNL | 476 | RG | PO51WNM |
| 472 | RG | PO51WNG | 474 | RG | PO51WNK | | | | | | |

| 501 | NB | RX07RKV | MAN ND243F A48 10.5m | East Lancs Kinetec + 4m | N50/29F | 2007 | |
| 599 | u | YN07LHD | Scania OmniCity N94 UD | Scania | N41/29F | 2007 | Scania demonstrator, 2008 |

601-604
Dennis Dart SLF 10.7m — SSC Compass — N37F — 2002 — Travel de Courcey, Coventry, '07

| 601 | NB | HX52WTC | 602 | NB | HX52WTD | 603 | NB | HX52WTE | 604 | NB | HX52WTF |

605	NB	GU52HKA	Dennis Dart SLF 10.7m	Plaxton Pointer 2	N38F	2002	Coastal, Newick, 2007
606	NB	GU52HKB	Dennis Dart SLF 10.7m	Plaxton Pointer 2	N38F	2002	Coastal, Newick, 2007
607	NB	NK55KBU	ADL Dart 10.7m	ADL Pointer	N37F	2005	Redby, Sunderland, 2007
608	NB	AE06ZBR	ADL Dart 10.7m	ADL Pointer	N37F	2005	E&M Horsburgh, Uphall, 2007
609	NB	AE06ZBT	ADL Dart 10.7m	ADL Pointer	N37F	2005	Cavalier, Long Sutton, 2007
610	NB	MX56HYA	ADL Dart 10.7m	ADL Pointer	N37F	2007	SM Coaches, Harlow, 2008
630	w	T553ADN	Optare MetroRider MR17	Optare	B31F	1999	
631	w	T554ADN	Optare MetroRider MR17	Optare	B29F	1999	
632	w	T556ADN	Optare MetroRider MR17	Optare	B29F	1999	

708-713
DAF DB250 — Optare Spectra — B48/29F — 1998 — Eastbourne Buses, 2001-02

| 708 | RG | S876BYJ | 710 | RG | S878BYJ | 712 | RG | S880BYJ | 713 | RG | S881BYJ |
| 709 | RG | S877BYJ | 711 | RG | S879BYJ | | | | | | |

721-739
DAF DB250 — Optare Spectra — N47/27F — 2001-03

721	NB	YJ51ZVE	726	RG	YE52FHF	731	RG	YJ51ZVH	736	RG	YJ51ZVO
722	RG	YJ51ZVF	727	RG	YE52FHG	732	RG	YJ51ZVK	737	RG	YG02FWA
723	RG	YJ51ZVG	728	RG	YJ03UMK	733	RG	YJ51ZVL	738	RG	YG02FWB
724	RG	YG02FWD	729	RG	YJ03UML	734	RG	YJ51ZVN	739	RG	YG02FWC
725	RG	YG02FWE				735	RG	YJ51ZVM			

English Bus Handbook: Notable Independents

Seen carrying the colours for routes 25/26 as it loads in Gun Street, Reading, is 825, YN06JWL. This is one of thirty-four Scania OmniDekka buses in the fleet. The OmniDekka is the East Lancs product, the similar integral unit built in Poland being known as an OmniCity double-deck. *Richard Godfrey*

801-817

Scania OmniDekka N94UD — East Lancs — N51/39F — 2004

801	RG	YN54AEP	806	RG	YN54AEX	810	RG	YN54AFE	814	RG	YN54AFO
802	RG	YN54AET	807	RG	YN54AEY	811	RG	YN54AFF	815	RG	YN54AFU
803	RG	YN54AEU	808	RG	YN54AEZ	812	RG	YN54AFJ	816	RG	YN54AFV
804	RG	YN54AEV	809	RG	YN54AFA	813	RG	YN54AFK	817	RG	YN54AFX
805	RG	YN54AEW									

818	RG	YN55NJZ	Scania OmniDekka N94UD — East Lancs — N47/33F — 2005

819-834

Scania OmniDekka N94UD — East Lancs — N47/33F — 2006

819	RG	YN06JWC	823	RG	YN06JWG	827	RG	YN06JWO	831	RG	YN06JWW
820	RG	YN06JWD	824	RG	YN06JWP	828	RG	MRD1	832	RG	YN06JWX
821	RG	YN06JWE	825	RG	YN06JWL	829	RG	YN06JWU	833	RG	YN06JWY
822	RG	YN06JWF	826	RG	YN06JWM	830	RG	YN06JWV	834	RG	YN06JWZ

835-861

Scania N230UD — East Lancs Olympus — N51/28F — 2007-08

835	GL	YN07LFA	842	RG	YN57FXB	849	RG	YN57FXK	856	RG	YN08HYM
836	GL	YN07LFB	843	RG	YN57FXC	850	RG	YN57FXL	857	RG	YN08HYT
837	GL	YN07LFD	844	RG	YN57FXD	851	RG	YN57FXM	858	RG	YN08HYU
838	GL	YN07LFE	845	RG	YN57FXE	852	RG	YN08HYO	859	RG	YN08HYW
839	GL	YN07LFF	846	RG	YN57FXF	853	RG	YN08HYP	860	RG	YN08HYX
840	GL	YN07LFG	847	RG	YN57FXH	854	RG	YN08HYR	861	RG	YN08HYY
841	RG	YN57FXA	848	RG	YN57FXJ	854	RG	YN08HYS			

925-931

Optare Excel L1150 — Optare — N39F — 1998-99

925	w	S925LBL	927	w	S927LBL	929	w	S929LBL	931	w	S931LBL
926	w	S926LBL	928	w	S928LBL	930	w	S930LBL			

961	RG	X961BPA	Optare Excel L1180	Optare	N39F	2000	
962	RG	X962BPA	Optare Excel L1180	Optare	N39F	2000	
963	RG	X963BPA	Optare Excel L1180	Optare	N39F	2000	
964	RG	X964BPA	Optare Excel L1180	Optare	N39F	2000	
974	NB	SK52USS	TransBus Enviro 300	TransBus	N44F	2002	TransBus demonstrator, 2004

Scania L94 1016, YN05GXT, is seen in the colours of route 9 as it heads for Shinfield Park. More recent single-deck arrivals feature the later K230 chassis. *Dave Heath*

1001-1020

Scania L94 UB   Wrightbus Solar   N42F   2005

1001	RG	YN05GXA	1006	RG	YN05GXF	1011	RG	YN05GXO	1016	RG	YN05GXT
1002	RG	YN05GXB	1007	RG	YN05GXG	1012	RG	YN05GXL	1017	RG	YN05GXU
1003	RG	YN05GXD	1008	RG	YN05GXH	1013	RG	YN05GXP	1018	RG	YN05GXV
1004	RG	YN05GXC	1009	RG	YN05GXJ	1014	RG	YN05GXR	1019	RG	YN05GXW
1005	RG	YN05GXE	1010	RG	YN05GXM	1015	RG	YN05GXS	1020	RG	YN05GXX

1021-1030

Scania L94 UB   Wrightbus Solar   N42F   2006

1021	RG	YN06NXP	1024	RG	YN06NXT	1027	RG	YN06NXW	1029	RG	YN06NXY
1022	RG	YN06NXR	1025	RG	YN06NXU	1028	RG	YN06NXX	1030	RG	YN06NXZ
1023	RG	YN06NXS	1026	RG	YN06NXV						

1031-1036

Scania K230 UB   Wrightbus Solar   N43F   2007

| 1031 | RG | YN57FWG | 1033 | RG | YN57FWJ | 1035 | RG | YN57FWL | 1036 | RG | YN57FWM |
| 1032 | RG | YN57FWH | 1034 | RG | YN57FWK | | | | | | |

| 1037 | RG | YR08SOU | Scania K230 UB | | Wrightbus Solar 2 | N43F | 2008 |

1101-1114

Scania K230 UB OmniDekka   Scania   N47/20F   2008

1101	RG	YN08MKM	1105	RG	YN08MKV	1109	RG	YN08MME	1112	RG	YN08MMK
1102	RG	YN08MKO	1106	RG	YN08MKX	1110	RG	YN08MMF	1113	RG	YN08MMO
1103	RG	YN08MKP	1107	RG	YN08MKZ	1111	RG	YN08MMJ	1114	RG	YN08MMU
1104	RG	YN08MKU	1108	RG	YN08MMA						

Ancillary vehicles:

| 310 | S516JJH | Dennis Dart SLF 10.7m | Plaxton Pointer 2 | N39F | 1998 | Tellings-Golden Miller, 2008 |
| 311 | S517JJH | Dennis Dart SLF 10.7m | Plaxton Pointer 2 | N39F | 1998 | Tellings-Golden Miller, 2008 |

Previous registrations:

| E247KCF | E475SON, MRD1 | | MRD1 | YN06JWJ |
| J257NLU | MRD1 | | | |

Depots: Great Knollys Street, Reading (RG) and Mill Lane, Newbury (NB) while the Goldline unit buses are shown (GL).

ROSSENDALE

Rossendale Transport Ltd, Knowsley Park Way, Haslingden, Rossendale, BB4 4RS

1-4   Mercedes-Benz Sprinter 411cdi   Koch   N15F   2004   *Operated for Lancashire CC*

1	HN	BK04MZU	2	HN	BK04MZV	3	HN	BK04MZW	4	HN	BK04MZZ

No.		Reg	Chassis	Body	Seating	Year	Notes
5	HN	PL52MZU	Optare Solo M850	Optare	N25F	2004	*Operated for Lancashire CC*
6	RE	L26FNE	Dennis Dart 9.8m	Marshall C37	B36F	1994	Mayne, Manchester, 1998
7	RE	L27FNE	Dennis Dart 9.8m	Marshall C37	B36F	1994	Mayne, Manchester, 1998
10	RE	R410XFL	Dennis Dart SLF 10.8m	Marshall Capital	N39F	1997	Halton, 2006
12	RE	R712MEW	Dennis Dart SLF 10.8m	Marshall Capital	N39F	1997	Halton, 2006

19-26   Volvo Olympian   Northern Counties Palatine   B47/29F   1998   Metrobus, Orpington, 2003

19	RE	S859DGX	21	RE	S861DGX	23	HN	S863DGX	25	HN	S865DGX
20	RE	S860DGX	22	RE	S862DGX	24	HN	S864DGX	26	HN	S866DGX

No.		Reg	Chassis	Body	Seating	Year	Notes
34	HN	G304UYK	Leyland Olympian ONCL10/1RZ	Leyland	B47/31F	1989	London United, 1999
35	HN	G307UYK	Leyland Olympian ON2R50C13Z4	Leyland	B47/31F	1989	London United, 1999
36	HN	G312UYK	Leyland Olympian ON2R50C13Z4	Leyland	B47/31F	1990	London United, 1999

41-53   Optare Solo M880   Optare   N25F   2004

41	RE	YJ54BUA	45	RE	YJ54BUO	48	RE	YJ54BUV	51	RE	YJ54UXU
42	RE	YJ54BUE	46	RE	YJ54BUP	49	RE	YJ54BUW	52	RE	YJ54UXV
43	RE	YJ54BUF	47	RE	YJ54BUU	50	RE	YJ54UXT	53	RE	YJ54UXW
44	RE	YJ54BUH									

Since the last edition of this Handbook the offices and main depot of Rossendale Transport have moved to Haslingden, once the home of a small municipal that with Rawtenstall Corporation made up Rossendale. Page 1 shows a Rawtenstall-coloured commemorative Olympian and here, 102, PN57LGF, is a Volvo B7TL with East Lancs Vyking bodywork. *Dave Heath*

Pictured in Burnley is 135, PO51WEC, the first of six Darts supplied from 2001 that carry East Lancs Spryte bodywork. *Alan Blagburn*

54-63

			Optare Solo M880 SL			Optare		N25F	2005
54	RE	YJ05JWC	57	HN	YJ05JWF	60	HN	YJ05JWL	62	HN	YJ05JWN
55	HN	YJ05JWD	58	HN	YJ05JWG	61	HN	YJ05JWM	63	HN	YJ05JWO
56	HN	YJ05JWE	59	HN	YJ05JWK						

100	HN	X2JPT	Volvo B7TL	East Lancs Vyking	N47/30F	2000	Moffat & Williamson, St Fort, 2007
101	HN	Y568LRN	Volvo B7TL	East Lancs Vyking	NC45/30F	2001	Cronins, Cork, 2007
102	HN	PN57LGF	Volvo B7TL	East Lancs Vyking	NC45/27F	2002	Circleline, Naas, 2007

106-113

			Dennis Dart SLF 10.1m			East Lancs Spryte	N28F	1996-97
106	HN	N106LCK	108	HN	N108LCK	110	HN	N110LCK	112	HN	P212DCK
107	HN	N107LCK	109	HN	N109LCK	111	HN	P211DCK	113	HN	P213DCK

114-123

			Dennis Dart SLF 11.3m			Plaxton Pointer SPD	N40F	1998
114	HN	S114KRN	117	HN	S117KRN	119	HN	S119KRN	121	HN	S121KRN
115	HN	S115KRN	118	HN	S118KRN	120	HN	S120KRN	123	RE	S123KRN
116	HN	S116KRN									

124	HN	R6BLU	Dennis Dart SLF 9.5m	Plaxton Pointer 2	N33F	1998	Blue Bus, Bolton, 2002
125	HN	R7BLU	Dennis Dart SLF 9.5m	Plaxton Pointer 2	N33F	1998	Blue Bus, Bolton, 2002
130	HN	R8BLU	Dennis Dart SLF 9.5m	Plaxton Pointer 2	N33F	1998	Blue Bus, Bolton, 2003
131	RE	X131JCW	Dennis Dart SLF 11.3m	Plaxton Pointer SPD	N40F	2000	
132	RE	X132JCW	Dennis Dart SLF 11.3m	Plaxton Pointer SPD	N40F	2000	
133	RE	X133JCW	Dennis Dart SLF 11.3m	Plaxton Pointer SPD	N40F	2000	
134	RE	X134JCW	Dennis Dart SLF 11.3m	Plaxton Pointer SPD	N40F	2000	

135-140

			Dennis Dart SLF 10.8m			East Lancs Spryte	N37F	2001-02
135	HN	PO51WEC	137	HN	PF51KMO	139	HN	PF51KMV	140	HN	PF51KMX
136	HN	PF51KMM	138	HN	PF51KMU						

141-148

			Dennis Dart SLF 11.3m			East Lancs Spryte	N41F	2002
141	HN	PF02XMX	143	HN	PN52WWK	145	HN	PN52WWM	147	HN	PN52WWP
142	HN	PF02XMW	144	HN	PN52WWL	146	HN	PN52WWO	148	HN	PN52WWR

Rochdale is now a constituent part of the Rossendale network. Pictured in Accrington while heading south is 133, X133JCW, one of four Super Pointer Darts allocated to Rochdale. *Richard Godfrey*

149	HN	S8BLU	Dennis Dart SLF 10.7m			Plaxton Pointer 2			N37F	1998	Blue Bus, Bolton, 2003
150-156			Volvo B7RLE			Wrightbus Eclipse Urban			N44F	2003	
150	RE	PO53OBU	**152**	HN	PO53OBN	**154**	HN	PO53OBR	**156**	RE	PJ53UHV
151	HN	PO53OBM	**153**	HN	PO53OBP	**155**	RE	PO53OBT			
157	RE	BX03BKU	Volvo B7RLE			Wrightbus Eclipse Urban			N44F	2003	Volvo demonstrator, 2003
158	RE	CU04AMX	MAN 14.220			MCV Evolution			N39F	2004	2Travel, Swansea, 2006
159	RE	CU04AOP	MAN 14.220			MCV Evolution			N39F	2004	2Travel, Swansea, 2006
160	RE	CU04AMV	MAN 14.220			MCV Evolution			N39F	2004	2Travel, Swansea, 2006
163-166			Dennis Dart SLF 10.7m			Alexander ALX200			N41F	2000	Munro's of Jedburgh, 2006
163	RE	X463UKS	**164**	RE	X464UKS	**165**	RE	X465UKS	**166**	RE	X466UKS
167-171			Dennis Dart SLF 10.1m			Caetano Compass			N34F	2002	Centra, Wandsworth, 2007
167	RE	HV52WSJ	**167**	RE	HV52WSL	**167**	RE	HV52WSN	**167**	RE	HV52WSO
167	RE	HV52WSK									
300	HN	N300EST	Dennis Javelin GX			Plaxton Première 350			C48FT	1996	Coachways, Rochdale, 2005
358	RE	P958YGG	Volvo B10M-62			Van Hool Alizée HE			C48FT	1997	Coachways, Rochdale, 2005
359	HN	P959YGG	Volvo B10M-62			Van Hool Alizée HE			C48FT	1997	Coachways, Rochdale, 2005

Ancillary vehicles:

366	HN	P466GUA	Bova Futura Club FLC12.280	Bova	TV	1997	Coachways, Rochdale, 2005

Previous registrations:

P958YGG	LSK878		PN57LGF	07KE5308
P959YGG	LSK877		Y568LRN	01D71334

Depots: Mandale Park, Rochdale (RE) and Knowsley Park Way, Haslingden (HN).
Web: www.rossendalebus.co.uk

THAMESDOWN

Thamesdown Transport Ltd, Barnfield Road, Swindon, SN2 2DJ

70-73

		Dennis Dominator DDA1033	East Lancs		B45/31F	1990	
70w	H970XHR	**71**	H971XHR	**72**	H972XHR	**73**	H973XHR

128	N128LMW	Dennis Dart 9.8m	Plaxton Pointer	B40F	1996	
129	XBZ7729	Dennis Dart 9.8m	Plaxton Pointer	B40F	1995	Isle of Man Transport, 2000
130	XBZ7730	Dennis Dart 9.8m	Plaxton Pointer	B40F	1995	Isle of Man Transport, 2000
131	XBZ7731	Dennis Dart 9.8m	Plaxton Pointer	B40F	1994	Isle of Man Transport, 2000
132	XBZ7732	Dennis Dart 9.8m	Plaxton Pointer	B40F	1994	Isle of Man Transport, 2000
141	R314NGM	Dennis Dart SLF	Plaxton Pointer 2	N37F	1997	The King's Ferry, Gillingham, 2000
142	R315NGM	Dennis Dart SLF	Plaxton Pointer 2	N37F	1997	The King's Ferry, Gillingham, 2000
143	R317NGM	Dennis Dart SLF	Plaxton Pointer 2	N37F	1997	The King's Ferry, Gillingham, 2000
144	R319NGM	Dennis Dart SLF	Plaxton Pointer 2	N37F	1997	The King's Ferry, Gillingham, 2000

151-158

		Dennis Dart SLF	Plaxton Pointer		B41F	1996	
151	P151SMW	**153**	P153SMW	**155**	P255SMW	**157**	P157SMW
152	P152SMW	**154**	P154SMW	**156**	P156SMW	**158**	P158SMW

159	P159VHR	Dennis Dart SLF	Plaxton Pointer 2	N41F	1997	
160	P160VHR	Dennis Dart SLF	Plaxton Pointer 2	N41F	1997	
161	P161VHR	Dennis Dart SLF	Plaxton Pointer 2	N41F	1997	
162	S162BMR	Dennis Dart SLF	Plaxton Pointer 2	N40F	1998	
163	T163RMR	Dennis Dart SLF	Plaxton Pointer 2	N40F	1999	
164	T164RMR	Dennis Dart SLF	Plaxton Pointer 2	N40F	1999	
165	T165RMR	Dennis Dart SLF	Plaxton Pointer 2	N40F	1999	
180	S838VAG	Dennis Dart SLF 11.3m	Plaxton Pointer SPD	N45F	1998	Plaxton demonstrator, 2000

181-191

		Dennis Dart SLF 11.3m	Plaxton Pointer SPD		N45F	1998-2000 *184-6 are N41F	
181	S181BMR	**184**	S184BMR	**187**	V187EAM	**190**	V190EAM
182	S182BMR	**185**	S185BMR	**188**	V188EAM	**191**	V191EAM
183	S183BMR	**186**	S186BMR	**189**	V189EAM		

The Super Pointer Dart forms a major element in the Thamesdown fleet. Illustrating the type is 208, WU52YWM, seen here passing through the Lawn district whlle heading for Pinehurst. *Richard Godfrey*

Pictured in the Eldene area of Swindon, 512, WX57TLU, illustrates the 2007 batch of Scania saloons to join the Thamesdown fleet. *Richard Godfrey*

192-197

						Dennis Dart SLF 11.3m		Plaxton Pointer SPD	N41F	2001	
192	Y192YMR	194	Y194YMR	196	Y196YMR	197	Y197YMR
193	Y193YMR	195	Y195YMR				

198-211

Dennis Dart SLF 11.3m — Plaxton Pointer SPD — N41F — 2002-03

198	WV02NNA	202	WU52YWF	206	WU52YWK	209	WX03YFD
199	WV02NNB	203	WU52YWG	207	WU52YWL	210	WX03YFE
200	WV02NNC	204	WU52YWH	208	WU52YWM	211	WX03ZNS
201	WU52YWE	205	WU62YWJ				

212-215

TransBus Dart SLF 11.3m — TransBus Super Pointer — N41F — 2004

212	WX04CZH	213	WX04CZJ	214	WX04CZK	215	WX04CZL

301-306

Volvo Olympian — Alexander RH — B51/33F — 1994 — Lothian Buses, 2007

301	L954MSC	303	L966MSC	305	L968MSC	306	L969MSC
302	L957MSC	304	L967MSC				

307-314

Leyland Olympian ON2R56C1 — Alexander RH — B51/33F — 1993 — Lothian Buses, 2007

307	K882CSF	309	K886CSF	311	K889CSF	313	K892CSF
308	K884CSF	310	K887CSF	312	K891CSF	314	K893CSF

315	M206VSX	Volvo Olympian		Alexander RH		B51/33F	1995	Lothian Buses, 2008

351-355

Volvo Olympian — Alexander RH — B51/33F — 1995 — Lothian Buses, 2008

351	M207VSX	353	M213VSX	354	M216VSX	355	M231VSX
352	M211VSX						

501-510

Scania L94 UB — Wrightbus Solar — N40F — 2006

501	WX55ZZR	504	WX06JXS	507	WX06JYC	509	WX06JYA
502	WX06JYD	505	WX06JXT	508	WX06JYB	510	WX06JXZ
503	WX06JXR	506	WX06JYE				

Many of the Thamesdown double-deck fleet are confined to school duties. Painted for that role is Volvo Olympian 305, L968MSC, which was new to Lothian Buses. *Richard Godfrey*

511-522

Scania K230 UB — Wrightbus Solar — N40F — 2007-08

511	WX57TLY	**514**	WX57TLZ	**517**	WX08SXD	**520**	XMW120
512	WX57TLU	**515**	WX57TLN	**518**	WX08SXE	**521**	WX08SXG
513	WX57TLV	**516**	WX57TLO	**519**	WX08SXF	**522**	WX08SXH

Special event vehicles:

145	JAM145E	Daimler CVG6/30	Northern Counties	B40/30F	1967
299	UMR199T	Leyland Fleetline FE30AGR	Eastern Coach Works	B43/31F	1978

Ancillary vehicle:

383	R189NFE	Dennis Dart SLF 10.7m	Plaxton Pointer 2	TV	1998	Hornsby, Ashly, 2007

Previous registrations:

XBZ7729	MAN14A, M410XTC		XBZ7731	CMN76X, M505XTC
XBZ7730	MAN15D, M409XTC		XBZ7732	CMN78X, M506XTC
			XMW120	WX08VDC

web: www.thamesdown-transport.co.uk
On order: 6 Scania K230 UBs with Wrightbus Solar bodywork

Named vehicles:- 70 *Western Explorer*; 71 *Western Pioneer*; 72 *Western Crusader*; 73, *Western Venturer*; 128 *Knight of the Grand Cross*; 129 *County of Gloucester*; 130 *County of Oxford*; 131 *County of Berks*; 132 *County of Wilts*; 141 *Sir Daniel Gooch*; 142 *Armstrong*; 143 *William Dean*; 144 *G J Churchward*; 151 *Saint Ambrose*; 152 *Saint Andrew*; 153 *Saint Augustine*; 154 *Saint Bartholomew*; 155 *Saint Benedict*; 156 *Saint Bernard*; 157 *Saint Cuthbert*; 158 *Saint David*; 159 *Saint Agatha*; 160 *Saint Catherine*; 161 *Saint Helena*; 162 *Saint Dunstan*; 163 *Saint Gabriel*; 164 *Saint George*; 165 *Saint Nicholas*; 180 *Caerphilly Castle*; 181 *Eclipse*; 182 *Vanguard*; 183 *Formidable*; 184 *Albion*; 185 *Avenger*; 186 *Benbow*; 187 *Caradoc*; 188 *Centaur*; 189 *Champion*; 190 *Cockade*; 191 *Daring*; 192 *Despatch*; 193 *Diadem*; 194 *Dragon*; 195 *Druid*; 196 *Glory*; 197 *Magnificent*; 198 *Cambrian*; 199 *Foxhound*; 200 *Goliath*; 201 *Salzgitter*; 202 *Ocotal*; 203 *Pontorson*; 204 *Grenville*; 205 *Greyhound*; 206 *Hercules*; 207 *Hermes*; 208 *Highflyer*; 209 *Intrepid*; 210 *Jupiter*; 211 *Torun - City of Copernicus*; 212 *Kelly*; 213 *Magpie*; 214 *Majestic*; 215 *Swindon*; 351 *Earl of Mount Edgcumbe*; 352 *Earl of Dunraven*; 353 *Earl of Dudley*; 354 *Earl Cawdor*; 355 *Earl of Dartmouth*; 501 *Brunel*; 502 *Iron Duke*; 503 *Great Britain*; 504 *Lightning*; 505 *Emperor*; 506 *Sultan*; 507 *Lord of the Isles*; 508 *Royal Sovereign* 509 *Tornado*; 510 *Great Western*; 511 *Pasha*; 512 *Courier*; 513 *Tartar*; 514 *Warlock*; 515 *Wizard*; 516 *Rougemont*; 517 *Hirondelle*; 518 *Swallow*; 519 *Timour*; 520 *Prometheus*; 520 *Perseus*; 521 *Estafette*.

ABUS

L C Munden & Sons Ltd, 6-7 Freestone Road, St Phillips, Bristol, BS2 0QH
A J Peters, 104 Winchester Road, Brislington, Bristol, BS4 3NL

XHK221X	Bristol VRT/SL3/6LXB	Eastern Coach Works	B39/31F	1981	First Cityline, 2000
RBO508Y	Leyland Olympian ONLXB/1R	East Lancs	B43/31F	1983	Cardiff Bus, 1998
B892UAS	Leyland Olympian ONLXB/1R	Alexander RL	B45/32F	1985	Stagecoach, 2004
B893UAS	Leyland Olympian ONLXB/1R	Alexander RL	B45/32F	1985	Stagecoach, 2005
B896UAS	Leyland Olympian ONLXB/1R	Alexander RL	B45/32F	1985	Stagecoach, 2004
B899UAS	Leyland Olympian ONLXB/1R	Alexander RL	B45/32F	1985	Stagecoach, 2004
J814NKK	Leyland Olympian ON2R50C13Z4	Northern Counties Palatine	B47/30F	1992	Stagecoach, 2007
K159PGO	DAF DB250	Optare Spectra	B44/27F	1993	Go North East, 2007
N646FKK	Volvo Olympian	Alexander Royale	B47/27D	1995	Dublin Bus, 2008
R222AJP	DAF DB250	Optare Spectra	N51/30F	1998	
S111AJP	DAF DB250	Optare Spectra	N50/27F	1998	
S333AJP	Mercedes-Benz Vario O814	Plaxton Beaver 2	B22F	1998	Firstgroup, 2006
V444AJP	DAF DB250	Alexander ALX400	N45/24F	1999	
X204NKR	Volvo B7TL	East Lancs Vyking	NC45/30F	2001	Dualway, Rathcoole, 2008
AP03BUS	Optare Solo M850	Optare	N29F	2003	
AP03BUZ	Optare Solo M850	Optare	N29F	2003	
AP53BUS	DAF DB250	Optare Spectra	N50/27F	2003	
AP04BUS	DAF DB250	Optare Spectra	N47/27F	2004	
SN04CPE	Scania OmniDekka N94UD	East Lancs	CO51/34F	2004	Anglian, Beccles, 2008
BU54AJP	Scania OmniDekka N94UD	East Lancs	N51/39F	2005	

Special event vehicles:

u	UHY384	Bristol KSW6G	Eastern Coach Works	B32/28RD	1955	Bristol OC
	969EHW	Bristol Lodekka LD6G	Eastern Coach Works	B33/25RD	1959	Bath Tramways
	OAE954M	Bristol RELL6L	Eastern Coach Works	B50F	1973	Bristol OC
	AFB592V	Bristol LH6L	Eastern Coach Works	B43F	1980	Bristol OC

Previous registrations:

K159PGO	K325FYG, WLT625	SN04CPE	SN04CPE, B16YBC
N646FKK	95D243	X204NKR	01D25209
S333AJP	S520RWP		

Depots: Kingsland Sidings, Bristol and Gas Lane, Bristol

An unusual combination is illustrated by the Alexander ALX400 body on a DAF DB250 chassis. Abus' V444AJP is such a combination and is soon leaving Bristol for Keynsham. *Alan Blagburn*

APL TRAVEL

APL Travel Ltd, 1 Peartree Cottage, The Street, Crudwell, SN16 3ES

Reg	Chassis	Body	Layout	Seating	Year	History
H882LOX	Mercedes-Benz 811D	Carlyle C17	B31F	1990		Weybus, Weymouth, 2006
K439DRW	Mercedes-Benz 609D	Mercedes-Benz	B7F	1992		Nightingale, Maidenhead, 2006
K50APL	Dennis Javelin 12m	Plaxton Paramount 3200 III	C49FT	1993		Southlands, Orpington, 2004
M711BMR	Dennis Dart 9.8m	Plaxton Pointer	B40F	1994		Thamesdown, 2008
M118BMR	Dennis Dart 9.8m	Plaxton Pointer	B40F	1994		Griffiths, Brinkworth, 2006
L666LMT	Bova Futura FHD 12.340	Bova	S70F	1995		Dailybus, Standish, 2007
N9LON	Dennis Javelin 12m	Berkhof Excellence 2000	C51FT	1997		Hodgson, Barnard Castle, 2004
P894FMO	Dennis Javelin 12m	Berkhof Excellence 2000	C53F	1997		Edwards, Soham, 2006
M50APL	Mercedes-Benz 814D	Autobus Classique	BC33F	1997		Bebb, Llantwit Fardre, 1999
S853PKH	Mercedes-Benz Vario 0814	Plaxton Beaver 2	B31F	1998		Turner, Bristol, 2004
S762XYA	Iveco TurboDaily 59-12	Marshall C19	B27F	1998		Bodman's. Worton, 2004
V78JKG	Optare Excel L1070	Optare	N38F	1999		Weavaway, Newbury, 2005
V81JKG	Optare Excel L1070	Optare	N38F	1999		Weavaway, Newbury, 2005
W2APL	Mercedes-Benz Vario 0814	Autobus Nouvelle	BC29F	2000		New Punjab, Southall, 2004
YP52BRF	Optare Alero	Optare	N13F	2002		*Operated for Wiltshire CC*
VO03DZD	MAN 14.220	Ikarus Polaris	N42F	2003		Bennetts, Gloucester, 2003
VO03DZE	MAN 14.220	Ikarus Polaris	N42F	2003		Bennetts, Gloucester, 2003
YJ54UCA	Optare Solo M850	Optare	N27F	2004		
YK07BFU	Optare Solo M780 SL	Optare	N21F	2007		

Previous registrations:

K50APL	K205GMX	P894FMO	P894FMO, 299SAE
L666LMT	M50FTG, GIB976, M367VBR	W2APL	W608KFE, MCH252
M50APL	P76VWO		

Autobus Classique bodywork is fitted to APL Travel's M50APL which is based on a Mercedes-Benz 814D chassis. The vehicle is seen in Cirencester. *Robert Edworthy*

AINTREE COACHLINE

Aintree Coaches (International) Ltd; Helms Coaches Ltd; J Cherry, 11 Clare Road, Bootle, L20 9LY

VRA124Y	Leyland Olympian ONLXB/1R	Northern Counties	B43/28F	1982	Chesterbus, 2007
A163HLV	Volvo-Ailsa B55 Mark III	Alexander RV	B44/37F	1984	Cardiff Bus, 2008
B201EFM	Leyland Olympian ONLXB/1R	Northern Counties	BC43/30F	1985	Chesterbus, 2007
E935CDS	Leyland Lion LDT11/1R	Alexander RH	BC49/37F	1987	Veolia England, 2008
F389GVO	Leyland-DAB LDT11/1R	East Lancs	B47/31F	1988	City of Nottingham, 2006
F406DUG	Volvo B10M-60	Plaxton Paramount 3500 III	C49FT	1989	Arriva Midlands, 2006
G864XDX	Leyland Olympian ONCL10/1RZ	Alexander RL	B45/32F	1989	Chambers, Bures, 2006
J54EDM	Dennis Dart 9m	Plaxton Pointer	B35F	1991	Chesterbus, 2007
J155EDM	Dennis Dart 9m	Plaxton Pointer	B35F	1991	Chesterbus, 2007
K100ACL	Leyland Olympian ON2R50C13Z4	Alexander RH	B47/31F	1992	Dublin Bus, 2006
L100ACL	Leyland Olympian ON2R50C13Z4	Alexander RH	B47/27D	1993	Dublin Bus, 2006
N349FKU	Kirn Mogul 232	East Lancs Spryte (2001)	NC49F	1993	Stagecoach, 2008
L202SKD	Volvo Olympian	Northern Counties	B47/29F	1993	Metroline, Harrow, 2007
L141VOM	Dennis Dart 9m	Plaxton Pointer	B35F	1993	Arriva NW & Wales, 2007
L86VDM	Dennis Dart 9m	Plaxton Pointer	B35F	1993	Arriva NW & Wales, 2007
L70ARK	Mercedes-Benz 709D	Plaxton Beaver 2	BC25F	1994	Ark, Wallasey, 2006
N401ARA	Dennis Arrow	Northern Counties Palatine	B44/36F	1995	City of Nottingham, 2005
N402ARA	Dennis Arrow	Northern Counties Palatine	B44/26F	1995	City of Nottingham, 2005
N403ARA	Dennis Arrow	Northern Counties Palatine	B49/35F	1995	City of Nottingham, 2005
N404ARA	Dennis Arrow	Northern Counties Palatine	B49/35F	1995	City of Nottingham, 2005
N414JBV	Volvo Olympian	Northern Counties Palatine	B43/30F	1995	Go-Ahead London, 2004
N415JBV	Volvo Olympian	Northern Counties Palatine	B43/30F	1995	Go-Ahead London, 2004
N419JBV	Volvo Olympian	Northern Counties Palatine	B43/30F	1995	Go-Ahead London, 2004

One of the entries at the 2008 Showbus event at Duxford was Aintree Coachline's new double-deck, SN58CDK, a Alexander Dennis Trident 2 with Enviro 400 bodywork. *Richard Godfrey*

Pictured in Chester, and lettered for the Cheshire Bus service from Chester to Whitchurch is Dart V549JBH, one of five of this model currently in service. *Richard Godfrey*

N421JBV	Volvo Olympian	Northern Counties Palatine	B43/30F	1995	Go-Ahead London, 2004
N422JBV	Volvo Olympian	Northern Counties Palatine	B43/30F	1995	Go-Ahead London, 2004
N734EOT	Bluebird Q	Bluebird	S60F	1996	Accord Southern, Chichester, 2007
P5ACL	Dennis Arrow	Northern Counties Palatine	B49/35F	1996	
P493MBY	Volvo Olympian	Alexander RH	B47/31F	1996	Metroline, Harrow, 2006
HKF151	Volvo B10M-62	Plaxton Première 350	C53F	1997	Wallace Arnold, Leeds, 1999
L1SLT	Dennis Dart SLF 10.1m	Plaxton Pointer 2	N35F	1997	Go-Ahead London, 2008
R55ACL	Dennis Arrow	East Lancs	B49/35F	1998	
S129RLE	Volvo Olympian	Alexander RH	B43/25D	1998	Metroline, Harrow, 2007
L2SLT	Dennis Dart SLF 10.1m	Plaxton Pointer 2	N35F	1999	Arriva NW&W, 2008
V544JBH	Dennis Dart SLF 10.1m	Plaxton Pointer 2	N33F	1999	Central Parking, Wandsworth, '08
V547JBH	Dennis Dart SLF 10.1m	Plaxton Pointer 2	N33F	1999	Centra, Heathrow, 2008
V549JBH	Dennis Dart SLF 10.1m	Plaxton Pointer 2	N33F	1999	Central Parking, Wandsworth, '08
W5ACL	Dennis Trident	East Lancs Lolyne	N51/36F	2000	
SK02TYS	Dennis Dart SLF 8.8m	Plaxton Pointer MPD	N29F	2002	Matthews, Heswall, 2008
YM52TPU	Optare Solo M850	Optare	N29F	2003	*Operates for Cheshire CC*
GB03ACL	Volvo B12B	Plaxton Paragon	C53F	2003	
SN58CDK	ADL Trident 2	ADL Enviro 400	N51/37F	2008	

Special event vehicles: (original owners shown)

XSL596A	AEC Routemaster R2RH	Park Royal	B36/28R	1962	London Transport
ALD978B	AEC Routemaster R2RH	Park Royal	B36/28R	1964	London Transport
TUB250R	Foden 04B016G800	Northern Counties	H43/31F	1977	West Yorkshire PTE

Previous registrations:

E935CDS	E164YGB, VLT204	L70ARK	L892VHT
HKF151	P331VWR	L100ACL	93D10175
K100ACL	92D151, K140KNO	L141VOM	L1SLT
L1SLT	R455LGH	N349PKU	1901HE
L2SLT	R465LGH	VIW4986	B975DWG
L86VDM	L2SLT	XSL596A	289CLT

Depots: Sefton Lane Industrial Estate, Maghull and Rivacre Road, Eastham.

ANDYBUS

Andybus & Coach Ltd, F6 Whitewalls, Easton Grey, Malmesbury, SN16 0RD

N3ARJ	Bova Futura Club FLC12.280	Bova	C53F	1995	
R30ARJ	Dennis Dart SLF	Marshall Capital	N37F	1997	Isle of Man Transport, 2003
M33ARJ	Dennis Javelin 12m	UVG Vanguard III	S70F	1997	MoD, 2008
R57JSG	Mercedes-Benz Vario 0814	Plaxton Beaver 2	B33F	1998	
R501SJM	Dennis Dart SLF 10.1m	Plaxton Pointer 2	N36F	1998	Tellings-Golden Miller, 2007
R502SJM	Dennis Dart SLF 10.1m	Plaxton Pointer 2	N36F	1998	Tellings-Golden Miller, 2007
R503SJM	Dennis Dart SLF 10.1m	Plaxton Pointer 2	N36F	1998	Tellings-Golden Miller, 2007
N222LFR	MAN NL222FR	East Lancs Spryte	N43F	1998	MAN demonstrator, 2001
T30ARJ	Dennis Dart SLF	Plaxton Pointer MPD	N29F	1999	Scarlet Band, West Cornforth, '05
WR02XXO	Bova Futura FHD12.340	Bova	C51FT	2002	
WP52YZA	Mercedes-Benz Vario 0814	Plaxton Beaver 2	BC33F	2002	Somerbus, Bristol, 2004
VU03ZPS	Mercedes-Benz Vario 0814	Plaxton Beaver 2	BC33F	2003	
VU03ZPY	Mercedes-Benz Vario 0814	Plaxton Beaver 2	BC33F	2003	
AJ05BUS	MAN 14.220	Ikarus Polaris	N40F	2005	
AJ55BUS	Toyota Coaster BB50R	Caetano Optimo V	C26F	2005	
AJ07BUS	VDL Bova Futura FHD127.365	VDL Bova	C55FT	2007	
YJ05WCC	Optare Solo M850	Optare	N23F	2008	Eastbourne Buses, 2008
AJ08BUS	Optare Versa V1100	Optare	N40F	2008	

Previous registrations:

R30ARJ	R822JHJ, DMN18R	T30ARJ	T76JBA	WP52YZA	620UKM

Web: www.andrew-james.co.uk

Representing the Andybus fleet is East Lancs-bodied MAN N222LFR, a number that reflects the chassis model. This bus was new as the MAN demonstrator and carries East Lancs bodywork. It is seen at the Great Western Outlet Centre. *Mark Lyons*

ANGLIAN

Anglian Bus & Coach Ltd, Beccles Business Park, Sandpit Lane, Beccles, NR34 7TH

103	HIJ6931	Mercedes-Benz O303	Mercedes-Benz	C49FT	1984	Fargo Coachlines, Rayne, 1998
107	CSK282	Mercedes-Benz O303	Mercedes-Benz	C53F	1988	Barry's, Moreton-in-Marsh, 2000
108	831HKA	Mercedes-Benz O303	Mercedes-Benz	C49FT	1985	Ron Lyles, Mirfield, 2001
109	J127LHC	Dennis Javelin 11m	Plaxton Derwent 2	BC53F	1991	Countywide, Basingstoke, 2001
110	G401DPD	Scania K93CRB	Plaxton Derwent 2	B57F	1989	Tim's Travel, Sheerness, 2003
111	784EYB	Dennis Javelin 11m	Wadham Stringer III	BC54F	1994	MoD (CX45AA), 2008
201	T81JBA	Mercedes-Benz Vario O814	Plaxton Beaver 2	B31F	1999	
203	T83JBA	Mercedes-Benz Vario O814	Plaxton Beaver 2	B31F	1999	
207	V227KAH	Mercedes-Benz Vario O814	Plaxton Beaver 2	B27F	1999	
210	T100CBC	Mercedes-Benz Vario O814	Plaxton Beaver 2	B31F	1999	Coakley Bus, Motherwell, 2002
211	T200CBC	Mercedes-Benz Vario O814	Plaxton Beaver 2	B27F	1999	Coakley Bus, Motherwell, 2002
213	T400CBC	Mercedes-Benz Vario O814	Plaxton Beaver 2	B27F	1999	Coakley Bus, Motherwell, 2002
217	V380HGG	Mercedes-Benz Vario O814	Plaxton Beaver 2	B27F	1999	Equalstorm, Laurencekirk, 2002
218	AO02LVC	Mercedes-Benz Vario O814	Plaxton Beaver 2	B31F	2002	
221	AO52LJF	Mercedes-Benz Vario O814	Plaxton Beaver 2	B31F	2002	
222	AU53GWC	Mercedes-Benz Vario O814	TransBus Beaver	B33F	2003	
226	Y784WHH	Mercedes-Benz Vario O814	Plaxton Beaver 2	B27F	2001	Reay's, Wigton, 2005
228	X228WRA	Optare Excel L1180	Optare	N44F	2000	Trent Barton, 2007
229	S195HOK	Mercedes-Benz Vario O814	Plaxton Beaver 2	B31F	1998	Thandi, Smethwick, 2001
302	MX53FDM	Optare Solo M920	Optare	N33F	2004	
303	MX53FDN	Optare Solo M920	Optare	N33F	2004	
304	MX53FDO	Optare Solo M920	Optare	N33F	2004	
305	MX53FDP	Optare Solo M920	Optare	N33F	2004	
306	AU04JKN	Optare Solo M920	Optare	N33F	2004	
307	AU54EOA	Optare Solo M780	Optare	N24F	2004	
308	AU54ENY	Optare Solo M780	Optare	N24F	2004	
309	AU07KMM	Optare Solo M950	Optare	N32F	2007	
310	AU07KMK	Optare Solo M950	Optare	N32F	2007	
311	AD57BDY	Optare Solo M950	Optare	N32F	2007	
312	YN57HPX	Optare Solo M950	Optare	N32F	2007	
313	YN57HRA	Optare Solo M950	Optare	N32F	2007	

2008 saw the arrival with Anglian of a pair of Optare Versa buses. With Great Yarmouth's Parish Church in the distant background, 411, AU08DKL is illustrated. *Steve Rice*

314	YN57HPV	Optare Solo M950	Optare	N32F	2007	
315	YN57HPU	Optare Solo M950	Optare	N32F	2007	
316	YN57HPZ	Optare Solo M950	Optare	N32F	2007	
317	AD57EXA	Optare Solo M950	Optare	N32F	2007	
318	AU08GLY	Optare Solo M950	Optare	N32F	2007	
319	AU58AKK	Optare Solo M950	Optare	N32F	2008	
320	AU58AKN	Optare Solo M950	Optare	N32F	2008	
401	AU06BOV	Volvo B7RLE	Wrightbus Eclipse Urban	N44F	2006	
402	AU06BPE	Volvo B7RLE	Wrightbus Eclipse Urban	N44F	2006	
403	AU06BPF	Volvo B7RLE	Wrightbus Eclipse Urban	N44F	2006	
404	AU06BPK	Volvo B7RLE	Wrightbus Eclipse Urban	N44F	2006	
405	AU06BPO	Volvo B7RLE	Wrightbus Eclipse Urban	N44F	2006	
406	YJ55BLX	Optare Tempo X1200	Optare	N42F	2005	Optare demonstrator, 2007
407	YJ06FZK	Optare Tempo X1200	Optare	N42F	2006	Optare demonstrator, 2007
408	R84EMB	Scania L113 CRL	Wight Access Ultralow	N42F	1997	Chesterbus, 2007
409	R85EMB	Scania L113 CRL	Wight Access Ultralow	N42F	1997	Chesterbus, 2007
410	R86EMB	Scania L113 CRL	Wight Access Ultralow	N42F	1997	Chesterbus, 2007
411	AU08DKL	Optare Versa V1100	Optare	N38F	2008	
412	AU08DKN	Optare Versa V1100	Optare	N38F	2008	
413	YN07LFU	Scania OmniLink K230 UB	Scania	N45F	2007	Scania demonstrator, 2008
414	YN07EZB	Scania OmniLink K230 UB	Scania	N45F	2007	Scania demonstrator, 2008
415	R81EMB	Scania L113 CRL	Wight Access Ultralow	N42F	1997	Aintree Coachline, Bootle, 2008
416	R82EMB	Scania L113 CRL	Wight Access Ultralow	N42F	1997	Aintree Coachline, Bootle, 2008
417	AU58AUV	Scania OmniLink CK230 UB	Scania	N45F	2008	
502	AU57EZM	Scania OmniCity CN230 UD	Scania	N47/29F	2008	
503	AU57EZL	Scania OmniCity CN230 UD	Scania	N47/29F	2008	
504	AO57HCD	Scania OmniCity CN230 UD	Scania	N47/29F	2008	
505	AO57HCC	Scania OmniCity CN230 UD	Scania	N47/29F	2008	
506	P476SWC	Volvo Olympian	Alexander RH	B47/27D	1996	Dublin Bus, 2008
507	P478SWC	Volvo Olympian	Alexander RH	B47/27D	1996	Dublin Bus, 2008

Previous registrations:

831HKA	B146MJU		
CSK282	E993KJF	P476SWC	96D305
HIJ6931	B152MJU	P478SWC	96D306

Depots: Sandpit Lane, Beccles and Green Lane West, Rackheath.

The village of Acle lies between Norwich and Great Yarmouth, and is the location for this view of Scania OmniCity 413, YN07LFU which was initially used as a demonstrator for Scania. *Richard Godfrey*

AVON BUSES

LW Smith & G Lewis, 10 Brookway, Prenton, CH43 3DT

272	A17AVN	Volvo B6	Alexander Dash	B40F	1994	Stagecoach, 2005
273	L273LHH	Volvo B6	Alexander Dash	B40F	1994	Stagecoach, 2005
274w	A19AVN	Volvo B6	Alexander Dash	B40F	1994	Stagecoach, 2005
9	A10AVN	Volvo B6LE	Wright Crusader	N39F	1995	Happy Al's, Birkenhead, 2007
10	A9AVN	Volvo B6LE	Wright Crusader	N39F	1995	Happy Al's, Birkenhead, 2007
4	B4AVN	Dennis Lance SLF 11m	Berkhof 2000	N37F	1995	Menzies, Heathrow, 2003
5	B5AVN	Dennis Lance SLF 11m	Berkhof 2000	N37F	1995	Menzies, Heathrow, 2003
185	M185UAN	Dennis Lance SLF 11m	Berkhof 2000	N37F	1995	McKindless, Wishaw, 2004
191	M191UAN	Dennis Lance SLF 11m	Berkhof 2000	N41F	1995	McKindless, Wishaw, 2004
197	M197UAN	Dennis Lance SLF 11m	Berkhof 2000	N37F	1995	McKindless, Wishaw, 2005
966	M966SDP	Dennis Lance SLF 11m	Berkhof 2000	N37F	1995	McKindless, Wishaw, 2005
509	S509NFR	Dennis Dart SLF 10.1m	East Lancs Spryte	N30F	1999	Express Travel, Speke, 2003
511	S511KFL	Dennis Dart SLF 10.7m	Marshall Capital	N37F	1998	Bleasdale, Liverpool, 2003
512	S512KFL	Dennis Dart SLF 10.7m	Marshall Capital	N37F	1998	Bleasdale, Liverpool, 2003
154	S54NCW	Dennis Dart SLF 10.7m	East Lancs Spryte	N37F	1999	
840	T840CCK	Dennis Dart SLF 10.1m	East Lancs Spryte	N30F	1999	Express Travel, Speke, 2003
841	T841CCK	Dennis Dart SLF 10.1m	East Lancs Spryte	N30F	1999	Express Travel, Speke, 2003
548	T548HNH	Dennis Dart SLF 8.8m	Plaxton Pointer MPD	N28F	1999	Norbus, Kirby, 2005
821	X821XCK	Dennis Dart SLF 10.7m	East Lancs Spryte	N36F	2000	
822	X822XCK	Dennis Dart SLF 10.7m	East Lancs Spryte	N36F	2000	
213	Y213BGB	Dennis Dart SLF 10.7m	Plaxton Pointer 2	N37F	2001	Beattie, Renfrew, 2005

Avon operates on the Wirral peninsula, an area once dominated by Crosville. The Dart forms the majority of the fleet, and an Alexander Dennis example, 308, AE08DKV, shows off its MCV Evolution body as it heads for Woodside, one of the ferry connections to Liverpool. *Richard Godfrey*

English Bus Handbook: Notable Independents

Added to the fleet in 2007, AE07DZT is one of the 9.2 metre Darts with MCV Evolution bodywork. It is seen leaving Birkenhead for Prenton. *Richard Godfrey*

486	Y486TSU	Dennis Dart SLF 10.7m	Plaxton Pointer 2	N37F	2001	Beattie, Renfrew, 2005
155	SN55DVF	ADL Dart 10.7m	ADL Pointer	N37F	2005	
255	SN55DVG	ADL Dart 10.7m	ADL Pointer	N37F	2005	
355	SN55DVP	ADL Dart 10.7m	ADL Pointer	N37F	2005	
156	AE56LWJ	ADL Dart 10.8m	MCV Evolution	N40F	2006	
256	AE56LWK	ADL Dart 10.8m	MCV Evolution	N40F	2006	
356	AE56LWL	ADL Dart 10.8m	MCV Evolution	N40F	2006	
456	YN56NHA	Enterprise Plasma EB01	Plaxton Primo	N28F	2006	
107	AE07DZS	ADL Dart 9.2m	MCV Evolution	N32F	2007	
207	AE07DZT	ADL Dart 9.2m	MCV Evolution	N32F	2007	
307	AE07NYS	ADL Dart 10.8m	MCV Evolution	N40F	2007	
407	AE07NYT	ADL Dart 9.2m	MCV Evolution	N32F	2007	
507	AE07NYU	ADL Dart 9.2m	MCV Evolution	N32F	2007	
108	AE08DKO	ADL Dart 4 11.3m	MCV Evolution	N40F	2008	
208	AE08DKU	ADL Dart 4 11.3m	MCV Evolution	N40F	2008	
308	AE08DKV	ADL Dart 4 11.3m	MCV Evolution	N40F	2008	

Previous registrations:

A9AVN	N902NNR, N900ALS	B4AVN	M961SDP
A10AVN	N901NNR, N100ALS	B5AVN	M184UAN
A17AVN	L272LHH	B17AVN	-
A18AVN	-	B19AVN	-
A19AVN	L274LHH	M197UAN	M197UAN, WIL9217
A20AVN	-	M966SDP	M966SDP, WIL9216

BAKERBUS

Bakerbus - Bakers Coaches - Niddries

Guideissue Ltd, The Coach Travel Centre, Prospect Way, Biddulph, Stoke-on-Trent, ST8 7PL

No	Reg	Chassis	Body	Seating	Year	Notes
3	7092RU	Volvo B10M-62	Plaxton Première 320	C53F	1999	
4	3601RU	Volvo B10M-62	Plaxton Première 320	C57F	1999	Linkline, London, 2001
5	4614RU	Volvo B10M-61	Plaxton Paramount 3200 III	C53F	1987	Excelsior, Bournemouth, 1990
6	3275RU	Volvo B10M-62	Plaxton Première 350	C53F	1999	
7	1513RU	Volvo B10M-62	Plaxton Première 320	C57F	2001	Londoners, Nunhead, 2001
8	8150RU	Volvo B10M-62	Plaxton Première 320	C53F	2000	
9	1497RU	Volvo B10M-62	Plaxton Première 320	C53F	2000	
16	4085RU	Volvo B10M-62	Plaxton Première 350	C49FT	1998	Tellings-Golden Miller, 2001
17	3353RU	Volvo B10M-62	Caetano Enigma	C51FT	2001	
19	8439RU	Mercedes-Benz Vario O814	Plaxton Cheetah	C33F	2002	
23	R578GDS	Mercedes-Benz Vario O814	Plaxton Beaver 2	B31F	1997	Superior Travel, Greenock, 1999
29	T129XVT	Mercedes-Benz Vario O810	Plaxton Beaver 2	B31F	1999	
163	R948AMB	Mercedes-Benz Vario O810	Plaxton Beaver 2	B31F	1998	Green Triangle, Atherton, 2003
164	5946RU	Volvo B10B	Plaxton Verde	B51F	1995	Countryliner, Guildford, 2003
172	3093RU	Volvo B12B	Caetano Enigma	C49FT	2004	
173	7025RU	Volvo B12B	Caetano Enigma	C49FT	2004	
175	X282MSP	Mercedes-Benz Vario O814	Plaxton Beaver 2	B31F	2000	Midland, Auchterarder, 2004
180	V116ESL	Mercedes-Benz Vario O814	Plaxton Beaver 2	B27F	2000	Moffat & Williamson, Gauldry, '04
181	T49JBA	Mercedes-Benz Vario O814	Plaxton Beaver 2	B33F	1999	McLennan, Laxy Lochs, 2004
182	V117ESL	Mercedes-Benz Vario O814	Plaxton Beaver 2	B29F	2000	Moffat & Williamson, Gauldry, '04
183	V723GGE	Mercedes-Benz Vario O814	Plaxton Beaver 2	B33F	1999	McLennan, Laxy Lochs, 2004
185	YJ54CEY	VDL Bus SB120	Wrightbus Cadet 2	N39F	2004	
186	YJ54CFA	VDL Bus SB120	Wrightbus Cadet 2	N39F	2004	
188	3566RU	Scania K114IB4	Irizar Century 12.35	C49FT	2005	
189	8399RU	Scania K114IB4	Irizar InterCentury 12.32	C57F	2005	
192	3102RU	Mercedes-Benz Vario O815	Sitcar Beluga	C27F	2002	Dawson, Heywood, 2005
193	YJ55KZT	VDL Bus SB120	Wrightbus Cadet 2	N39F	2005	
194	9595RU	Scania L113CRL	Wright Axcess-ultralow	N47F	1995	White, Bridge of Walls, 2006
195	Y783WHH	Mercedes-Benz Vario O814	Plaxton Beaver 2	B27F	2001	Belle Vue, Heaton Chapel, 2006
197	MX54KXY	Optare Solo M920	Optare	N33F	2005	Mistral, Knutsford, 2006
198	MX54KXO	Optare Solo M920	Optare	N33F	2005	Mistral, Knutsford, 2006
199	MX05OTN	Optare Solo M920	Optare	N33F	2005	Mistral, Knutsford, 2006
200	MX05OTP	Optare Solo M920	Optare	N33F	2005	Mistral, Knutsford, 2006
201	3563RU	Scania K114 EB4 12.9m	VDL Berkhof Axial 50	C55FT	2007	
202	6280RU	Scania K340 EB4 12.9m	VDL Berkhof Axial 50	C55FT	2007	

Bakerbus operates services for Staffordshire and Cheshire authorities, and here VDL Bus 185, YJ54CEY, is seen leaving Hanley for Stafford rail station.
Alan Blagburn

Congleton's town bear was a major attraction at the town's Wakes, but in 1662, the bear died. That is the background to the name carried by the town's bus service. Plaxton Primo 205, DX07WEF, is seen on service 93, one of the five routes in the network. *Alan Blagburn*

205	DX07WEF	Enterprise Plasma EB03	Plaxton Primo	N28F	2007	
206	DX07WEH	Enterprise Plasma EB03	Plaxton Primo	N28F	2007	
207	DX07WEJ	Enterprise Plasma EB03	Plaxton Primo	N28F	2007	
208	ESK697	DAF MB230	Plaxton Paramount 3500 III	C57F	1988	Niddrie's, Middlewich, 2007
209	ESK841	DAF SB2300	Van Hool Alizée	C51FT	1988	Niddrie's, Middlewich, 2007
210	MX07BAV	Enterprise Plasma EB03	Plaxton Primo	N28F	2007	Princess, West End, 2007
211	MX07BBE	Enterprise Plasma EB01	Plaxton Primo	N28F	2007	Princess, West End, 2007
215	3471RU	Volvo B6LE	Alexander ALX200	N35F	1997	Transdev Blazefield, 2008
216	5621RU	Volvo B6LE	Alexander ALX200	N35F	1997	Transdev Blazefield, 2008
217	5777RU	Volvo B6LE	Alexander ALX200	N35F	1997	Transdev Blazefield, 2008
218	9423RU	Volvo B6LE	Alexander ALX200	N35F	1997	Transdev Blazefield, 2008
219	BIG5013	Volvo B10M	Van Hool Alizée	C57F	1992	Charlton, Services, 2008
220	R107GNW	Mercedes-Benz Vario 0810	Plaxton Beaver 2	B27F	1998	TM Travel, Sheffield, 2008
222	KX58GVC	ADL Dart 4	ADL Enviro 200	N29F	2009	
223	KX58GSZ	ADL Dart 4	ADL Enviro 200	N29F	2009	

Previous registrations:

1497RU	-	5777RU	P344JND
1513RU	T632JWB	6280RU	From new
3093RU	From new	6663RU	-
3102RU	C8DWA, SL02OLB	7025RU	From new
3275RU	From new	7092RU	From new
3353RU	From new	8150RU	From new
3471RU	P338JND	8399RU	From new
3563RU	From new	8439RU	From new
3566RU	From new	8830RU	-
3601RU	T633JWB	9423RU	P346JND
4085RU	T30TGM	9530RU	-
4614RU	D255HFX, 4614RU, D780FVT	9595RU	N162KPS
5621RU	P342JND	BIG5013	J223NNC, 6709PO
		ESK697	F219RJX
5946RU	M875NWK	ESK841	E356EVH

web: www.bakerbus.co.uk; www.bakerscoaches.com

BEESTONS

Beestons - Constables

Beeston's (Hadleigh) Ltd; Constable Coaches Ltd, 21 Long Bessels, Hadleigh, Ipswich, IP7 5DB

B	FIL4166	Leyland Tiger TRCTL11/3R	Van Hool Alizée	C57F	1984	Travellers, Hounslow, 1988
B	FIL8615	Leyland Tiger TRCTL11/3ARZ	Van Hool Alizée H	C57F	1987	Travellers, Hounslow, 1989
B	SJI8098	Leyland Tiger TRCTL11/3RZ	Van Hool Alizée H	C57F	1987	Travellers, Hounslow, 1989
B	NIW6517	Volvo B10M-61	Van Hool Alizée H	C49F	1987	Smith, Groton, 2008
B	SJI9321	Volvo B10M-61	Van Hool Alizée H	C57F	1988	Shearings, 1993
B	SJI9320	Volvo B10M-61	Van Hool Alizée H	C57F	1988	Shearings, 1993
B	SJI9319	Volvo B10M-61	Van Hool Alizée H	C57F	1988	Shearings, 1993
B	131ASV	Volvo B10M-60	Van Hool Alizée H	C57F	1990	First Eastern Counties, 1999
C	L212TWM	Volvo Olympian	Northern Counties Palatine	B47/29F	1994	Metroline, Harrow, 2007
B	532FN	Volvo Olympian	Alexander RH	B57/45F	1995	Citybus, Hong Kong, 2006
B	LSU413	Volvo Olympian	Alexander RH	B51/31F	1995	East Yorkshire, 2007
B	MAZ6907	Ford Transit	Ford	M14	1996	Silver Shield, Layham, 2000
C	R279RAU	Mercedes-Benz Vario O810	Plaxton Beaver 2	B31F	1997	Paul S Winson, Loughborough, '08
B	P366UUG	Volvo B10B	Wright Endurance	BC49F	1997	Stagecoach, 2009
B	229LRB	Volvo B10M-62	Van Hool T9 Alizée H	C53F	1998	Irving, Carlisle, 2004
B	221GRA	Volvo B10M-62	Van Hool T9 Alizée H	C53F	1998	Andrew's, Whiteness, 2004
B	524FN	Volvo B12(T)	Van Hool Astrobel	C57/14FT	1998	Berry's, Taunton, 2007
C	EIG1357	Mercedes-Benz Vario O814	Plaxton Beaver 2	B31F	1998	
C	EIG1358	Mercedes-Benz Vario O814	Plaxton Beaver 2	B33F	1999	PC Coaches, Lincoln, 2006
B	217OPO	Mercedes-Benz Vito 112cdi	Mercedes-Benz	M7	1999	private owner, 2006
C	AY51EFG	Mercedes-Benz Vario O814	Plaxton Beaver 2	B31F	2001	
C	KV51KZD	Mercedes-Benz Vario O814	Plaxton Beaver 2	B33F	2001	D&G, Adderley Green, 2006
C	KV51KZF	Mercedes-Benz Vario O814	Plaxton Beaver 2	B33F	2001	D&G, Adderley Green, 2006
C	AF51JZU	LDV Convoy	Olympus	C16F	2001	Grant Palmer, Dunstable, 2008
B	FHJ565	Scania K114EB4	Van Hool T9 Alizée	C53F	2002	Leon's, Stafford, 2006
B	219GRA	Scania K114IB4	Van Hool T9 Alizée	C53F	2002	Leon's, Stafford, 2006
C	YS02XDO	Scania L94UB	Wrightbus Solar	N43F	2002	
C	AV52GHZ	Mercedes-Benz Vario O814	Plaxton Beaver 2	B31F	2002	

Beeston's OmniDekka YN54OBX is one of five currently operated. It is seen at rest between journeys.
Alan Blagburn

Mayday 2008 in Colchester and Beestons' YN56EZX is seen heading for Dedham. This vehicle has since left the fleet, being replaced by another similar model from Scania. *Richard Godfrey*

C	YN03DDA	Scania OmniDekka N94UD	East Lancs	N47/31F	2003	
B	222GRA	Scania K114EB4	Van Hool T9 Alizée	C53F	2004	
C	YN04ANX	Scania L94UB	Wrightbus Solar	N43F	2004	
C	YN04OBX	Scania OmniDekka N94UD	East Lancs	N47/33F	2004	Scania demonstrator, 2006
C	YN04OCJ	Scania OmniDekka N94UD	East Lancs	N47/33F	2004	Scania demonstrator, 2006
C	YN54OBB	Scania OmniDekka N94UD	East Lancs	N45/33F	2005	
C	YN06CJE	Scania OmniCity CN94UB	Scania	N42F	2006	Menzies, Heathrow, 2006
C	YJ08EBZ	Van Hool Astron T916	Van Hool	C53FT	2008	
C	YN08HXZ	Scania OmniLink CN230 UB	Wrightbus Solar	N43F	2008	
C	YN55NDY	Scania OmniDekka N94UD	East Lancs	N47/33F	2005	Thames Travel, Wallingford, 2008
C	YN07EXP	Scania OmniCity N94UB	Scania	N42F	2007	Scania demonstrator, 2008
C		Volvo B9TL	Optare Olympus	N(100)F	On order	

Special event vehicle:

C	ATV11B	Bedford SB5	Plaxton Embassy	C41F	1964	Mulleys, Ixworth, 1981

Previous registrations:

131ASV	G260RNS	FHJ565	BT02LCT
217OPO	T128DNW	FIL4166	A145RMJ
219GRA	CT02LCT	FIL8615	D283HMT
221GRA	LSK509, R427EOS, HIL8280	LSU413	N597BRH
222GRA	YN04ANV	NIW6517	D357MVR
229LRB	S11BUS, S914RHH	SJI8098	D284HMT, FIL8617
524FN	R380XYD	SJI9319	E621UNE, WSV528, E489CDB, PJI6391
532FN	GE5021(HK), M129BNO	SJI9320	E620UNE, SPR124, E683CDB, PJI6392, RJI7972
EIG1357	S617URT	SJI9321	E619UNE, XTW359, E684CDB, PJI6393, RJI7973
EIG1358	T7PCC, T108BFW		

Web: www.beestons.co.uk
Depot: Ipswich Road, Hadleigh

BENNETTS

P, R A, & D Bennett and P A Lane, Eastern Avenue, Gloucester, GL4 7BU

L950MSC	Volvo Olympian	Alexander RH	B51/30D	1994	Lothian Buses, 2007
L951MSC	Volvo Olympian	Alexander RH	B51/30D	1994	Lothian Buses, 2007
L952MSC	Volvo Olympian	Alexander RH	B51/30D	1994	Lothian Buses, 2007
L958MSC	Volvo Olympian	Alexander RH	B51/30D	1994	Lothian Buses, 2007
L519EHD	DAF SB2700	Van Hool Alizée	C53F	1994	Enfield Transport, Co. Meath, 1998
L463RDN	DAF SB2700	Van Hool Alizée	C51F	1994	Deros, Killarney, 1995
M225VSX	Volvo Olympian	Alexander RH	B51/30D	1995	Lothian Buses, 2008
R179GNW	DAF SB3000	Van Hool T9 Alizée	C53F	1998	North Kent Express, 2003
T419PDG	DAF SB220	Ikarus CitiBus	N43F	1999	
Y200BCC	Neoplan Euroliner N316SHD	Neoplan	C49FT	2001	
Y300BCC	Neoplan Euroliner N316SHD	Neoplan	C49FT	2001	
Y400BCC	Neoplan Euroliner N316SHD	Neoplan	C49FT	2001	
YR02UMU	Neoplan Euroliner N316SHD	Neoplan	C49FT	2002	
VO03DZC	MAN 14.220	Ikarus Polaris	B42F	2003	
BU53AYB	Mercedes-Benz Touro OC500	Mercedes-Benz	C49FT	2003	Eurorider, Offerton, 2005
YK04KWA	Optare Solo M1020	Optare	N37F	2004	Operated for Gloucestershire CC
YK04KWB	Optare Solo M1020	Optare	N37F	2004	Operated for Gloucestershire CC
YK04KWC	Optare Solo M1020	Optare	N37F	2004	Operated for Gloucestershire CC
YK04KWD	Optare Solo M1020	Optare	N37F	2004	Operated for Gloucestershire CC
BX54EBD	Mercedes-Benz Citaro O530	Mercedes-Benz	N42F	2004	
BX54EBF	Mercedes-Benz Citaro O530	Mercedes-Benz	N42F	2004	
BX54EBG	Mercedes-Benz Citaro O530	Mercedes-Benz	N42F	2004	
BX54ECV	Mercedes-Benz Citaro O530	Mercedes-Benz	N42F	2004	Evobus demonstrator, 2005
GL05BUS	Mercedes-Benz Touro OC500	Mercedes-Benz	C49FT	2005	
BU06CVA	Mercedes-Benz Touro OC500	Mercedes-Benz	C49FT	2006	
BX56VTP	Mercedes-Benz Citaro O530	Mercedes-Benz	N40F	2006	
BT08SZR	Mercedes-Benz Tourismo	Mercedes-Benz	C49FT	2008	
	Mercedes-Benz Tourismo	Mercedes-Benz	C32F	2009	
	Mercedes-Benz Tourismo	Mercedes-Benz	C49FT	2009	

Previous registrations:

L463RDN	94KY1609		
		L519EHD	L519EHD, 94MH3061

Oldham bus station is this location for this view of Bluebird 71, FF56BLU, an Alexander Dennis Enviro 200. Seating twenty-nine, this version replaced the Mini Pointer Dart in the ADL catalogue. *Richard Godfrey*

BLUEBIRD

Bluebird Coaches Ltd, Alexander House, Greengate, Middleton, Rochdale, M24 1RU

1	N126XEG	MAN 11.220	Marshall	B37F	1996	Far East Travel, Ipswich, 2006
3	MX54BLU	Dennis Dart SLF 10.7m	Caetano Nimbus	N37F	2004	
8	Y8BLU	Dennis Dart SLF 10.7m	East Lancs Spryte	N37F	2001	
9	MD02BLU	Dennis Dart SLF 10.7m	East Lancs Spryte	N37F	2002	
10	MV54BLU	ADL Dart 10.7m	ADL Pointer	N37F	2004	
11	MW54BLU	ADL Dart 10.7m	ADL Pointer	N37F	2004	
12	AJ54AMJ	ADL Dart 10.7m	ADL Pointer	N37F	2004	
15	MA02BLU	Dennis Dart SLF 10.7m	Caetano Nimbus	N37F	2002	
16	MB02BLU	Dennis Dart SLF 10.7m	Caetano Nimbus	N37F	2002	
17	DD56BLU	ADL Dart 10.7m	ADL Pointer	N37F	2006	
18	196BLU	Dennis Dart SLF 10.7m	Alexander ALX200	N42F	2001	Ambassador, Gt Yarmouth, 2006
19	SS55BLU	ADL Dart 8.8m	ADL Mini Pointer	N29F	2006	
20	OO06BLU	ADL Dart 8.8m	ADL Mini Pointer	N29F	2006	
23	ME52BLU	Dennis Dart SLF 8.8m	Plaxton Pointer MPD	N29F	2002	
24	MF03BLU	Dennis Dart SLF 8.8m	Plaxton Pointer MPD	N29F	2003	
25	MG53BLU	Dennis Dart SLF 8.8m	Plaxton Pointer MPD	N29F	2003	
26	MH53BLU	Dennis Dart SLF 8.8m	Plaxton Pointer MPD	N29F	2003	
27	MP04BLU	ADL Dart 8.8m	ADL Mini Pointer	N29F	2004	
28	AA05BLU	ADL Dart 8.8m	ADL Mini Pointer	N29F	2005	
29	BB05BLU	ADL Dart 8.8m	ADL Mini Pointer	N29F	2005	
30	CC05BLU	ADL Dart 8.8m	ADL Mini Pointer	N29F	2005	
31	VX05TZO	ADL Dart 8.8m	ADL Mini Pointer	N29F	2005	
41	270BLU	MAN 11.190	Optare Vecta	B40F	1994	Trent, 2005

A surprise in 2008 was the repainting of one of Bluebird's Mini Pointers into the old North Western colours. 25, MG53BLU, is seen operating the Middleton to Oldham route 159. *John Young*

During 2007 a pair of VDL Bus SB120s with Plaxton Centro bodywork joined the Bluebird fleet. Seen arriving in Manchester's Piccadilly is 81, KK07BLU. *John Young*

51	Y251KNB	Dennis Dart SLF 8.9m	Alexander ALX200	N29F	2001	Compass, Worthing, 2006
52	Y252KNB	Dennis Dart SLF 8.9m	Alexander ALX200	N29F	2001	Compass, Worthing, 2006
53	Y253KNB	Dennis Dart SLF 8.9m	Alexander ALX200	N29F	2001	Compass, Worthing, 2006
54	Y254KNB	Dennis Dart SLF 8.9m	Alexander ALX200	N29F	2001	Compass, Worthing, 2006
61	360BLU	ADL E300	ADL Enviro300	N44F	2004	Go North East, 2006
62	1BLU	ADL E300	ADL Enviro300	N44F	2007	
63	MX57OEL	ADL E300	ADL Enviro300	N44F	2007	
71	EE56BLU	ADL Dart 4 8.9m	ADL Enviro200	N29F	2007	
72	FF56BLU	ADL Dart 4 8.9m	ADL Enviro200	N29F	2007	
73	MX56WWA	ADL Dart 4 8.9m	ADL Enviro200	N29F	2007	
74	MX56WWB	ADL Dart 4 8.9m	ADL Enviro200	N29F	2007	
75	MX56WWC	ADL Dart 4 8.9m	ADL Enviro200	N29F	2007	
76	MX58SGU	ADL Dart 4 8.9m	ADL Enviro200	N29F	2008	
77	MX58SGV	ADL Dart 4 8.9m	ADL Enviro200	N29F	2008	
81	KK07BLU	VDL Bus SB120	Plaxton Centro	N40F	2007	
82	LL07BLU	VDL Bus SB120	Plaxton Centro	N40F	2007	
91	PP57BLU	ADL Dart 4 10.7m	MCV Evolution	N40F	2007	
92	RR57BLU	ADL Dart 4 10.7m	MCV Evolution	N37F	2007	
93	SS57BLU	ADL Dart 4 10.7m	MCV Evolution	N37F	2007	
94	TT57BLU	ADL Dart 4 10.7m	MCV Evolution	N40F	2007	
95	JJ58BLU	ADL Dart 4 10.7m	MCV Evolution	N40F	2008	
96	WW58BLU	ADL Dart 4 10.7m	MCV Evolution	N40F	2008	

Previous registrations:

1BLU	MX57KXR		
196BLU	VX51RCO	360BLU	SN04EGF
270BLU	M805PRA	N126XEG	N126XEG, 1BLU

Web: www.bluebirdbus.co.uk

English Bus Handbook: Notable Independents

BODMANS

RJ CJ DW Bodman & KG Heath, The Old Forge Garage, 88 High Street,
Worton, Devizes, SN10 5RU

JIL3959	DAF SB3000	Van Hool Alizee	C53F	1993	Hatts, Foxham, 2008
L780GMJ	Dennis Javelin 12m	Plaxton Première 320	C57F	1993	Scancoaches, Battersea, 1997
M112BMR	Dennis Dart 9.8m	Plaxton Pointer	B40F	1994	Thamesdown, 2008
M115BMR	Dennis Dart 9.8m	Plaxton Pointer	B40F	1994	Thamesdown, 2008
M203EGF	Dennis Dart 9m	Plaxton Pointer	B34F	1995	London General, 2004
M112UWY	Volvo B10M-62	Plaxton Première 320	C53F	1995	Taylor's, Sutton Scotney, 2005
M153XHW	Dennis Javelin 12m	Wadham Stringer Vanguard III	S70F	1995	Hatts, Foxham, 2008
M241XWS	Dennis Javelin 12m	Wadham Stringer Vanguard III	S70F	1995	MoD (16RN62), 2003
M774XHW	Dennis Javelin 12m	Wadham Stringer Vanguard III	S70F	1995	MoD (16RN61), 2003
N890SBB	Mercedes-Benz 811D	Plaxton Beaver	B30F	1996	Docklands Minibus, 2000
N653THO	Volvo B10M-62	Plaxton Première 320	C53F	1996	Jim Bevan, Lydney, 2005
N131XEG	MAN 11.220	Marshall	B38F	1996	Metroline, Harrow, 2005
P292MLD	Dennis Dart 10.1m	Plaxton Pointer	B39F	1996	Metroline, Harrow, 2005
P302MLD	Dennis Dart 10.1m	Plaxton Pointer	B38F	1996	Metroline, Harrow, 2005
P304MLD	Dennis Dart 10.1m	Plaxton Pointer	B35F	1996	Metroline, Harrow, 2005
P306MLD	Dennis Dart 10.1m	Plaxton Pointer	B39F	1996	Metroline, Harrow, 2005
XUD367	Dennis Javelin GX 12m	Berkhof Axial 50	C53FT	1997	Voel, Dyserth, 2003
R708NJH	Dennis Javelin 12m	Berkhof Axial 50	C49FT	1997	Voel, Dyserth, 2003
P214RUM	DAF SB3000	Ikarus Blue Danube	S67F	1997	Alfa, Leyland, 2005
R576NFX	Volvo B10M-62	Plaxton Première 320	C57F	1997	Brook, Chadderton, 2002
R54AWO	Volvo B10M-62	Plaxton Première 320	C57F	1997	Wickson, Walsall Wood, 2003
R436FWT	Volvo B10M-62	Plaxton Première 350	C53F	1998	Wallace Arnold, Leeds, 2002
R50TGM	Volvo B10M-62	Plaxton Première 350	C57F	1998	Tellings-Golden Miller, 2007

Bodmans' Dart P292MLD, seen awaiting time in Warminster from where the operator runs several services to nearby towns. *Robert Edworthy*

Devizes is the location for this view of Bodmans' Dart P304MLD, seen operating route X72. The vehicle was new to Metroline in 1996 and shows a variant of Bodmans' colours. *Richard Godfrey*

X291ABU	Dennis Dart SLF 8.8m	Alexander ALX200	N29F	2000	Southdown PSV, Copthorne, 2007
X292ABU	Dennis Dart SLF 8.8m	Alexander ALX200	N29F	2000	Quantock MS, 2007
X46CNY	Volvo B10M-62	Plaxton Paragon	C49FT	2001	Beeline, Warminster, 2008
MM51XVB	Dennis Dart SLF 8.8m	Plaxton Pointer MPD	N29F	2002	South Lancs, Atherton, 2007
MM51YCB	Dennis Dart SLF 8.8m	Plaxton Pointer MPD	N29F	2002	South Lancs, Atherton, 2007
YR52MBU	Volvo B7R	Plaxton Prima	C57F	2002	Glenvic, Stanton Wick, 2006
YK53HLV	Mercedes-Benz Vario O814	TransBus Beaver	BC33F	2004	Hutchinson, Easingwold, 2005
SF05FNW	Mercedes-Benz Vario O814	Plaxton Beaver 2	B32F	2005	Weavaway, Newbury, 2007
SN56AXM	ADL E300	ADL Enviro 300	S60F	2007	ADL demonstrator, 2008
SF07DLO	Mercedes-Benz Vario O814	Plaxton Beaver 2	B32F	2007	Weavaway, Newbury, 2007
SK07DYA	ADL E300	ADL Enviro 300	N49F	2007	ADL demonstrator, 2007
RX57GOE	ADL E300	ADL Enviro 300	S60F	2007	ADL demonstrator, 2008
WA57JZW	Volvo B12B	Van Hool Alicron	C53FT	2008	
SN08AAF	ADL E300	ADL Enviro 300	N49F	2008	
WX08RZJ	ADL E300	ADL Enviro 300	S56F	2008	
AJ58WAA	MAN 14.240	MCV Evolution	N43F	2008	

Previous registrations:

JIL3959	K104TCP	R576NFX	R576NFX, A4XCL
MM51YCB	M88SLT	SF05FNW	SN05FNW, OO02TEN
MM51XVB	M99SLT	XUD367	P883FMO, 1760VC
N653THO	XEL254	YR52MBU	YR52MBU, H10BOD

BRYLAINE

Brylaine Travel Ltd, 291 London Road, Boston, PE21 7DD

XAZ1314	MCW Metrobus DR115/5	MCW	B63/44F	1988	New World FB, Hong Kong, 2001
XAZ1408	MCW Metrobus DR115/5	MCW	B63/44F	1988	New World FB, Hong Kong, 2001
XAZ1312	Leyland Olympian ONCL10/1RZ	Alexander RH	B45/30F	1989	Bullock, Cheadle, 2004
H908PTW	Leyland Olympian ON2R50C13Z4	Alexander RH	B47/33F	1990	Chambers, Bures, 2006
J717CEV	Volvo Olympian	Alexander RH	B47/31F	1991	Dublin Bus, 2006
K132TCP	DAF SB220	Ikarus CitiBus	B48F	1993	Arriva NW & Wales, 2008
XAZ1310	Dennis Dart 9.8m	Marshall C37	B40F	1995	Halton, 2002
XAZ1316	Dennis Dart 9.8m	Marshall C37	B40F	1995	Halton, 2002
N720FKK	Volvo Olympian	Alexander RH	B47/31F	1995	Dublin Bus, 2008
N721FKK	Volvo Olympian	Alexander RH	B47/31F	1995	Dublin Bus, 2008
N722FKK	Volvo Olympian	Alexander RH	B47/31F	1995	Dublin Bus, 2008
N723FKK	Volvo Olympian	Alexander RH	B47/31F	1995	Dublin Bus, 2008
N724FKK	Volvo Olympian	Alexander RH	B47/31F	1995	Dublin Bus, 2008
N725FKK	Volvo Olympian	Alexander RH	B47/31F	1995	Dublin Bus, 2008
N607JNO	Volvo Olympian	Alexander RH	B47/31F	1996	Dublin Bus, 2008
P201RUM	DAF SB220	Ikarus CitiBus	B40F	1997	BCP, Crawley, 2004
P205RUM	DAF SB220	Ikarus CitiBus	B40F	1997	BCP, Crawley, 2004
P128RWR	DAF SB220	Ikarus CitiBus	B40F	1997	Central Parking, Luton, 2003
P129RWR	DAF SB220	Ikarus CitiBus	B40F	1997	Central Parking, Luton, 2003
R686MEW	Dennis Dart SLF	Marshall Capital	N35F	1998	
R687MEW	Dennis Dart SLF	Marshall Capital	N35F	1998	
R23GNW	DAF SB220	Ikarus CitiBus	B40F	1998	Hallmark, Luton, 2003
S401JUA	DAF SB220	Ikarus CitiBus	B40F	1998	BlueBus, Horwich, 2003
W967JNF	Mercedes-Benz Vario O814	Plaxton Beaver 2	B31F	2000	
W829UMJ	Ford Transit	Ford	M8	2000	private owner, 2005

Many British independent fleets have taken advantage of the large numbers of Olympians being displaced by Dublin Bus and Lothian Buses. Now converted to single-door, N724FKK is one from Dublin. *Richard Godfrey*

Many of the newest buses with Brylaine are Optare Solos. Spalding is the location for this view of MX08DHF a slimline version of the Solo seen here with the *IntoTown* livery. *Richard Godfrey*

FY02VHF	Optare Solo M850	Optare	N29F	2002	*Operated for Lincolnshire CC*
FY02VHG	Optare Solo M850	Optare	N29F	2002	*Operated for Lincolnshire CC*
VU02TSZ	Optare Solo M920	Optare	N33F	2002	
YJ05JXD	Optare Solo M1020	Optare	N37F	2005	
YJ05JXE	Optare Solo M1020	Optare	N37F	2005	
YJ05JXF	Optare Solo M1020	Optare	N37F	2005	
YJ05JXG	Optare Solo M1020	Optare	N37F	2005	
YJ05JXH	Optare Solo M1020	Optare	N37F	2005	
YJ05JXK	Optare Solo M1020	Optare	N37F	2005	
YJ05JXL	Optare Solo M1020	Optare	N37F	2005	
BU05EEO	BMC Falcon 1100FE	BMC	S44F	2005	
BU05EEP	BMC Falcon 1100FE	BMC	S44F	2005	
BU05EER	BMC Falcon 1100FE	BMC	S44F	2005	
YJ57BBE	VDL Bus DB250	East Lancs Olympus	N51/29F	2007	
YJ57BNB	VDL Bus DB250	East Lancs Olympus	N51/29F	2007	
MX08DHF	Optare Solo M880 SL	Optare	N29F	2008	
YJ08PHV	Optare Solo M880 SL	Optare	N29F	2008	
YJ08PHX	Optare Solo M880 SL	Optare	N29F	2008	
YJ08PHY	Optare Solo M880 SL	Optare	N29F	2008	

Previous registrations:

H908PTW	90D1044	N725FKK	95D244
J717CEV	91D1122	XAZ1310	M579WLV
N607JNO	96D289	XAZ1312	F236YTJ
N720FKK	95D238	XAZ1314	F165UJN
N721FKK	95D239	XAZ1316	M582WLV
N722FKK	95D240	XAZ1320	E223WBG
N723FKK	95D241	XAZ1408	F166UJN
N724FKK	95D242		

Depots: London Road, Boston; Old Boston Lane, Coningsby and Hassall Road, Skegness.

CAROUSEL

Carousel Buses Ltd, 134 Desborough Road, High Wycombe, HP11 2PU

	WLT896	AEC Routemaster R2RH	Park Royal	B40/32R	1965	Arriva London, 2005
M336	EYE336V	MCW Metrobus DR101/12	MCW	B43/28F	1980	Arriva Southern Counties, 2002
M524	GYE524W	MCW Metrobus DR101/14	MCW	B43/28D	1981	Metroline, 2002
M703	KYV703X	MCW Metrobus DR101/14	MCW	B43/30F	1981	Aventa, Crawley, 2002
M737	KYV737X	MCW Metrobus DR101/14	MCW	B43/28D	1982	Abbey Cars, High Wycombe, 2002
M758	KYV758X	MCW Metrobus DR101/14	MCW	B43/28D	1982	Abbey Cars, High Wycombe, 2002
M1345	C345BUV	MCW Metrobus DR101/17	MCW	B43/28D	1985	London United, 2004
M1356	C356BUV	MCW Metrobus DR101/17	MCW	B43/28D	1985	London United, 2004
M1386	C386BUV	MCW Metrobus DR101/17	MCW	B43/28D	1985	Abbey Cars, High Wycombe, 2002
M1432	C432BUV	MCW Metrobus DR101/17	MCW	B43/28D	1985	Abbey Cars, High Wycombe, 2002
L510	F510NJE	Leyland Olympian ONLXB/1R	Northern Counties	B45/30F	1989	Stagecoach, 2007
L511	F511NJE	Leyland Olympian ONLXB/1R	Northern Counties	B45/30F	1989	Stagecoach, 2007
L530	G530VBB	Leyland Olympian ON2R50C13Z4	Northern Counties	B47/27D	1990	Arriva London, 2006
L534	G534VBB	Leyland Olympian ON2R50C13Z4	Northern Counties	B47/27D	1990	Arriva London, 2006
L554	H554GKX	Leyland Olympian ON2R50C13Z4	Leyland	B47/31F	1991	Armchair, 2003
L556	H556GKX	Leyland Olympian ON2R50C13Z4	Leyland	B47/31F	1991	Armchair, 2003
L563	H563GKX	Leyland Olympian ON2R50C13Z4	Leyland	B47/31F	1991	Armchair, 2002
L564	H564GKX	Leyland Olympian ON2R50C13Z4	Leyland	B47/31F	1991	Armchair, 2003
V431	A431VNY	Volvo-Ailsa B55 Mk III	Northern Counties	B39/33F	1984	Cardiff Bus, 2007
V451	A971YSX	Volvo-Ailsa B55 Mk III	Alexander RV	B44/37F	1984	Cardiff Bus, 2007
P7	RX07BNF	Enterprise Plasma EB01	Plaxton Primo	N28F	2007	
XL202	P202BNR	Optare Excel L1150	Optare	N42F	1996	Timetrak, Shepperton, 2006
DPL423	P423MLE	Dennis Dart 10.1m	Plaxton Pointer 2	N37F	1997	Airlinks, Heathrow, 2006
DPL479	P479MLE	Dennis Dart 9.5m	Plaxton Pointer 2	N29F	1997	Airlinks, Heathrow, 2006
DPL480	P480MLE	Dennis Dart 9.5m	Plaxton Pointer 2	N29F	1997	Airlinks, Heathrow, 2006

Carousel's purple route links High Wycombe with Flackwell Heath and DMS6, R706MEW is seen here returning to the town centre. *Richard Godfrey*

Carousel's DAF984, R984FNW carries dedicated livery for route 336 and is seen crossing Chorleywood Common heading for Watford. New as a dual-doored vehicle, the conversion to single door is hardly noticeable.
Mark Lyons

DMS6	R706MEW	Dennis Dart SLF	Marshall Capital	N27F	1998	Metroline, Harrow, 2007
DMS8	R708MEW	Dennis Dart SLF	Marshall Capital	N27F	1998	Metroline, Harrow, 2007
DAF976	R976FNW	DAF SB220	Plaxton Prestige	B30D	1998	NCP Birmingham, 2005
DAF981	R981FNW	DAF SB220	Plaxton Prestige	BC40F	1998	Heathrow Express, 2005
DAF984	R984FNW	DAF SB220	Plaxton Prestige	BC40F	1998	Heathrow Express, 2005
DAF986	R986FNW	DAF SB220	Plaxton Prestige	BC40F	1998	Heathrow Express, 2005
DAF988	R988FNW	DAF SB220	Plaxton Prestige	DC40F	1998	NCP Birmingham, 2005
DMS13	S513KFL	Dennis Dart SLF	Marshall Capital	N27F	1998	Metroline, Harrow, 2008
DMS14	S514KFL	Dennis Dart SLF	Marshall Capital	N27F	1998	Metroline, Harrow, 2008
DMS23	S523KFL	Dennis Dart SLF	Marshall Capital	N27F	1998	Metroline, Harrow, 2008
MB51	CB51BUS	Mercedes-Benz Citaro O530	Mercedes-Benz	N42F	2003	
MB52	CB52BUS	Mercedes-Benz Citaro O530	Mercedes-Benz	N42F	2003	
MB53	CB53BUS	Mercedes-Benz Citaro O530	Mercedes-Benz	N42F	2003	
DAF54	CB54BUS	VDL Bus SB120	Wrightbus Cadet 2	N30F	2005	
AL1	C1WYC	Irisbus AgoraLine	Irisbus	B41F	2006	
AL2	C2WYC	Irisbus AgoraLine	Irisbus	B41F	2006	
AL3	C3WYC	Irisbus AgoraLine	Irisbus	B41F	2006	

Previous registration:
C432BUV	C432BUV, WLT432	CSL498	JXN9

Depot: Halifax Road, Cressex Business Park, High Wycombe and Abbey Business Centre, Westbourne Street, High Wycombe.

CEDAR COACHES

Cedar Coaches Ltd, Arkwright Road, Bedford, MK42 0LE

2	L979MSC	Volvo Olympian	Alexander RH	B51/30D	1994	Lothian Buses, 2007
3	L981MSC	Volvo Olympian	Alexander RH	B51/30D	1994	Lothian Buses, 2007
4	A711YFS	Leyland Olympian ONTL11/2R	Eastern Coach Works	B51/32D	1983	Lothian Buses, 2000
6	H805AHA	Dennis Dominator DDA1031	East Lancs	B47/29F	1990	Arriva NW & Wales, 2006
7	E889CDS	Leyland Lion LDTL11/2R	Alexander RH	BC49/37F	1987	Holmeswood Coaches, 2003
8	F44YHB	Scania N113 DRB	Alexander RH	B47/33F	1988	Travelspeed, Burnley, 2008
9	J702HMY	Leyland Olympian ON3R49C18Z4	Alexander RH	BC53/41F	1991	Stagecoach, 2008
10	JSK492	Leyland Olympian ON3R49C18Z4	Alexander RH	BC53/41F	1987	Stagecoach, 2008
11	G641BPH	Volvo Citybus B10M-50	Northern Counties	B45/31F	1989	Sun Fun, Earith, 2006
12	F440AKB	Leyland Olympian ONCL10/1RZ	Northern Counties Palatine	B47/33F	1988	Liverpool Motor Services, 2004
13	G372RTO	Scania N113 DRB	Alexander RH	B47/33F	1990	Dunn-Line, Nottingham, 2003
14	P493FRR	Volvo Olympian	East Lancs	B49/35D	1997	City of Nottingham, 2008
15	S333HEB	Scania N113 DRB	East Lancs Cityzen	B46/31F	1998	Beeston's, Hadleigh, 2006
16	H811WKH	Scania N113 DRB	East Lancs	B47/37F	1991	Black Prince, Morley, 2003
17	G376RTO	Scania N113 DRB	Alexander RH	B47/33F	1991	City of Nottingham, 2004
18	H212LOM	Scania N113 DRB	Alexander RH	B45/31F	1990	Ausden Clark, Leicester, 2006
19	H206LOM	Scania N113 DRB	Alexander RH	B45/31F	1990	Ausden Clark, Leicester, 2006
20	H221LOM	Scania N113 DRB	Alexander RH	B45/31F	1990	Ausden Clark, Leicester, 2006
21	J697CEV	Leyland Olympian ON3R56C18Z4	Alexander RH	B57/45F	1992	Citybus, Hong Kong, 2006

Cedar Coaches uses this former Hong Kong Citybus Olympian, J697CEV, which was specially purchased to meet a high capacity demand on school services. When new, and photographed by your editor while in Hong Kong, it was operating commuter services for China Light & Power in a white and blue livery. *Ken Lansdowne*

When photographed, shortly after delivery, W31COM was carrying its initial mark PN08CPO. This modern coach is a MAN with Marcopolo Viaggio 330 bodywork. Marcopolo being a Brazilian coachbuilder which also has assembly facilities in Portugal where this coach was constructed. *Ken Landsdowne*

BEZ7262	DAF MB220	Van Hool Alizée	C57F	1993	Ludlows, Halesowen, 2005
N682YAV	Iveco TurboDaily 59.12	Marshall C31	B27F	1996	Pete's Travel, West Bromwich, '01
R59EDW	Bova Futura Cub FLC12.280	Bova	C53F	1998	Reliant, Heather, 2003
R9CCC	Scania L94 IB	Irizar InterCentury 12.32	C55F	1998	
S285UAL	Mercedes-Benz Vario O814	Plaxton Beaver 2	B31F	1998	Trent Barton, 2006
S289UAL	Mercedes-Benz Vario O814	Plaxton Beaver 2	B31F	1998	Trent Barton, 2006
V901LOH	Mercedes-Benz Vario O814	Autobus Nouvelle 2	C29F	1999	Isleworth, Heston, 2005
V208EAL	Iveco TurboDaily 59.12	Mellor	BC25F	1999	Goodwin, Witheridge, 2003
Y4CCC	Ayats Bravo A3E/BR1	Ayats	C57/18CT	2001	
YJ51ZWA	Optare Solo M850	Optare	NC29F	2001	
713WAF	Ayats Atlantis A18-12/AT	Ayats	C53FT	2003	Sun Fun, Earith, 2004
YN53SVW	Optare Solo M850	Optare	N27F	2003	Expresslines, Bedford, 2005
MX54KXN	Optare Solo M850	Optare	N29F	2004	Vale, Manchester, 2008
YJ54UXH	Optare Solo M780	Optare	N23F	2005	
BX55OFL	BMC Probus 850	BMC	C35F	2006	
W31COM	MAN 18.350	Marcopolo Viaggio III 330	C57F	2008	

Special event vehicles:

HOD55	Bedford OB	Duple Vista	C29F	1949	Western National
MPP747	Dennis Lancet III	Yeates	C35F	1950	Red Rover, Chesham

Previous registrations:

713WAF	AJ03LZA		J697CEV	FC1067 (HK)
BEZ7262	K116TCP, YWD687		JSK492	EW9698 (HK)
E889CDS	E889CDS, A4HWD		W31COM	PN08CPO

Depots: Arkwright Road, Bedford and Manor Farm, Harrowden, Bedford

CEDRIC

Cedric Coaches Ltd, A120 North, Ardleigh, Colchester, CO7 7SL

2	H552GKX	Leyland Olympian ON2R50C13Z4	Leyland	B47/31F	1991	Holmeswood Coaches, 2005
3	G756UYT	Leyland Olympian ONCL10/1RV	Northern Counties	B45/30F	1989	Arriva North East, 2005
4	K814HMV	Leyland Olympian ON2R50C13Z4	Leyland	B47/31F	1992	Metrobus, Orpington, 2006
5	K815HMV	Leyland Olympian ON2R50C13Z4	Leyland	B47/31F	1992	Metrobus, Orpington, 2006
6	K816HMV	Leyland Olympian ON2R50C13Z4	Leyland	B47/31F	1992	Metrobus, Orpington, 2006
7	H547GKX	Leyland Olympian ON2R50C13Z4	Leyland	B47/31F	1991	Armchair, Brentford, 2005
8	H548GKX	Leyland Olympian ON2R50C13Z4	Leyland	B47/31F	1991	Armchair, Brentford, 2005
9	G285UMJ	Leyland Olympian ONCL10/1RV	Leyland	B47/31F	1989	Beeston, Hadleigh, 2007
10	K360DWJ	Leyland Olympian ON2R50G13Z4	Alexander RL	BC43/37F	1992	Stagecoach, 2008
11	YN51MHO	Scania K114 EB4	Irizar Century	C49FT	2001	Chambers, Moneymore, 2004
12	YD02RCO	DAF SB4000	Van Hool T9 Alizée	C32FT	2002	Eavesway, Wigan, 2007
13	FIG6292	Scania K113 TRB	Irizar Century 12.37	C51FT	1997	APT, Rayleigh, 2008
14	FSK868	Bova Futura FHD12.340	Bova	C49FT	1997	Gibson, Moffat, 2006
15	R15CED	Volvo B10M-62	Jonckheere Mistral 50	C51FT	1998	
16	P960DNR	Toyota Coaster BB50R	Caetano Optimo IV	C18F	1997	Princess, West End, 2005
17	YIL7758	Volvo B12(T)	Van Hool Astrobel	C53/14CT	1995	Trathens, Plymouth, 2001
18	OUI8418	Bova Futura FHD12.340	Bova	C49FT	1996	Godson, Crossgates, 2005
19	FJ03AAZ	Volvo B12M	Berkhof Axial 50	C49FT	2003	Harry Shaw, Coventry, 2008
20	R20CED	Volvo B10M-62	Berkhof Axial 50	C49FT	1998	Scancoaches, Battersea, 1998

Previous registrations:

BW04ZGU	04CN809		
FIG6292	P103GHE, 827APT	P960DNR	P960DNR, SAZ4959
FSK868	R50TPB	R20CED	R924ULA
OUI8418	N972EUG	YIL7758	N316BYA

New to Metrobus of Orpington, 4, K814HMV is one of three Leyland Olympians that have, so far, stayed together. It is seen with fleet number 24 when pictured in Colchester during May 2008. *Richard Godfrey*

CENTRAL BUSES

Central Buses Ltd, 177 New Town Row, Newtown, Birmingham, B6 4QZ

1001	J117DUV	Dennis Dart 8.5m	Plaxton Pointer	B24F	1992	2Travel, Swansea, 2003
1004	K100SDC	Dennis Dart 9.8m	Plaxton Pointer	B35F	1992	Excel, Stansted, 2003
1014	V944DNB	Dennis Dart SLF 8.8m	Plaxton Pointer MPD	N29F	1999	Bluebird, Middleton, 2006
1015	V946DNB	Dennis Dart SLF 8.8m	Plaxton Pointer MPD	N29F	2000	Bluebird, Middleton, 2006
1016	T103KCC	Dennis Dart SLF 8.8m	Plaxton Pointer MPD	N29F	1998	Supertravel, Speke, 2006
1019	S290NRB	Optare Solo M920	Optare	N33F	1998	City of Nottingham, 2008
1020	S291NRB	Optare Solo M920	Optare	N33F	1998	City of Nottingham, 2008
1021	X713CCA	Dennis Dart SLF 8.8m	Plaxton Pointer MPD	N29F	1998	Supertravel, Speke, 2008
1022	S515KFL	Dennis Dart SLF	Marshall Capital	N27F	1998	Metroline, Harrow, 2008
1023	S527KFL	Dennis Dart SLF	Marshall Capital	N27F	1998	Metroline, Harrow, 2008
	S105LBL	Optare Solo M850	Optare	N29F	1999	Reading Buses, 2008

Previous registrations:

K100SDC	K712KGU	T103KCC	T6STM

The small Central Buses fleet is based by the main A34 artery to the north of Birmingham. Illustrating the fleet is Dart T103KCC which was new to Supertravel on Merseyside. *Alan Blagburn*

CHALKWELL

Chalkwell Garage & Coach Hire Ltd, 195 Chalkwell Road, Sittingbourne, ME10 1BJ

VIB5237	Leyland Olympian ONLXB/2R	East Lancs	B45/29F	1982	Warrington BT, 2005
YIL4540	Leyland-DAF 400	Jubilee (1993)	M16L	1991	van, 1993
CAZ6603	Dennis Dart 9.8m	Wright HandyBus	BC39F	1994	Ulsterbus, 2006
CAZ6604	Dennis Dart 9.8m	Wright HandyBus	BC39F	1994	Ulsterbus, 2006
XIL7250	Mercedes-Benz 814D	Wadham Stringer Wessex	BC25F	1995	MoD (79RN01), 2004
YIL9717	Dennis Javelin 12m	Wadham Stringer Vanguard III	S70F	1995	MoD (EC45AA), 2006
YIL8824	Dennis Javelin 12m	Wadham Stringer Vanguard III	S70F	1995	MoD (EC41AA), 2005
YIL8825	Dennis Javelin 12m	Wadham Stringer Vanguard III	S70F	1995	MoD (EC42AA), 2005
XIL4697	Mercedes-Benz 811D	Alexander Sprint	B31F	1995	Arriva Midlands, 2006
XIL4698	Mercedes-Benz 811D	Alexander Sprint	B31F	1995	Arriva Midlands, 2006
N355OBC	Mercedes-Benz 709D	Alexander Sprint	B27F	1995	Arriva Fox County, 2003
INZ2296	Mercedes-Benz 711D	Plaxton Beaver	BC25F	1996	Bassetts, Tittensor, 2003
XIL9100	Mercedes-Benz Vario 0814	Wadham Stringer Wessex II	BC33F	1996	Autopoint, Herstmonceux, 2006
N624JNO	Volvo Olympian	Alexander RH	B41/37F	1996	Dublin Bus, 2008
N644JNO	Volvo Olympian	Alexander RH	B41/37F	1996	Dublin Bus, 2008
577HLU	Volvo B10M-62	Plaxton Première	C49FT	1997	Steel's, Skipton, 2004
187NKN	Volvo B10M-62	Plaxton Excalibur	C49FT	1997	Reg's, Hertford, 2004
P848YGB	Mercedes-Benz Vario 0810	Plaxton Beaver 2	B31F	1997	NCP, Glasgow Airport, 2005
P850YGB	Mercedes-Benz Vario 0810	Plaxton Beaver 2	B31F	1997	NCP, Glasgow Airport, 2005
XIL9300	Mercedes-Benz Vario 0814	UVG CitiStar	B33F	1997	Autopoint, Herstmonceux, 2006
R105CKN	Dennis Javelin 12m	UVG S320	C53F	1997	
R107GKN	Dennis Javelin 12m	UVG S320	C53F	1997	
XIL9400	Mercedes-Benz Vario 0814D	Autobus Nouvelle 2	C25F	1997	
R108AKP	Dennis Javelin 12m	Plaxton Première 320	C53F	1998	
R109AKP	Dennis Javelin 12m	Plaxton Première 320	C53F	1998	
R110JKP	Dennis Javelin 12m	Plaxton Première 320	C53F	1998	
YIL8826	Dennis Javelin 12m	UVG S320	C57F	1998	Lewis, Llanrhystyd, 2005
T504RPN	Volvo B10M-62	Plaxton Excalibur	C49FT	1999	Brighton & Hove, 2008
V362HKG	Mercedes-Benz Sprinter 412	Mercedes-Benz	M16	2000	Shiptonslink, 2007
Y402GKJ	Renault Master	Rohill	N17F	2001	*Operated for Kent CC*
Y403GKJ	Renault Master	Rohill	N17F	2001	*Operated for Kent CC*

Many of the fleets in the Notable Independents Handbook that operate Olympians from Dublin Bus and Lothian Buses which have been replaced by new low-floor buses. Initially dual-door, many are now single-doored including N624JNO, seen here in Chalkwell's colours. *Ken Lansdowne*

Pictured arriving in Sittingbourne is of Chalkwell's Mercedes-Benz Vario GN54SVG. Bodywork is a Plaxton Beaver 2, the most popular body on the Vario for the British market. *Richard Godfrey*

GJ02LVT	Renault Master	Rohill	N15F	2002	Classic, Annfield Plain, 2007
FC02DRX	Volvo B12M	Jonckheere Mistral 50	C49FT	2002	Collinson, Claregalway, 2006
FC02DRZ	Volvo B12M	Jonckheere Mistral 50	C49FT	2002	John Pike, Andover, 2006
CSU960	Mercedes-Benz Sprinter 413 cdi	Mercedes-Benz	M16	2002	Andy Jones, Dauntsey, 2007
GJ52LUY	Mercedes-Benz 0814	Plaxton Beaver 2	BC33F	2003	
WX03ENM	Optare Solo M920	Optare	N33F	2003	Hopleys, Mount Hawke, 2007
YM03EOY	Volvo B12M	Plaxton Panther	C49FT	2003	Proctor, Leeming Bar, 2007
YN53EJG	Mercedes-Benz Vario 0814	Plaxton Beaver 2	BC33F	2003	Parry, Malton, 2006
YO53OUG	Optare Solo M920	Optare	N33F	2004	Silverline, Solihull, 2008
YO53OVC	Optare Solo M920	Optare	N33F	2004	Silverline, Solihull, 2008
SJ04KEU	Volvo B12M	Jonckheere Mistral 50	C49FT	2004	Park's of Hamilton, 2007
MX04YLR	Optare Solo M850	Optare	N29F	2004	Darwen Coach Services, 2007
GX04AWR	Optare Solo M850	Optare	N29F	2004	Autopoint, Herstmonceux, 2006
GN54SVF	Mercedes-Benz Vario 0814	Plaxton Beaver 2	BC31F	2004	
GN54SVG	Mercedes-Benz Vario 0814	Plaxton Beaver 2	BC31F	2004	
SF05XDC	Volvo B12M	Jonckheere Mistral 50	C49FT	2005	Park's of Hamilton, 2007
YN07LJU	Scania K114 IB4	Irizar Century	C53FT	2007	Scania demonstrator, 2008

Previous registrations:

187NKN	R929LAA, A13XEL		
577HLU	R131XWF	XIL4698	N471EHA
CSU960	NF02BKK	XIL7250	M613USL, M291UKM
FC02DRX	02G3614	XIL9100	5752AP, N133RAP
FC02DRZ	02G4036	XIL9300	P953CJK
GX04AWR	GX04AWR, 3069AP	XIL9400	R26TKO
INZ2296	N755GBF, XIL7321	YIL4540	H651ENK
N624JNO	96D281	YIL8824	M741UKM
N644JNO	96D259	YIL8825	M751UKM
SJ04KEU	LSK875	YIL8826	R884HEJ
VIB5237	A210DTO	YIL9717	M170UKN
XIL4697	N469EHA		

Web: www.chalkwell.co.uk

Depots: Chalkwell Road, Sittingbourne; Eurolink Commercial Park, Sittingbourne and Church Road, Murston.

English Bus Handbook: Notable Independents

CHAMBERS

H C Chambers & Son Ltd, Knowle House, High Street, Bures, CO8 5AB

Reg	Chassis	Body	Type	Year	Notes
F246HNE	Talbot Pullman	Talbot	B22F	1988	Cooper, Ashton-u-Lyne, 1992
M655KVU	Volvo B10M-62	Van Hool Alizée	C49FTL	1995	Shearings, 2003
M81MYM	Volvo Olympian	Alexander RH	B45/29F	1995	London Central, 2004
N952KBJ	Volvo Olympian	Northern Counties Palatine	B47/30F	1995	
N529LHG	Volvo Olympian	Northern Counties Palatine	B48/31F	1996	London General, 2006
N531LHG	Volvo Olympian	Northern Counties Palatine	B48/31F	1996	London General, 2006
N532LHG	Volvo Olympian	Northern Counties Palatine	B48/31F	1996	London General, 2006
P475MBY	Volvo Olympian	Alexander RH	B43/29F	1996	Metroline, Harrow, 2006
P484MBY	Volvo Olympian	Alexander RH	B43/29F	1996	Metroline, Harrow, 2006
P488MBY	Volvo Olympian	Alexander RH	B43/29F	1996	Metroline, Harrow, 2006
P515UUG	Volvo B10B	Wright Endurance	BC47F	1997	Stagecoach, 2008
P901RYO	Volvo Olympian	Northern Counties Palatine	B47/27F	1997	London Central, 2005
P902RYO	Volvo Olympian	Northern Counties Palatine	B47/27F	1997	London Central, 2005
P549WGT	Volvo Olympian	Northern Counties Palatine	B47/29F	1997	London Central, 2004
P225EJW	Mercedes-Benz 811D	Marshall C16	B27F	1997	Travel West Midlands, 2005
P236EJW	Mercedes-Benz 811D	Marshall C16	B27F	1997	Travel West Midlands, 2005
P240EJW	Mercedes-Benz 811D	Marshall C16	B27F	1997	Travel West Midlands, 2005
R415XFL	MAN 11.220	Marshall City	B40F	1997	Dart Buses, Paisley, 2002
R995YJO	Mercedes-Benz 208D	Mercedes-Benz	M8	1998	private owner, 2003
R112NTA	Mercedes-Benz Vario O814	Alexander ALX100	B29F	1998	Stagecoach, 2008
R361GDX	Volvo B10M-62	Plaxton Première 320	C55F	1998	Halfpenny, Blackrock, 2002
R941YOV	Volvo Olympian	Alexander RH	B47/29F	1998	UK North, Manchester, 2007
S233RLH	Volvo Olympian	Alexander RH	B43/29F	1998	Metroline, Harrow, 2007
S131RLE	Volvo Olympian	Alexander RH	B43/25D	1998	Metroline, Harrow, 2008
S848DGX	Volvo Olympian	East Lancs Pyoneer	B47/29F	1998	Courtney, Bracknell, 2005
S851DGX	Volvo Olympian	East Lancs Pyoneer	B47/29F	1998	Metrobus, Crawley, 2005
S855DGX	Volvo Olympian	East Lancs Pyoneer	B47/29F	1998	Metrobus, Crawley, 2005
S856DGX	Volvo Olympian	East Lancs Pyoneer	B47/29F	1998	Metrobus, Crawley, 2005
S48RGA	MAN 11.220	Marshall City	B40F	1999	Dart, Paisley, 2002
BX55FYH	Mercedes-Benz Touro OC500	Mercedes-Benz	C49FT	2006	
BX55FYJ	Mercedes-Benz Touro OC500	Mercedes-Benz	C49FT	2006	

Previous registration:
R361GDX   98LH859

Chambers' fleet comprises mostly double-deck Olympians. S848DGX, seen here at Newbury Park, features East Lancs Pyoneer bodywork.
Alan Blagburn

CLARIBELS

Claribel Coaches Ltd, B10 Fortnum Close, Tile Cross, Birmingham, B33 0JT

T290ROF	Volvo B7BLE	Wrightbus Crusader	N43F	1999	Whitelaw, Stonehouse, 2008
V385KVY	Optare Excel L1150	Optare	N45F	1999	
V386KVY	Optare Excel L1150	Optare	N45F	1999	
V387KVY	Optare Excel L1150	Optare	N45F	1999	
X92FOR	Dennis Dart SLF	Caetano Compass	N37F	2000	
YD02RCX	DAF SB200	Wrightbus Commander	N44F	2002	
YD02RCY	DAF SB200	Wrightbus Commander	N44F	2002	
YD02RCZ	DAF SB200	Wrightbus Commander	N44F	2002	
YJ03PKA	DAF SB120	Wrightbus Cadet 2	N39F	2003	
YJ53VDL	VDL Bus SB200	Wrightbus Commander	N44F	2003	
YJ54CKL	VDL Bus DB250	East Lancs Lowlander	N51/29F	2004	
YJ54CEA	VDL Bus SB200	Wrightbus Commander	N44F	2005	
YJ54CEF	VDL Bus SB200	Wrightbus Commander	N44F	2005	
YJ54CEK	VDL Bus SB200	Wrightbus Commander	N44F	2005	
YJ54CEN	VDL Bus SB200	Wrightbus Commander	N44F	2005	
YJ54CEO	VDL Bus SB200	Wrightbus Commander	N44F	2005	
YJ55KZX	VDL Bus SB200	Wrightbus Commander	N44F	2005	
YJ56JXX	VDL Bus SB200	Wrightbus Commander	N44F	2006	
YJ56KCV	VDL Bus SB200	Plaxton Centro	N40F	2006	
YJ57BOF	VDL Bus SB200	Wrightbus Commander	N44F	2007	
YJ57BOH	VDL Bus SB200	Wrightbus Commander	N44F	2007	

Claribels has built an extensive fleet of Wrightbus Commanders, the body based on the low floor SB200 from VDL Bus, the successor to DAF. The latest to join the fleet is YJ57BOH, seen here in Birmingham. *Alan Blagburn*

COASTAL

PH Jenkins, 18 West Point, Newick, Lewes, BN8 4NU

115	GU52HAO	Dennis Dart SLF 8.5m	Plaxton MPD	N29F	2002
116	GU52HAX	Dennis Dart SLF 8.5m	Plaxton MPD	N29F	2002
117	XS2210	Dennis Dart SLF 10.7m	Plaxton Pointer 2	N38F	2002
118	GU52HJY	Dennis Dart SLF 10.7m	Plaxton Pointer 2	N38F	2002
121	GU52HKC	Dennis Dart SLF 10.7m	Plaxton Pointer 2	N38F	2002
122	GU52HKD	Dennis Dart SLF 10.7m	Plaxton Pointer 2	N38F	2002
123	GX04ASU	TransBus Dart 8.8m	TransBus Mini Pointer	N29F	2004
124	GX05AOP	ADL Dart 10.7m	ADL Pointer	N37F	2005
125	GX07AVO	ADL Dart 4 10.7m	ADL Enviro 200	N38F	2007
126	GX07BYO	ADL Dart 4 10.7m	ADL Enviro 200	N37F	2007

Previous registration:
XS2210   GU52HJX

Depot: Coppards Lane, Northiam

Following on from the fleet of Darts at Coastal Coaches, the latest arrivals are Dart 4s with Alexander Dennis Enviro 200 bodies. Pictured near Westfield is GX07AVO. *Richard Godfrey*

COMPASS BUS

Compass Travel (Sussex) Ltd, Faraday Close, Durrington, Worthing, BN13 3RB

YXI5860	Volvo B10M-61	Van Hool Alizée	C53F	1989	Foster, Aylesbury, 2007
KAZ4127	DAF SB2305	Van Hool Alizée	C53F	1991	Kirby Lonsdale Coaches, 2007
J811KHD	DAF SB3000	Van Hool Alizée	C53F	1992	Jackson, Bicknacre, 2008
J599DUV	Dennis Dart 8.5m	Plaxton Pointer	B28F	1992	London United, 2003
J100SOU	DAF MB230	Van Hool Alizée	C51FT	1992	Galloway, Mendlesham, 2005
L287MJH	DAF SB3000	Van Hool Alizée	C55F	1993	Weavaway, Newbury, 2007
L476CFT	Dennis Lance 11.5m	Optare Sigma	B47F	1994	Go North East, 2007
L478CFT	Dennis Lance 11.5m	Optare Sigma	B47F	1994	Go North East, 2007
L481CFT	Dennis Lance 11.5m	Optare Sigma	B47F	1994	Go North East, 2007
L488CFT	Dennis Lance 11.5m	Optare Sigma	B47F	1994	Go North East, 2007
N506LUA	DAF SB3000	Ikarus Blue Danube 350	C53F	1996	Holmeswood Coaches, 2004
A18HOF	Scania K113 CRB	Van Hool Alizée	C49FT	1996	Hellyers, Fareham, 2007
R93HUA	Optare MetroRider MR15	Optare	B31F	1997	Alansway, Heathfield, 2007
R449PRH	Mercedes-Benz Vario O814	Plaxton Beaver 2	B27F	1997	East Yorkshire, 2008
R81EDW	Mercedes-Benz Vario O814	Autobus Nouvelle	BC33F	1998	Bebb, Llantwit Fardre, 1999
R84EDW	Mercedes-Benz Vario O814	Autobus Nouvelle	BC33F	1998	Clintona, Brentwood, 2007
S73TNM	Optare MetroRider MR17	Optare	B29F	1998	UNÖ, Hatfield, 2007
T421ADN	Optare MetroRider	Optare	BC33F	1999	Optare demonstrator, 2001
T422ADN	Optare MetroRider	Optare	BC33F	1999	Optare demonstrator, 2001
V896LOH	Mercedes-Benz Vario O814	Plaxton Beaver 2	B29F	1999	Judge, Corby, 2007
W426CWX	Optare MetroRider MR17	Optare	B29F	2000	Henderson, Hamllton, 2008
W821RJT	Ford Transit	Ford	M16	2001	private owner, 2008
X228AWB	Mercedes-Benz Vario O814	Plaxton Beaver 2	BC33F	2001	Hearn, Harrow Weald, 2007
GX02CGY	Mercedes-Benz Sprinter 411cdi	Mercedes-Benz	M14	2002	

Another Enviro 200-bodied Dart is Compass Bus' GX57AFV which joined the fleet in 2007. It is seen arriving in South Ferrins, its normal service. *Richard Godfrey*

Carrying Tesco corporate colours is Compass Bus' J599DUV. New as one of the early Darts supplied to London Buses, it features the initial Pointer body designed for Reeve Burgess coachworks. *Richard Godfrey*

SA52MYR	Dennis Dart SLF 8.8m	Plaxton Pointer MPD	N29F	2003	A1A, Birkenhead, 2008
SA52MYS	Dennis Dart SLF 8.8m	Plaxton Pointer MPD	N29F	2003	A1A, Birkenhead, 2008
GX03AZJ	Dennis Dart SLF	Plaxton Pointer 2	N37F	2003	
GX03AZL	Dennis Dart SLF	Plaxton Pointer 2	N37F	2003	
DF03NTE	TransBus Dart SLF 8.8m	TransBus Pointer MPD	N29F	2003	Huyton Travel, 2007
SN53ETJ	TransBus Dart 8.8m	TransBus Mini Pointer	N29F	2003	
SN53ETK	TransBus Dart 8.8m	TransBus Mini Pointer	N29F	2003	
SN53ETL	TransBus Dart 8.8m	TransBus Mini Pointer	N29F	2003	
SN53ETO	TransBus Dart 8.8m	TransBus Mini Pointer	N29F	2003	
SN53FTR	TransBus Dart 8.0m	TransBus Mini Pointer	N29F	2003	
GX04BXN	Mercedes-Benz Sprinter 411cdi	Mercedes-Benz	M14	2004	
GX54AWH	ADL Dart 8.8m	ADL Mini Pointer	N29F	2004	
VX05UHS	Dennis Dart SLF 8.8m	Plaxton Pointer MPD	N29F	2005	Aston's, Worcester, 2007
MX06ACY	Optare Solo M950	Optare	N31F	2006	Bakerbus, Biddulph, 2008
GX57AFV	ADL Dart 4 8.9m	ADL Enviro 200	N29F	2007	

Previous registrations:

A18HOF	N92WVC, 3KOV, N605WVC		N506LUA	N506LUA, SBV703
J100SOU	J100SOU, 5611PP		W821RJT	00D51482
J811KHD	J811KHD, 5516PP, MJI3084		YXI5860	F740ENE
KAZ4127	H526YCX, B10MPT			
L287MJH	L551EHD, LUI9691, C1OXF			

Web: www.compass-travel.co.uk

COUNTRYLINER

Countryliner Coach Hire Ltd; Countryliner Sussex Ltd; Countryliner Buses Ltd, GB House,
Slyfield Industrial Estate, Guildford, GU1 1RR

DLA116 NM	J116WSC	Dennis Lance 11m	Alexander PS	B47F	1992	Blue Bus, Bolton, 2005
DLO411 GD	N411MPN	Dennis Lance 11m	Optare Sigma	B47F	1996	Brighton & Hove, 2007
DLO414 PN	N414MPN	Dennis Lance 11m	Optare Sigma	B47F	1996	Brighton & Hove, 2006
DLO416 NM	N416MPN	Dennis Lance 11m	Optare Sigma	B47F	1996	Brighton & Hove, 2006
DP1 GD	RL51CXB	Dennis Dart SLF 8.5m	Plaxton Pointer MPD	N26F	2002	
DP2 GD	RL51CXC	Dennis Dart SLF 8.5m	Plaxton Pointer MPD	N26F	2002	
DP3 GD	RL51CXD	Dennis Dart SLF 8.5m	Plaxton Pointer MPD	N26F	2002	
DP7 GD	KU52RXX	Dennis Dart SLF 8.5m	Plaxton Pointer MPD	N26F	2002	
DP9 BH	W937JNF	Dennis Dart SLF 8.5m	Alexander ALX200	N26F	2000	Go-West, King's Lynn, 2002
DP10 GD	W921JNF	Dennis Dart SLF 8.5m	Plaxton Pointer MPD	N26F	2000	Liskeard & District, 2002
DP11 BH	T78JBA	Dennis Dart SLF 8.5m	Plaxton Pointer MPD	N26F	1999	M R Travel, Rochdale, 2003
DP12 BH	T550HNH	Dennis Dart SLF 8.5m	Plaxton Pointer MPD	N26F	1999	Pink Elephant Parking, 2003
DP13 BH	V247BNV	Dennis Dart SLF 8.5m	Plaxton Pointer MPD	N26F	1999	Nostalgiabus, Mitcham, 2003
DP14 GD	P682RWU	Dennis Dart SLF 9.8m	Plaxton Pointer	N34F	1997	First PMT, 2003
DP15 BH	P686RWU	Dennis Dart SLF 9.8m	Plaxton Pointer	N34F	1997	Tellings-Golden Miller, 2003
DP16 BH	P688RWU	Dennis Dart SLF 9.8m	Plaxton Pointer	N34F	1997	First PMT, 2003
DP17 PN	V257BNV	Dennis Dart SLF 8.8m	Plaxton Pointer MPD	N26F	1999	Cardiff Bus, 2003
DP18 GD	KV51KZJ	Dennis Dart SLF 8.8m	Plaxton Pointer MPD	N29F	2001	Menzies, Heathrow, 2003
DP19 GD	W364ABD	Dennis Dart SLF 8.8m	Plaxton Pointer MPD	N26F	2000	
DP20 GD	KP51SXV	Dennis Dart SLF 8.8m	Plaxton Pointer MPD	N26F	2001	Courtney, Bracknell, 2004
DP21 GD	W773URP	Dennis Dart SLF 8.8m	Plaxton Pointer MPD	N26F	2000	
DP22 GD	KP51SYC	Dennis Dart SLF 8.8m	Plaxton Pointer MPD	N26F	2001	Supertravel, Speke, 2006
DP23 GD	KP51SYA	Dennis Dart SLF 8.8m	Plaxton Pointer MPD	N26F	2001	Supertravel, Speke, 2006
DP24 BH	KX51UDG	Dennis Dart SLF 10.7m	Plaxton Pointer 2	N37F	2001	Buzzlines, Hythe, 2006
DP25 BH	X601AHE	Dennis Dart SLF 10.7m	Plaxton Pointer 2	N37F	2001	RDH Services, Ditchling, 2006
DP26 BH	X602AHE	Dennis Dart SLF 10.7m	Plaxton Pointer 2	N37F	2001	RDH Services, Ditchling, 2006
DP27 BH	X214ONH	Dennis Dart SLF 10.7m	Plaxton Pointer 2	N37F	2001	Flights-Hallmark, Hounslow, 2006
DP28 BH	YT51EAC	Dennis Dart SLF 8.8m	Plaxton Pointer MPD	N26F	2001	Connex, 2006
DP29 BH	YT51EAE	Dennis Dart SLF 8.8m	Plaxton Pointer MPD	N26F	2001	Connex, 2006
DP30 BH	YT51EAF	Dennis Dart SLF 8.8m	Plaxton Pointer MPD	N26F	2001	Compass, Worthing, 2006
DP31 BH	YT51DZY	Dennis Dart SLF 8.8m	Plaxton Pointer MPD	N26F	2001	Compass, Worthing, 2006
DP32 BH	GX04AZA	Dennis Dart SLF 8.8m	Plaxton Pointer MPD	N26F	2004	Compass, Worthing, 2006
DM33 BH	F196FFM	Dennis Dart SLF 10.8m	Marshall Capital	N39F	1998	Halton, 2006
DP34 GD	RX56DWE	ADL Dart 8.8m	ADL Mini Pointer	N26F	2006	
DP35 GD	T337TVM	Dennis Dart SLF 8.8m	Plaxton Pointer MPD	N26F	1999	Shuttle Buses, Kilwinning, 2006
DP36 GD	T314SMV	Dennis Dart SLF 8.8m	Plaxton Pointer MPD	N26F	1999	Menzies, Heathrow, 2007
DC37 u	KU02YBG	Dennis Dart SLF	Caetano Compass	N31F	2002	Centra, Heathrow, 2007
DC38 u	KM51BFN	Dennis Dart SLF	Caetano Compass	N31F	2002	Centra, Heathrow, 2008
DP39 u	P307HDP	Dennis Dart SLF 9.8m	Plaxton Pointer 2	N33F	1997	Metrobus, Crawley, 2008
DP40 u	P724RYL	Dennis Dart SLF 10.1m	Plaxton Pointer 2	N36F	1997	Metrobus, Crawley, 2008

Mostly used on
school services,
Countryliner
WDL696Y illustrates
the standard body
for the Olympian
supplied by Eastern
Coach Works.
Alan Blagburn

Preparing to leave Tunbridge Wells for Hastings is Countryliner DP24, KX51UDG, a Dennis Dart with Plaxton Pointer 2 bodywork. *Mark Lyons*

DP41	-	P725RYL	Dennis Dart SLF 10.1m	Plaxton Pointer 2	N36F	1997	Metrobus, Crawley, 2008
DM	-	V361DLH	Dennis Dart SLF	Marshall Capital	N25F	1999	Metrobus, Crawley, 2008
DS2	BH	H712LOL	Dennis Dart 9.8m	Carlyle Dartline	B40F	1991	RDH Services, Ditchling, 2006
DS6	BH	K430OKH	Dennis Dart 9m	Plaxton Pointer	B35F	1993	RDH Services, Ditchling, 2006
DS7	BH	K479JHJ	Dennis Dart 9m	Plaxton Pointer	B34F	1992	RDH Services, Ditchling, 2006
DS8	PN	K864LGN	Dennis Dart 9m	Plaxton Pointer	B34F	1993	RDH Services, Ditchling, 2006
DS13	BH	L717OMV	Dennis Dart 9.8m	Plaxton Pointer	B35F	1994	Compass, Worthing , 2006
DS14	BH	L718OMV	Dennis Dart 9.8m	Plaxton Pointer	B35F	1994	Compass, Worthing , 2006
DS15	BH	L719OMV	Dennis Dart 9.8m	Plaxton Pointer	B35F	1994	Compass, Worthing , 2006
DS16	u	N720KGF	Dennis Dart 10m	Plaxton Pointer	B35F	1995	Metrobus, Crawley, 2007
DS17	u	P895PWW	Dennis Dart 10m	Plaxton Pointer	B39F	1997	Metrobus, Crawley, 2007
DS18	u	R58GNW	Dennis Dart 10m	Plaxton Pointer	B40F	1997	Metrobus, Crawley, 2007
DFC1	GD	OYD693	DAF MB230	Van Hool Alizée SH	C53F	1988	Horsham Buses, 1998
DFC3	GD	776WME	DAF MB230	Van Hool Alizée SH	C53F	1990	Horsham Buses, 1998
ELD906	BH	X906RHG	DAF SB220	East Lancs Myllennium	N47F	2001	Courtney, Bracknell, 2006
MB653	u	F653OFG	Mercedes-Benz 811D	Wright	BC31F	1989	Autopoint, Herstmonceux, 2006
MB	u	BU04UTP	Mercedes-Benz Sprinter 413cdi	Koch	N16F	2004	Metrobus, Crawley, 2007
MB	u	R101NTA	Mercedes-Benz Vario 0814	Alexander ALX100	B29F	1998	Stagecoach, 2007
MC2	NM	BU03LXV	Mercedes-Benz Touro OC500	Mercedes-Benz	C53F	2003	Compass, Worthing, 2006
MC3	NM	BU03LXW	Mercedes-Benz Touro OC500	Mercedes-Benz	C49FT	2003	Compass, Worthing, 2006
MC5	NM	BX54EBP	Mercedes-Benz Touro OC500	Mercedes-Benz	C49FT	2004	
MC6	NM	LX03KPE	Mercedes-Benz Touro OC500	Mercedes-Benz	C49FT	2003	Redwing, Herne Hill, 2005
MM412	NM	R412XFL	MAN 11.220	Marshall C37	B40F	1997	Northumbria Coaches, 2005
MRD9	-	AE56MDK	ADL Dart 4	MCV Evolution	N40F	2007	
MRD10	-	AE56MDO	ADL Dart 4	MCV Evolution	N40F	2007	
MRM1	BH	AE06VPY	MAN 14.220	MCV Evolution	N40F	2006	
MRM2	BH	AE06VPZ	MAN 14.220	MCV Evolution	N40F	2006	Compass, Worthing , 2006
MRM3	GD	AE56MDF	MAN 14.220	MCV Evolution	N34F	2006	Compass, Worthing , 2006
MRM4	GD	AE56MDJ	MAN 14.220	MCV Evolution	N34F	2006	
MRM5	GD	AE56MBX	MAN 14.220	MCV Evolution	N34F	2006	
MRM6	GD	AE56MBY	MAN 14.220	MCV Evolution	N34F	2006	
MRM7	-	AE07DZD	MAN 18.220	MCV Evolution	N35F	2007	
MRM8	-	AE07DZE	MAN 18.220	MCV Evolution	N35F	2007	
MRM9	-	AE07DZF	MAN 18.220	MCV Evolution	N35F	2007	
NSD1	GD	FSK598	Neoplan Skyliner N122/3	Neoplan	C55/22DT	1992	Moffat & Williamson, Gauldry, '05

Nine MAN single-deck buses with MCV Evolution bodies are now operated by Countryliner. Pictured in Brighton was MRM2, AE06VPZ. *Richard Godfrey*

NSD2	NM	W892WDT	Neoplan Skyliner N122/3	Neoplan	C55/22DT	2000	Ferris, Nantgarw, 2006
OE	-	R100PAR	Optare Excel L1150	Optare	N40F	1997	Swanbrook, Cheltenham, 2008
OE	-	R200PAR	Optare Excel L1150	Optare	N40F	1997	Swanbrook, Cheltenham, 2008
OS	-	YK04KVU	Optare Solo M850 SL	Optare	N25F	2004	Hampshire CC, 2008
OS	-	YK04KVW	Optare Solo M780 SL	Optare	N23F	2004	Hampshire CC, 2008
OT904	PN	N904HWY	Optare MetroRider MR13	Optare	B26F	1996	Metrobus, Crawley, 2003
OT425	PN	P425VRG	Optare MetroRider	Optare	B26F	1997	Go North East, 2006
	PN	YK05CDU	Optare Solo M850 SL	Optare	N25F	2005	*operated for East Sussex CC*
	PN	YK05CDX	Optare Solo M850 SL	Optare	N25F	2005	*operated for East Sussex CC*
PP1	GD	KX06LYP	Enterprise Plasma EB01	Plaxton Primo	N28F	2006	
PP2	GD	MX56NLZ	Enterprise Plasma EB01	Plaxton Primo	N28F	2006	
PP3	GD	KX57FMM	Enterprise Plasma EB01	Plaxton Primo	N28F	2007	Silverjet, Luton, 2008
VP40	GD	R40TGM	Volvo B10M-62	Plaxton Première 350	C53F	1998	Tellings-Golden Miller, 2005
VP511	NM	R511WDC	Volvo B10M-62	Plaxton Première 350	C49FT	1998	Compass Royston, Stockton, '04
V710	GD	P710OOA	Volvo B10M-62	Caetano Enigma	C53F	1997	Pat Kavanagh, Urlingford, 2004
V	-	S365VKW	Volvo B10M-62	Plaxton Première 350	C44FT	1999	Stagecoach, 2008
V	-	CN04NBY	Volvo B12B	Berkhof Axial 100	C(83)FT	2004	Ferns, Nantgarw, 2008
RLH521	u	C521LJR	Leyland Olympian ONCL10/1RV	Eastern Coach Works	BC45/27F	1985	Renown, Bexhill, 2006
RLH681	GD	A681KDV	Leyland Olympian ONLXB/1R	Eastern Coach Works	B45/32F	1983	Stagecoach, 2004
RLH693	GD	WDL693Y	Leyland Olympian ONLXB/1R	Eastern Coach Works	BC40/30F	1983	Southern Vectis, 2002
RLH695	GD	WDL695Y	Leyland Olympian ONLXB/1R	Eastern Coach Works	BC40/30F	1983	Southern Vectis, 2002
RLH696	GD	WDL696Y	Leyland Olympian ONLXB/1R	Eastern Coach Works	BC40/30F	1983	Southern Vectis, 2002

Previous registrations:

776WME	G974KJX		
F653OFG	HDZ2609, 1108AP	P710OOA	97CW1
FSK598	J41XHE	S365VKW	S365VKW, 1619HE
K479JHJ	OMN71P	T337TVM	BUS1N
OYD693	F618HGO	W773URP	W444APS

Depots: Fairbridge Way, Burgess Hill (BH); Westfield Road, Slyfield Industrial Estate, Guildford (GD); Heathfield (HE); Oak Ferrars Farm, Piltdown, Uckfield (PN) and Surrey CC yard, Merrow, Guildford. **Web:** www.countryliner-coaches.com

COURTNEY

Courtney Coaches Ltd, 1 Berkshire Business Centre, Downmill Road, Bracknell, RG12 1QS

S847DGX	Volvo Olympian	East Lancs Pyoneer	B47/29F	1998	Metrobus, Crawley, 2004
S853DGX	Volvo Olympian	East Lancs Pyoneer	B47/29F	1998	Metrobus, Crawley, 2004
S857DGX	Volvo Olympian	East Lancs Pyoneer	B47/29F	1998	Metrobus, Crawley, 2004
YE52FGX	Optare Solo M920	Optare	N33F	2002	Zak's, Birmingham, 2005
GU52HXM	Optare Solo M850	Optare	N29F	2003	
RO03JVA	Renault Master	Frank Guy	M8	2003	RB Windsor & Maidenhead, 2005
KX04HRA	Optare Solo M920	Optare	N33F	2004	
KX54NLA	DAF SB220	East Lancs Myllennium	N50F	2004	
KX54NLC	Optare Solo M920	Optare	N33F	2004	
KX54NLD	Optare Solo M920	Optare	N33F	2004	
YJ54UXA	Optare Solo M920	Optare	N31F	2005	
YJ05XMT	Optare Solo M920	Optare	N33F	2005	
YJ05XNA	Optare Solo M920	Optare	N33F	2005	
YJ55BLV	Optare Solo M920	Optare	N33F	2005	
RX55AOT	Optare Solo M920	Optare	N33F	2005	
SN55DVA	ADL E300	ADL Enviro300	S60F	2005	
RX06XFD	ADL E300	East Lancs Myllennium	N51F	2006	
RX06XFE	ADL E300	East Lancs Myllennium	N51F	2006	
RX06WRU	VDL Bus DB250	East Lancs Lowlander	N47/33F	2006	
YJ56KCK	VDL Bus DB250	East Lancs Lowlander	N47/33F	2006	
RK07KDA	Optare Solo M710 SE	Optare	N24F	2007	
RX57MDZ	Optare Solo M710 SE	Optare	N23F	2007	
YJ57EKH	Optare Solo M950 SL	Optare	N29F	2007	
YJ57EKK	Optare Solo M880 SL	Optare	N31F	2007	
YJ57XWM	Optare Solo M950 SL	Optare	N33F	2007	
YJ57XWN	Optare Solo M950 SL	Optare	N33F	2007	
YJ57XWO	Optare Solo M950 SL	Optare	N33F	2007	
YJ57XWP	Optare Solo M950 SL	Optare	N33F	2007	
YJ57XWR	Optare Solo M950 SL	Optare	N33F	2007	
YJ57XWS	Optare Solo M950 SL	Optare	N33F	2007	
YJ57XWU	Optare Solo M950 SL	Optare	N33F	2007	
YJ57XWY	Optare Solo M950 SL	Optare	N33F	2007	
YJ57XXE	Optare Solo M1020	Optare	N37F	2007	
YJ08PGU	Optare Solo M710 SE	Optare	N23F	2008	
YJ08PGV	Optare Solo M710 SE	Optare	N23F	2008	
KX08HMD	ADL Dart 4 8.9m	ADL Enviro 200	N29F	2008	
KX08HME	ADL Dart 4 8.9m	ADL Enviro 200	N29F	2008	
RX58HVJ	ADL Trident II	Optare Olympus	N51/31F	2008	

Web: www.courtneycoaches.com

Newly into service with Courtney are two Enviro 200 Darts to the former Mini Pointer length. KX08HMD is seen on Basingstoke's Park & Ride service.
Mark Lyons

D&G BUS COMPANY

D&G - Choice

D&G Coach & Bus Ltd, Mossfield Road, Adderley Green, Stoke-on-Trent, ST3 5BW
D&G Coach & Bus Ltd, 26 The Meadows, Kingstone, Uttoxeter, ST14 8QE

1	AG	YN53EMF	Optare Solo M850	Optare	N27F	2003	Wardle Transport, Norton, 2006
3	AG	YN53EMK	Optare Solo M850	Optare	N27F	2003	
4	AG	YN53EMV	Optare Solo M850	Optare	N27F	2003	
5	AG	YN53EMX	Optare Solo M850	Optare	N27F	2003	
6	w	FJZ4196	Dennis Dart 9.8m	East Lancs	B31F	1996	Quantock MS, Taunton, 2003
7	AG	ENZ4635	Dennis Dart 9.5m	East Lancs	B34F	1994	London General, 2003
8	w	LUI9649	Dennis Dart 9.5m	East Lancs	B31F	1996	Metropolitan Police, 2002
9	AG	W82NDW	Optare Solo M850	Optare	N29F	2000	Addison, Steeton, 2007
10	AG	CHZ8960	ADL Dart 8.8m	ADL Mini Pointer	N29F	2004	
11	AG	CNZ2250	ADL Dart 8.8m	ADL Mini Pointer	N29F	2004	
12	AG	M67HHB	Dennis Dart 9.8m	Wright HandyBus	B39F	1995	Stagecoach, 2006
14	CR	R713MEW	Dennis Dart SLF 10.7m	Marshall Capital	N39F	1998	Halton, 2006
15	CR	R714MEW	Dennis Dart SLF 10.7m	Marshall Capital	N39F	1998	Halton, 2006
18	w	EJZ2291	Mercedes-Benz 811D	Wright NimBus	B33F	1993	Arriva Cymru, 2002
19	w	ENZ2127	Mercedes-Benz 709D	Plaxton Beaver	B27F	1994	Cooper, Dukinfield, 2003
20	AG	YM52TPV	Optare Solo M850	Optare	N28F	2003	Chesterbus, 2008
21	AG	YM52TPX	Optare Solo M850	Optare	N28F	2003	First, 2008
22	AG	YM52TPY	Optare Solo M850	Optare	N28F	2003	First, 2008
28	w	G541TBD	Mercedes-Benz 811D	Wright NimBus	B33F	1989	
30	CR	YD02RGY	Dennis Dart SLF 10.7m	Plaxton Pointer 2	N38F	2002	Quantock, Wiveliscombe, 2008
34	w	LUI9633	Mercedes-Benz 711D	Plaxton Beaver	B27F	1993	Gibson Direct, Renfrew, 2002
37	WN	BHZ8675	Mercedes-Benz 709D	Dormobile Routemaker	B25F	1991	Orion, Wemyss Bay, 2001
40	AG	VIL8577	Mercedes-Benz 709D	Plaxton Beaver	B27F	1995	Cooper, Dukinfield, 2002

Since the last edition of this publication D&G, along with its sister operation, Choice, has expanded in Cheshire, Shropshire and Staffordshire. JJZ5291 is an East Lancs-bodied Dart seen in Walsall. *Alan Blagburn*

Carrying the blue of Shropshire Bus is Optare Solo 171, DX05OMB, Choice operates several tendered services for the county, including the main link north from Wellington to Market Drayton. *Alan Blagburn*

41	AG	L330CHB	Mercedes-Benz 811D	Marshall C16	B33F	1993	Stagecoach, 2007
42	CR	VIL8677	Mercedes-Benz Vario O814	Marshall Master	B31F	1999	Leven Valley, South Bank, 2002
44	CR	P688KCC	Mercedes-Benz 711D	Plaxton Beaver	B27F	1997	Arriva NW & Wales, 2007
48	CR	LLZ3249	Mercedes-Benz 709D	Plaxton Beaver	B27F	1995	Cooper, Dukinfield, 2003
61	AG	YJ54BSX	Optare Solo M780 SL	Optare	N23F	2004	
62	AG	YJ54BSY	Optare Solo M780 SL	Optare	N23F	2004	
63	AG	YJ54BSZ	Optare Solo M780 SL	Optare	N23F	2004	
64	CR	YJ54ZYA	Optare Solo M850	Optare	N28F	2005	
65	CR	YJ54ZYB	Optare Solo M050	Optare	N28F	2005	
66	CR	YJ54ZYC	Optare Solo M850	Optare	N28F	2005	
67	CR	YJ54ZYD	Optare Solo M850	Optare	N28F	2005	
68	CR	YJ54ZYE	Optare Solo M850	Optare	N28F	2005	
69	CR	YJ54ZYF	Optare Solo M850	Optare	N28F	2005	
71	AG	YJ05JXR	Optare Solo M850	Optare	N27F	2005	
72	AG	YJ05JXS	Optare Solo M850	Optare	N27F	2005	
75	CR	YJ54UBD	Optare Solo M780	Optare	N27F	2005	Malbank, Nantwich, 2005
77	w	N808PDS	Mercedes-Benz 811D	Marshall C19	B29F	1996	Arriva Scotland, 2005
79	w	H367XGC	Dennis Dart 8.5m	Wright HandyBus	B29F	1991	Arriva The Shires, 2005
84	w	J545GCD	Dennis Dart 9.8m	Alexander Dash	B40F	1992	Stagecoach, 2006
85	AG	R54OCK	Dennis Dart SLF 10.1m	East Lancs Spryte	N33F	1997	Sadler, Gloucester, 2006
86	WN	T466HNH	Dennis Dart SLF 8.8m	Plaxton Pointer MPD	N29F	1999	City Nippy, Middleton, 2006
87	WN	T467HNH	Dennis Dart SLF 8.8m	Plaxton Pointer MPD	N29F	1999	City Nippy, Middleton, 2006
88	WN	T551HNH	Dennis Dart SLF 8.8m	Plaxton Pointer MPD	N29F	1999	City Nippy, Middleton, 2006
89	WN	T552HNH	Dennis Dart SLF 8.8m	Plaxton Pointer MPD	N29F	1999	City Nippy, Middleton, 2006
90	CR	W586YDM	Dennis Dart SLF 8.8m	Plaxton Pointer MPD	N29F	2000	Swans, Chadderton, 2006
91	CR	DG02WYB	Dennis Dart SLF 8.8m	Plaxton Pointer MPD	N26F	2002	Al's Coaches, Birkenhead, 2006
92	CR	V261BNV	Dennis Dart SLF 8.8m	Plaxton Pointer MPD	N28F	2001	Dawson Rentals, 2006
93	WN	N472EHA	Mercedes-Benz 709D	Plaxton Beaver	B31F	1995	Arriva Midlands, 2007
94	WN	N189EMJ	Mercedes-Benz 811D	Plaxton Beaver	B27F	1995	Faresaver, Chippenham, 2007
95	WN	R726EGD	Mercedes-Benz Vario O810	Plaxton Beaver 2	B31F	1997	Travel Wright, Newark, 2007
96	WN	N784EUA	Mercedes-Benz 811D	Plaxton Beaver	B35F	1995	Veolia England, 2007
100	WN	GNZ3561	Dennis Dart SLF 8.8m	Plaxton Pointer MPD	N29F	1999	
101	WN	GNZ3462	Dennis Dart SLF 8.8m	Plaxton Pointer MPD	N29F	1999	
102	WN	JJZ5368	Dennis Dart SLF 11.3	Plaxton Pointer SPD	N41F	1198	Duke's Travel, Berry Hill, 2005
103	WN	OUI9143	Dennis Dart SLF 10.7m	Plaxton Pointer 2	N41F	1999	People's Exp, W Bromwich, 2005

104	WN	OUI2376	Dennis Dart SLF 10.7m	Plaxton Pointer 2	N41F	1999	People's Exp, W Bromwich, 2005
105	WN	P513UUG	Volvo B10B	Wright Endurance	BC47F	1997	Stagecoach, 2008
106	WN	JJZ5289	TransBus Dart SLF 8.8m	TransBus Min Pointer	N29F	2002	Stuart's, Carluke, 2006
107	WN	JJZ5250	TransBus Dart SLF 8.8m	TransBus Min Pointer	N29F	2002	Stuart's, Carluke, 2006
108	WN	JJZ5278	Dennis Dart SLF 8.8m	Plaxton Pointer MPD	N29F	2001	
109	WN	JJZ3437	Dennis Dart SLF 8.8m	Plaxton Pointer MPD	N29F	2001	
110	WN	R110VNT	Mercedes-Benz Vario 0810	Plaxton Beaver 2	B27F	1997	
111	WN	R211VNT	Mercedes-Benz Vario 0810	Plaxton Beaver 2	B27F	1997	
112	AG	JJZ5248	Dennis Dart SLF 8.8m	Plaxton Pointer MPD	N29F	2001	
114	WN	K405FHJ	Dennis Dart 9.8m	Plaxton Pointer	B40F	1993	
115	WN	OUI7148	Dennis Dart SLF 10.8m	Marshall Capital	N39F	1997	Halton Bus, 2005
116	WN	OUI9120	Dennis Dart SLF 10.8m	Marshall Capital	N39F	1997	Halton Bus, 2005
117	WN	JJZ5291	Dennis Dart SLF 10.1m	East Lancs Spryte	N40F	1997	Durham Travel, London, 2005
118	WN	JJZ5312	Dennis Dart SLF 10.1m	East Lancs Spryte	N40F	1997	Durham Travel, London, 2005
119	AG	M456EDH	Mercedes-Benz 811D	Marshall C16	B31F	1999	Veolia England, 2007
120	WN	Y381HKE	Dennis Dart SLF 8.8m	Plaxton Pointer MPD	N29F	2001	Metrobus, Crawley, 2006
121	WN	Y382HKE	Dennis Dart SLF 8.8m	Plaxton Pointer MPD	N29F	2001	Metrobus, Crawley, 2006
122	WN	Y383HKE	Dennis Dart SLF 8.8m	Plaxton Pointer MPD	N29F	2001	Metrobus, Crawley, 2006
123	WN	Y384HKE	Dennis Dart SLF 8.8m	Plaxton Pointer MPD	N29F	2001	Metrobus, Crawley, 2006
124	WN	Y385HKE	Dennis Dart SLF 8.8m	Plaxton Pointer MPD	N29F	2001	Metrobus, Crawley, 2006
125	WN	Y386HKE	Dennis Dart SLF 8.8m	Plaxton Pointer MPD	N29F	2001	Metrobus, Crawley, 2006
126	WN	Y387HKE	Dennis Dart SLF 8.8m	Plaxton Pointer MPD	N29F	2001	Metrobus, Crawley, 2006
127	WN	Y388HKE	Dennis Dart SLF 8.8m	Plaxton Pointer MPD	N29F	2001	Metrobus, Crawley, 2006
128	WN	Y389HKE	Dennis Dart SLF 8.8m	Plaxton Pointer MPD	N29F	2001	Metrobus, Crawley, 2006
129	WN	Y391HKE	Dennis Dart SLF 8.8m	Plaxton Pointer MPD	N29F	2001	Metrobus, Crawley, 2006
130	WN	Y392HKE	Dennis Dart SLF 8.8m	Plaxton Pointer MPD	N29F	2001	Metrobus, Crawley, 2006
131	WN	Y393HKE	Dennis Dart SLF 8.8m	Plaxton Pointer MPD	N29F	2001	Metrobus, Crawley, 2006
132	WN	T310MBU	Dennis Dart SLF 8.8m	Marshall Capital	N28F	1999	Nip-on, St Helens, 2006
133	AG	M718WUD	Dennis Dart 9.8m	Marshall C37	B40F	1995	Central Parking, Heathrow, 2007
134	WN	P134MEH	Mercedes-Benz 709D	Plaxton Beaver	B27F	1996	
135	WN	Y301KNB	Dennis Dart SLF 8.8m	Alexander ALX200	N29F	2001	Al's Coaches, Birkenhead, 2006
136	WN	Y302KNB	Dennis Dart SLF 8.8m	Alexander ALX200	N29F	2001	Al's Coaches, Birkenhead, 2006
137	WN	S194FFM	Dennis Dart SLF 10.7m	Marshall Capital	N39F	1998	Halton Transport, 2007
138	WN	S197FFM	Dennis Dart SLF 10.7m	Marshall Capital	N39F	1998	Halton Transport, 2007
139	WN	S195FFM	Dennis Dart SLF 10.7m	Marshall Capital	N39F	1998	Halton Transport, 2007
140	WN	S42FWY	Optare Solo M850	Optare	N29F	1999	A1A, Birkenhead, 2008
141	WN	R741BUJ	Optare Excel L1150	Optare	N39F	1998	
142	w	R742BUJ	Optare Excel L1150	Optare	N39F	1998	
143	AG	R743BUJ	Optare Excel L1150	Optare	N39F	1998	
144	AG	R744BUJ	Optare Excel L1150	Optare	N39F	1998	
145	WN	S119RCS	Dennis Dart SLF 10.7m	Marshall Capital	N43F	1998	Gibson Direct, Renfrew, 2007
146	WN	S582PGB	Dennis Dart SLF 8.8m	Plaxton Pointer MPD	N29F	1998	Gibson Direct, Renfrew, 2007
147	WN	W895AGA	Dennis Dart SLF 8.8m	Plaxton Pointer MPD	N29F	2000	Gibson Driect, Renfrew, 2007

Crewe has seen several additonal services operated by D&G in recent months. Pulling away from the bus station is 179, K574NHC, an Alexander-bodied Dart new to Sussex Coastline.
Alan Blagburn

148	WN	X332ABU	Optare Solo M850	Optare	N29F	2000	Henderson, Hamilton, 2007
149	WN	X349AUX	Optare Solo M850	Optare	N28F	2000	
150	WN	X939NUB	Optare Solo M850	Optare	N29F	2000	Dawsonrentals, 2007
151	WN	X943NUB	Optare Solo M850	Optare	N29F	2000	Dawsonrentals, 2007
152	AG	S2CLA	Optare Excel L1150	Optare	N39F	1998	Classic, Annfield Plain, 1999
153	AG	S3CLA	Optare Excel L1150	Optare	N39F	1998	Classic, Annfield Plain, 1999
154	AG	R987EWU	Optare Excel L1000	Optare	N33F	1998	TGM Logistics, 2000
155	WN	WL03AUL	Optare Solo M920	Optare	N33F	2003	Rowe, Kilmarnock, 2007
156	WN	G122PGT	Mercedes-Benz 811D	Alexander Sprint	B28F	1990	MK Metro, 2001
157	WN	V3JPT	Optare Solo M920	Optare	N37F	1999	J P Travel, Middleton, 2002
158	WN	G114PGT	Mercedes-Benz 811D	Alexander Sprint	B28F	1990	MK Metro, 2003
160	WN	X209ONH	Dennis Dart SLF 8.8m	Plaxton Pointer MPD	N29F	2001	Birmingham Motor Traction, 2007
161	WN	V332CVV	Dennis Dart SLF 8.8m	Plaxton Pointer MPD	N29F	2001	Birmingham Motor Traction, 2007
162	WN	YN53ELV	Optare Solo M850	Optare	N26F	2003	
163	WN	YN53SVT	Optare Solo M920	Optare	N33F	2003	
164	WN	X731FPO	Optare Solo M920	Optare	N27F	2000	Firstbus, 2008
168	WN	DX04MVR	Optare Solo M920	Optare	N33F	2004	*Operated on behalf of Shropshire CC*
169	WN	DX04XMS	Optare Solo M920	Optare	N31F	2004	*Operated on behalf of Shropshire CC*
171	WN	DX05OMB	Optare Solo M920	Optare	N33F	2005	*Operated on behalf of Shropshire CC*
172	WN	W304EYG	Optare Solo M850	Optare	N23F	2000	Dawsonrentals, 2007
173	WN	W282EYG	Optare Solo M850	Optare	N23F	2000	Dawsonrentals, 2007
174	WN	W287EYG	Optare Solo M850	Optare	N23F	2000	Dawsonrentals, 2007
175	WN	W291EYG	Optare Solo M850	Optare	N23F	2000	Dawsonrentals, 2007
176	WN	W298EYG	Optare Solo M850	Optare	N23F	2000	Dawsonrentals, 2007
177	WN	W299EYG	Optare Solo M850	Optare	N23F	2000	Dawsonrentals, 2007
179	CR	K574NHC	Dennis Dart 9.8m	Alexander Dash	B40F	1992	Sussex Coastline, 2008

Previous registrations:

CHZ8960	DK54KKZ	JJZ5368	S781RNE, 98D70555
CNZ2250	DK54KKY	JJZ5312	R521YRP
EJZ2291	L38OKV	LLZ2349	N322YNC
ENZ2127	M728MBU	LUI9633	L655MYG
ENZ4635	L903JRN	LUI9649	N754OYR
FJZ4196	N248OYR	OUI2376	T429EBD
JJZ3437	X706UKS	OUI7148	R408XFL
JJZ5248	DE51EWJ	OUI9120	R409XFL
JJZ5250	SL02GYE	OUI9143	T428EBD
JJZ5278	X702UKS	S119RCS	S400CBC
JJZ5289	SL02GYD	VIL8577	N321YNC
JJZ5291	R519YRP, 97D83740, 98D70818	VIL8677	V719GGE

Depots: Mossfield Road, Adderley Green, Longton (AG - D&G); Lockett Street, Crewe(CR - D&G) and Planetary Road, Wednesfield, Wolverhampton (WN - Choice). **Web:** www.dgbus.co.uk

D&G's CityRider livery is shown on Optare Solo 62, YJ54BSY. Many of the D&G services operate from Hanley and the Stoke-on-Trent area while Choice concentrates further south.
Alan Blagburn

DRM

D R Morris, The Coach Garage, Bromyard, HR7 4NT

MOI9565	Leyland National 11351/1R	East Lancs Greenway (1994)	B49F	1976	Arriva Kent & Sussex, 2003
MOI5055	Volvo B10M-56	Alexander P	BC53F	1987	First, 2007
MOI4000	Volvo B10M-61	East Lancashire 2000 (1995)	BC53F	1990	Classic, Paignton, 1995
PJI9172	Leyland Lynx LX2R11C15Z4R	Leyland	B49F	1990	First, 2007
W1DRM	Volvo B10BLE	Alexander ALX300	N44F	2000	
DM51BUS	Volvo B6BLE	East Lancashire Spryte	N33F	2001	
UK02DRM	Scania OmniCity CN94 UB	Scania	N42F	2002	Scania demonstrator, 2004
UK54DRM	Scania OmniCity CN94 UB	Scania	N42F	2004	Scania demonstrator, 2005
DM55DRM	Scania OmniCity CN94 UB	Scania	N42F	2005	Scania demonstrator, 2006
YN56EZV	Scania OmniCity CN94 UB	Scania	N42F	2006	Scania demonstrator, 2007
UK56DRM	Scania OmniCity CN230 UB	Scania	N42F	2006	

Previous registrations:

DM55DRM	YN55RCY		MOI9565	JOX490P, PDZ6273
			PJI9172	G146HND
MOI4000	G68RGG		UK02DRM	YU02GGZ
MOI5055	D112GHY, TPR354		UK54DRM	YN54OBA

DRM now operates five Scania OmniCity buses. UK54DRM is shown. *David Longbottom*

English Bus Handbook: Notable Independents

DELAINE

Delaine Buses Ltd, 8 Spalding Road, Bourne, PE10 9LE

116	M1OCT	Volvo Olympian	East Lancs	B51/35F	1995	
117	M2OCT	Volvo Olympian	East Lancs	B51/35F	1995	
118	N3OCT	Volvo Olympian	East Lancs	B51/35F	1995	
121	P1OTL	Volvo Citybus B10M-55	East Lancs	B53F	1996	
122	P2OTL	Volvo Citybus B10M-55	East Lancs	B53F	1996	
127	R4OCT	Volvo Olympian	East Lancs	B51/35F	1997	
128	S5OCT	Volvo Olympian	East Lancs	B47/33F	1998	
129	T6OCT	Volvo Olympian	East Lancs	B47/33F	1999	
130	X7OCT	Volvo B7TL	East Lancs Vyking	N47/31F	2000	
135	Y8OCT	Volvo B7TL	East Lancs Vyking	N47/31F	2001	
136	AD03OCT	Volvo B7TL	East Lancs Vyking	N47/31F	2002	
137	P87SAF	Volvo B10B	Wright Renown	B51F	1997	Hopley, Mount Hawke, 2003
138	P112SGS	Volvo B10B	Wright Renown	B51F	1997	Sovereign, Stevenage, 2003
139	AD04OCT	Volvo B7TL	East Lancs Vyking	N45/31F	2004	
140	AD05OCT	Volvo B7TL	East Lancs Vyking	N45/31F	2005	
141	AD56DBL	Volvo B9TL	East Lancs Olympus	N47/31F	2006	
142	AD07DBL	Volvo B9TL	East Lancs Olympus	N47/31F	2007	
143	AD08DBL	Volvo B9TL	East Lancs Olympus	N47/31F	2008	
144	AD58DBL	Volvo B9TL	East Lancs Olympus	N47/31F	2008	
145	AD09DBL	Volvo B9TL	Optare Olympus	N47/31F	2009	

Special event vehicles:

45	KTL780	Leyland Titan PD2/20	Willowbrook	B35/28R	1956
50	RCT3	Leyland Titan PD3/1	Yeates	B39/34R	1960
72	ACT540L	Leyland Atlantean AN68/2R	Northern Counties	B47/35F	1973
100	E100AFW	Leyland Tiger TRCTL11/2RZ	Duple Dominant	S59F	1987

Named vehicles: 141 *Hugh Delaine-Smith MBE*; 142 *Thomas Arthur Smith*; 143 *Emma Jane Smith*; 144 *Derek Vickers Tilley*
Web: www.delainebuses.com

Delaine was the first company to operate the East Lancs Olympus body and has recently placed an order for a fifth. Number 144, AD58DBL, is seen in Bourne in September 2008. *Richard Godfrey*

EMSWORTH & DISTRICT

Emsworth & District Motor Services Ltd, Bus Garage, Clovelly Road, Southbourne, PO10 8PE

Reg	Chassis	Body	Seating	Year	History
OUC45R	Leyland Fleetline FE30AGR	MCW	B45/32F	1976	London Buses, 1999
LUA255V	Volvo B58-56	Plaxton Supreme IV	C53F	1980	preservation, 2007
TND409X	DAF MB200	Plaxton Supreme IV	C45DL	1982	Hatts, Foxham, 2002
NYH161Y	Bedford YNT	Plaxton Supreme V Express	C45DL	1983	Netley Waterside Home, 1996
A883SUL	Leyland Titan TNTL11/2RR	Leyland	B44/32F	1983	Stagecoach London, 2001
UOI2679	Van Hool T815	Van Hool Alicron	C49FT	1984	Morgan, Bristol, 1999
D602RGJ	Bedford YMT	Plaxton Derwent 2	B53F	1987	Metrobus, 1999
G505XLO	Leyland Swift LBM6T/2RA	Reeve Burgess Harrier	B41F	1989	Parfitts, Rhondda, 1998
G727RGA	Leyland Swift LBM6T/2RA	Reeve Burgess Harrier	B39F	1989	Arriva Cymru, 2001
G767CDU	Leyland Swift LBM6T/2RA	Reeve Burgess Harrier	B39F	1989	Arriva Cymru, 2001
G516VYE	Dennis Dart 8.5m	Duple Dartline	B31F	1990	Go South Coast, 2007
G39TGW	Dennis Dart 8.5m	Carlyle Dartline	B34F	1990	Tilley, Wainhouse Corner, 2003
H36YCW	Leyland Swift ST2R44C97A4	Reeve Burgess Harrier	B39F	1990	Stagecoach Ribble, 1998
H37YCW	Leyland Swift ST2R44C97A4	Reeve Burgess Harrier	B39F	1990	Stagecoach Ribble, 1998
H201DVM	Van Hool T815	Van Hool Alizée	C53F	1991	Warners, Tewkesbuy, 2005
H204DVM	Van Hool T815	Van Hool Alizée	C45FT	1991	Chambers, Bures, 2002
J502GCD	Dennis Dart 9.8m	Alexander Dash	B41F	1992	Stagecoach, 2006
J503GCD	Dennis Dart 9.8m	Alexander Dash	B41F	1992	Stagecoach, 2006
J504GCD	Dennis Dart 9.8m	Alexander Dash	B41F	1992	Stagecoach South, 2004
J539GCD	Dennis Dart 9.8m	Alexander Dash	B41F	1992	Stagecoach South, 2004
J548GCD	Dennis Dart 9.8m	Alexander Dash	B41F	1992	Stagecoach, 2007
L452UEB	Dennis Dart 9.8m	Marshall C27	B40F	1993	Town & Around, Folkestone, 2003
M3KFC	MAN 11.190	Berkhof Excellence 1000L	C33FT	1994	AR, Hemel Hempstead, 2003
M452LLJ	Dennis Dart 9.8m	East Lancs	B40F	1995	Transdev Yellow Buses, 2008
M456LLJ	Dennis Dart 9.8m	East Lancs	B40F	1995	Transdev Yellow Buses, 2008
N731RDD	Mercedes-Benz 709D	Alexander Sprint	BC25F	1996	Stagecoach, 2007
C8LEA	Mercedes-Benz Vario 0814	Autobus Nouvelle 2	C29F	1998	Palmer, Southall, 2001
W195CDN	DAF SB3000	Van Hool T9 Alizée	C48FT	2000	Eavesway, Wigan, 2008
Y297HUA	DAF SB3000	Van Hool T9 Alizée	C48FT	2000	Eavesway, Wigan, 2008

Previous registrations:

C8LEA	R749ECT	LUA255V	LUA255V, AJF405A, SIB2633, 3927TR, KAO221V
H201DVM	H201DVM, 315ASV	UOI2679	A102OYG, WPT43

Web: www.emsworth&district.co.uk

Stagecoach has been the source for several recent purchases by Emsworth & District, many being Darts with Alexander Dash bodywork. J504GCD is seen in its new colours.
Richard Godfrey

ENSIGNBUS

Ensignbus - City Sightseeing

Bath Bus Co Ltd; Ensign Bus Company Ltd, Juliette Close, Purfleet Industrial Park, Purfleet, RM15 4YF

107	P	YJ51ZVO	DAF DB250	Optare Spectra	N47/27F	2001	Reading Buses, 2007
111	P	PO58NPG	Volvo B9TL	East Lancs Olympus	N51/31F	2008	
112	P	PO58NPJ	Volvo B9TL	East Lancs Olympus	N51/31F	2008	
113	P	PO58NPK	Volvo B9TL	East Lancs Olympus	N51/31F	2008	
114	P	PO58NPN	Volvo B9TL	East Lancs Olympus	N51/31F	2008	
115	P	PO58NPP	Volvo B9TL	East Lancs Olympus	N51/31F	2008	
116	P	PO58NPU	Volvo B9TL	East Lancs Olympus	N51/31F	2008	
117	P	PO58NPV	Volvo B9TL	East Lancs Olympus	N51/31F	2008	
118	P	PO58NPX	Volvo B9TL	East Lancs Olympus	N51/31F	2008	
119	P	PO58NPY	Volvo B9TL	East Lancs Olympus	N51/31F	2008	
120	P	PO58NRE	Volvo B9TL	East Lancs Olympus	N51/31F	2008	
124	P	S124RLE	Volvo Olympian	Alexander RH	B43/29F	1998	Metroline, Harrow, 2008
125	P	S125RLE	Volvo Olympian	Alexander RH	B43/29F	1998	Metroline, Harrow, 2008
126	P	P426UUG	Volvo Olympian	Alexander Royale	BC45/27F	1997	Lancashire United, 2005
127	P	P427UUG	Volvo Olympian	Alexander Royale	BC45/27F	1997	Lancashire United, 2005
128	P	WLT428	Volvo Olympian	Alexander Royale	BC45/27F	1997	Lancashire United, 2005
129	P	P429UUG	Volvo Olympian	Alexander Royale	BC45/27F	1997	Lancashire United, 2005
132	P	S132RLE	Volvo Olympian	Alexander RH	B43/29F	1998	Metroline, Harrow, 2008
136	P	S136RLE	Volvo Olympian	Alexander RH	B43/29F	1998	Metroline, Harrow, 2008
138	P	S138RLE	Volvo Olympian	Alexander RH	B43/29F	1998	Metroline, Harrow, 2008
141	BA	A441UUV	MCW Metrobus DR102/45	MCW	O43/30F	1984	Stagecoach, 2007
156	P	S126RLE	Volvo Olympian	Alexander RH	B43/29F	1998	Metroline, Harrow, 2008
159	P	WJY759	Volvo Olympian	Alexander RH	B43/29F	1998	Metroline, Harrow, 2008
195	P	P495SWC	Volvo Olympian	Alexander RH	BC47/31F	1997	Dublin Bus, 2008
200	CF	F815YLV	MCW Metrobus DR132/15	MCW	PO46/31F	1989	Arriva NW & Wales, 2003
214	CH	BOK68V	MCW Metrobus DR101/12	MCW	PO43/30F	1980	Midland (Stevensons), 1997
220	NR	G520VBB	Leyland Olympian ON2R50C13Z4	Northen Counties	O47/31F	1990	Arriva London, 2007
233	P	KYV633X	MCW Metrobus DR101/14	MCW	PO31/14F	1981	OLST, 2005
238	CH	D238FYM	Leyland Olympian ONLXB/1R	Eastern Coach Works	PO42/26D	1987	Arriva London, 2004
248	EB	N548LHG	Volvo Olympian	Northen Counties Palatine	O48/27D	1995	London Central, 2004
270	GY	KYV670X	MCW Metrobus DR101/14	MCW	PO43/28D	1981	London General, 2003
272	BA	EU05VBG	Volvo B7L	Ayats Bravo	PO55/24F	2005	
273	BA	EU05VBJ	Volvo B7L	Ayats Bravo	PO55/24F	2005	
274	DA	EU05VBK	Volvo B7L	Ayats Bravo	PO55/24F	2005	
275	WI	EU05VBP	Volvo B7L	Ayats Bravo	PO55/24F	2005	
276	CF	EU05VBT	Volvo B7L	Ayats Bravo	PO55/24F	2005	
303	WI	K703BBL	DAF DB250	Optare Spectra	O47/27F	1992	Reading Buses, 2007
307	WI	WLT307	Volvo Olympian	Northen Counties Palatine	PO48/27D	1995	London Central, 2004
313	EB	J813HMC	Scania N113DRB	Alexander RH	O47/23D	1991	Metroline, 2003
325	EB	H225LOM	Scania N113DRB	Alexander RL	O45/29D	1990	Travel West Midlands, 2005
331	CF	F31XOF	MCW Metrobus DR102/64	MCW	O43/30F	1989	Travel West Midlands, 2002
335	CH	D235FYM	Leyland Olympian ONLXB/1R	Eastern Coach Works	PO42/26D	1986	Arriva London, 2004
345	BL	F45YHB	Scania N113DRB	Alexander RH	O47/33F	1989	Newport, 2003
346	CF	M646RCP	DAF DB250	Northen Counties Palatine 2	O47/30F	1995	City Sightseeing, Florence, 2004
347	AB	M647RCP	DAF DB250	Northen Counties Palatine 2	O47/30F	1995	City Sightseeing, Florence, 2004
351	CH	D251FYM	Leyland Olympian ONLXB/1R	Eastern Coach Works	PO42/26D	1987	Arriva London, 2004
352	CF	F52XOF	MCW Metrobus DR102/64	MCW	O43/30F	1989	Travel West Midlands, 2003
367	WI	E767LBT	Leyland Olympian ONCL10/1RZ	Northern Counties	O43/28F	1988	Arriva Yorkshire, 2002
372	BA	EU05BZM	Volvo B7L	Ayats Bravo City	O55/24F	2005	
373	BA	EU05VBL	Volvo B7L	Ayats Bravo City	O55/24F	2005	
374	BA	EU05VBM	Volvo B7L	Ayats Bravo City	O55/24F	2005	
375	BA	EU05VBN	Volvo B7L	Ayats Bravo City	O55/24F	2005	
376	WI	EU05VBO	Volvo B7L	Ayats Bravo City	O55/24F	2005	
381	BA	EU04CPV	Volvo B7L	Ayats Bravo City	O59/26F	2004	
382	P	EU04CUW	Volvo B7L	Ayats Bravo City	O59/26F	2004	
383	BA	EU04CZS	Volvo B7L	Ayats Bravo City	O59/26F	2004	
384	BA	EU04CZR	Volvo B7L	Ayats Bravo City	O59/26F	2004	
385	BL	B285WUL	MCW Metrobus DR101/17	MCW	O43/28D	1983	Arriva London, 2003
393	CH	D183FYM	Leyland Olympian ONLXB/1R	Eastern Coach Works	PO42/26D	1986	Arriva London, 2004
404	P	304CLT	MCW Metroliner DR140/3	MCW	C47/16DT	1988	Stagecoach, 2007
464	P	864DYE	MCW Metroliner DR140/4	MCW	C47/18DT	1988	Stagecoach, 2007

Ensign is one of Britain's premier bus dealerships as well as operating local services under the Ensignbus name. Seen at Lakeside is Dart 792, R692MEW, which has a Marshall Capital body. *Richard Godfrey*

702	P	P82MOR	Dennis Dart SLF 10.7m	UVG UrbanStar	N39F	1997	National Express, 2004
703	P	P503MOT	Dennis Dart SLF 10.7m	UVG UrbanStar	N39F	1997	National Express, 2004
704	P	P504MOT	Dennis Dart SLF 10.7m	UVG UrbanStar	N39F	1997	National Express, 2004
705	P	P505MOT	Dennis Dart SLF 10.7m	UVG UrbanStar	N39F	1997	National Express, 2004
710	P	P910MOR	Dennis Dart SLF 10.7m	UVG UrbanStar	N39F	1997	National Express, 2004
714	P	P514CVO	Dennis Dart SLF 10.7m	East Lancs Flyte	N44F	1996	City of Nottingham, 2008
719	P	R619VEG	Dennis Dart SLF 10.2m	Marshall Capital	N37F	1998	Metroline, Harrow, 2006
720	P	R620VEG	Dennis Dart SLF 10.2m	Marshall Capital	N37F	1998	Metroline, Harrow, 2006
721	P	R621VEG	Dennis Dart SLF 10.2m	Marshall Capital	N37F	1998	Metroline, Harrow, 2007
722	P	R622VEG	Dennis Dart SLF 10.2m	Marshall Capital	N37F	1998	Metroline, Harrow, 2006
723	P	R623VEG	Dennis Dart SLF 10.2m	Marshall Capital	N37F	1998	Metroline, Harrow, 2007
724	P	R624VEG	Dennis Dart SLF 10.2m	Marshall Capital	N37F	1998	Metroline, Harrow, 2006
725	P	R625VEG	Dennis Dart SLF 10.2m	Marshall Capital	N37F	1998	Metroline, Harrow, 2007
726	P	R626VEG	Dennis Dart SLF 10.2m	Marshall Capital	N37F	1998	Metroline, Harrow, 2007
727	P	R627VEG	Dennis Dart SLF 10.2m	Marshall Capital	N37F	1998	Metroline, Harrow, 2007
729	P	R629VEG	Dennis Dart SLF 10.2m	Marshall Capital	N37F	1998	Metroline, Harrow, 2007
730	P	R630VEG	Dennis Dart SLF 10.2m	Marshall Capital	N37F	1998	Metroline, Harrow, 2006
731	P	R631VEG	Dennis Dart SLF 10.2m	Marshall Capital	N37F	1998	Metroline, Harrow, 2006
732	P	R632VEG	Dennis Dart SLF 10.2m	Marshall Capital	N37F	1998	Metroline, Harrow, 2007
738	P	R638VEG	Dennis Dart SLF 10.2m	Marshall Capital	N37F	1998	Metroline, Harrow, 2006
761	P	R561UOT	Dennis Dart SLF 10.7m	UVG UrbanStar	N44F	1997	TM Travel, Sheffield, 2007
764	P	R864MCE	Dennis Dart SLF 10.2m	Marshall Capital	N28D	1998	Metroline, Harrow, 2008
776	P	R876MCE	Dennis Dart SLF 10.2m	Marshall Capital	N28D	1998	Metroline, Harrow, 2008
783	P	R83GNW	Dennis Dart SLF 10.7m	UVG UrbanStar	N37F	1998	TM Travel, Sheffield, 2007
792	P	R692MEW	Dennis Dart SLF 10.2m	Marshall Capital	N37F	1998	Metroline, Harrow, 2007

Ensignbus now runs ten Volvo B9TLs with East Lancs Olympus bodywork. Illustrating the batch as it passes through Chadwell St Mary is 117, PO58NPV. *Richard Godfrey*

Special event vehicles - museum unit (* fully licenced)

T499	P	ELP223	AEC Regal III 0662	LPTB	C33F	1938	London Transport
RT8	P*	FXT183	AEC Regent III 0661	LPTB	B30/26R	1939	London Transport
RT624	P*	JXC432	AEC Regent III 0961	Weymann	B30/26R	1949	London Transport
RT1239	P	KGK708	AEC Regent III 0961	Saunders	B30/26R	1949	London Transport
RT1431	P*	JXC194	AEC Regent III 0961	Cravens	B30/26R	1949	London Transport
RT1499	P*	KGK758	AEC Regent III 0961	Cravens	B30/26R	1949	London Transport
TD695	P*	HLJ44	Bristol K6A	Eastern Coach Works	L27/28R	1949	Hants & Dorset
SP665	P	KR8385	Leyland Tiger TS2	Burlingham	B30F		
TD161	P	KR1728	Leyland Titan TD1	Short	B30/26R	1950	Maidstone & District
RT3232	P*	KYY961	AEC Regent III 0961	Weymann	B30/26R	1950	London Transport
RTW335	P	KXW435	Leyland Titan PD2	Leyland	B30/26R	1950	London Transport
RLH61	P*	MXX261	AEC Regent III 9613	Weymann	B27/26R	1952	London Transport
MLL735	P	MLL735	AEC Regal IV	Park Royal	C18C	1953	London Transport
RT4421	P*	NXP775	AEC Regent III 0961	Weymann	B30/26R	1954	London Transport
5280	P*	5280NW	Leyland Titan PD3/5	Roe	B38/32R	1959	Leeds
RM25	CF*	VLT25	AEC Routemaster R2RH	Park Royal	B36/28R	1959	London Transport
RM54	P*	LDS279A	AEC Routemaster R2RH	Park Royal	B36/28R	1959	London Transport
RM1361	P*	VYJ808	AEC Routemaster R2RH	Park Royal	B36/28R	1962	London Transport
TC710	P*	BCJ710B	Leyland Tiger Cub PSUC1/12	Harrington Cavalier	C45F	1963	Wye Valley Motors, Hereford
204	P*	KTJ204C	Leyland Titan PD2/37	East Lancashire	B37/28F	1965	Lancaster City Transport
DMO3	P	CRU184C	Daimler Fleetline CRG6LX	Weymann	O43/31F	1965	Bournemouth
RCL2220	P*	CUV220C	AEC Routemaster R2RH/3	Park Royal	CO40/27R	1965	London Transport
RCL2226	P	CUV226C	AEC Routemaster R2RH/3	Park Royal	B40/26R	1965	London Transport
RML2405	P*	JJD405D	AEC Routemaster R2RH/3	Park Royal	B40/26R	1966	London Transport
RMA58	P*	NMY655E	AEC Routemaster R2RH	Park Royal	B32/24F	1967	London Transport (BEA)
RML2665	BA*	SMK665F	AEC Routemaster R2RH/3	Park Royal	B40/26R	1968	London Transport
333	P	EGP33J	Daimler Fleetline CRG6LXB	Park Royal	O45/23F	1970	London Transport
DM2646	P*	THX646S	Leyland Fleetline FE30AGR	Park Royal	B44/28D	1978	London Transport
M1	P	THX101S	MCW Metrobus DR101/3	MCW	B43/28D	1978	London Transport
549	P	A249SVW	Leyland Tiger TRCTL11/3RP	Duple Laser	C57F	1984	Southend
192	P*	F292NHJ	MCW Metrobus DR102/71	MCW	B46/31F	1988	London Pride, 2001

One of the many restored buses retained by Ensign is RT1499, KGK758, seen here on a special in Greenhithe. It is one of a pair of rare survivors from the non-standard Cravens-built batch of RTs that were early withdrawals from service in London. *Richard Godfrey*

Previous registrations:

864DYE	E906TOJ, YTC858	LAT508V	LAT508V, WLT428
304CLT	E99AAK, HE8899	P495SWC	97D334
E767LBT	TWY7	VYJ808	361CLT
GYE420W	GYE420W, 304CLT	WJY759	S128RLE
H648PVW	EV2009(HK)	WLT307	N537LHG
KGY158V	BYX171V, VLT71	WLT428	P428UUG

Web: www.ensignbus.com; www.city-sightseeing.com

Depots and locations for Ensign's buses: Purfleet Industrial Park, Purfleet (P); Aberdeen (AB); Bath (BA); Bristol (BL); Cardiff (CF); Chester (CH); Eastbourne (EB); Great Yarmouth (GY); Norwich (NR) and Windsor (WI).

Citysightseeing tours are operated in partnership with local operators at the following locations: Cambridge, Colchester, Edinburgh, Felixstowe, Glasgow, Ipswich, Leicester, Llandudno, London, Newcastle, Newport, Oxford, South Tyneside, Stratford-upon-Avon, Thanet and York.

FELIX

Felix Bus Services Ltd, 157 Station Road, Stanley, Ilkeston, DE7 6FJ

J564URW	Leyland Lynx LX2R11C15245	Leyland Lynx 2	B49F	1992	Volvo demonstrator, 1992
M301KRY	Volvo B10B-58	Alexander Strider	B51F	1995	
V708GRY	Volvo B10BLE	Alexander ALX300	N43F	2000	
W709PTO	Volvo B10M-62	Plaxton Paragon	C53F	2000	
FE02LWD	Optare Solo M920	Optare	N33F	2002	
FG52WUC	Optare Solo M920	Optare	N33F	2003	
YN03WRA	Scania L94UB	Wrightbus Solar	N43F	2003	
YN04AGY	Scania L94UB	Wrightbus Solar	N43F	2004	
YN05GZB	Scania L94UB	Wrightbus Solar	N43F	2005	
YN55YSE	Irisbus EuroRider 397E.12.31	Plaxton Paragon	C53F	2006	
YN55YSF	Irisbus EuroRider 397E.12.31	Plaxton Paragon	C53F	2006	
YJ56KBF	VDL Bus SB200	Plaxton Centro	N45F	2007	
YJ08EFL	VDL Bus SB200	Plaxton Centro	N45F	2008	

Continuing to provide services between Derby and Ilkeston is Felix using the black cat motif to good effect. One of three Scania buses operated is YN03WRA, seen here in Derby. *Dave Heath*

FISHWICK

John Fishwick & Sons Ltd, Golden Hill Garage, Golden Hill Lane, Leyland, PR5 2LE

1	L1JFS	DAF SB120	Wright Cadet	N42F	2000	
2w	A462LFV	Leyland Atlantean AN69/2L	Eastern Coach Works	B47/35F	1983	
3	J7JFS	Leyland Lynx LX2R11C15Z4R	Leyland Lynx 2	B47F	1991	
4	H64CCK	Leyland Lynx LX2R11C15Z4R	Leyland Lynx 2	B47F	1991	
5	H65CCK	Leyland Lynx LX2R11C15Z4R	Leyland Lynx 2	B47F	1991	
6	YJ54CFN	VDL Bus SB120	Wrightbus Cadet 2	N39F	2005	
7	YJ08EFS	VDL Bus SB200	Wrightbus Commander	N44F	2008	
8	YJ08EFW	VDL Bus SB200	Wrightbus Commander	N44F	2008	
9	YJ54CFM	VDL Bus SB120	Wrightbus Cadet 2	N39F	2005	
10	YJ53VDT	DAF SB200	Wrightbus Commander	N44F	2004	
11	YJ53VDV	DAF SB200	Wrightbus Commander	N44F	2004	
14	J14JFS	Leyland Lynx LX2R11C15Z4R	Leyland Lynx 2	B47F	1992	
15	R845VEC	Dennis Dart SLF	Wright Crusader	N41F	1997	
16	OFV620X	Leyland National 2 NL116AL11/1R		B49F	1981	
17	R846VEC	Dennis Dart SLF	Wright Crusader	N41F	1997	
18	R847VEC	Dennis Dart SLF	Wright Crusader	N41F	1997	
19	R848VEC	Dennis Dart SLF	Wright Crusader	N41F	1997	
20	UHG150V	Leyland Atlantean AN68A/2R	East Lancs	B49/33F	1980	Preston Bus, 2000
22	UHG149V	Leyland Atlantean AN68A/2R	East Lancs	B49/33F	1980	Preston Bus, 2000
23w	GRN895W	Leyland Atlantean AN69/1L	Eastern Coach Works	B43/31F	1981	
24	YJ07JWD	VDL Bus SB200	Plaxton Centro	N44F	2007	
25	YJ07JWE	VDL Bus SB200	Plaxton Centro	N44F	2007	
26	YJ07JDZ	VDL Bus SB200	Plaxton Centro	N44F	2007	
27	YJ57BLF	VDL Bus SB200	Plaxton Centro	N45F	2007	
28	YJ55KZP	VDL Bus SB200	Wrightbus Commander	N44F	2005	
29	YJ55KZR	VDL Bus SB200	Wrightbus Commander	N44F	2005	

John Fishwick & Sons continue their progression from Leyland products to DAF/VDL buses with the purchase of two further Plaxton Centro-bodied SB200s. From the 2007 intake, 25, YJ07JWE, is seen at Preston rail station.
Mark Bailey

Pictured leaving Preston bus station on a return leg of route 111 to Leyland is Fishwick's 4, H64CCK, one of only four Leyland Lynx that remain in the fleet. It was back in 1907 that John Fishwick started a haulage business in Leyland using a steam lorry. He later obtained a charabanc body that replaced the flat deck of the lorry and he used it for passenger service. Some of his original routes continue to the present day. *Richard Godfrey*

30	G802GSX	Leyland Olympian ONCL10/2R	Alexander RH	B51/34F	1989	Lothian Buses, 2006
31	G806GSX	Leyland Olympian ONCL10/2R	Alexander RH	B51/34F	1989	Lothian Buses, 2006
32	F352WSC	Leyland Olympian ONCL10/2R	Alexander RH	B51/30F	1989	Lothian Buses, 2004
33	F353WSC	Leyland Olympian ONCL10/2R	Alexander RH	B51/30F	1989	Lothian Buses, 2004
34	F355WSC	Leyland Olympian ONCL10/2R	Alexander RH	B51/30F	1989	Lothian Buses, 2004
36	X821NWX	DAF SB120	Wrightbus Cadet	N42F	2001	
37	X822NWX	DAF SB120	Wrightbus Cadet	N42F	2001	
38	X823NWX	DAF SB120	Wrightbus Cadet	N42F	2001	
39	YG52EVY	DAF SB200	Wrightbus Commander	N42F	2002	
40	YG52CFY	DAF SB200	Wrightbus Commander	N44F	2002	
41	YJ03PFG	DAF SB200	Wrightbus Commander	N44F	2003	
42	YJ03PFF	DAF SB200	Wrightbus Commander	N44F	2003	
	YJ04BJF	VDL Bus SB4000	Van Hool T9 Alizée	C42FT	2004	
	YJ05PWN	VDL Bus SB4000	Van Hool T9 Alizée	C42FT	2005	
	YJ06LGA	VDL Bus SB4000	Van Hool T9 Alizée	C42FT	2006	
	YJ07JWF	VDL Bus SB4000	Van Hool T9 Alizée	C42FT	2007	
	YJ07JWG	VDL Bus SB4000	Van Hool T9 Alizée	C42FT	2007	

Depots: Golden Hill Lane, Leyland and Chapel Brow, Leyland.

GELDARDS

Geldards Coaches Ltd, 1 Chapel Lane, Armley, Leeds, LS12 2DJ

-	J992XKU	Ford Transit	Advanced	M14	1992	Kenning, 1994
-	YE06FVH	Vauxhall Vivaro	Vauxhall	M8	2006	
-	SMK702F	AEC Routemaster	Park Royal	B40/32R	1967	London United, 2004
309	E215WBG	Leyland Olympian ONCL10/1RZ	Northern Counties	B41/30F	1988	Arriva Merseyside, 2001
314	J815HMC	Scania N113 DRB	Alexander RH	B47/31F	1991	Metroline, Harrow, 2003
315	J816HMC	Scania N113 DRB	Alexander RH	B47/31F	1991	Metroline, Harrow, 2003
316	J811HMC	Scania N113 DRB	Alexander RH	B47/31F	1991	Metroline, Harrow, 2003
317	J817HMC	Scania N113 DRB	Alexander RH	B47/31F	1991	Metroline, Harrow, 2003
318	F123PHM	Volvo Citybus B10M-50	Alexander RV	B46/33F	1988	Arriva London, 2003
319	F129PHM	Volvo Citybus B10M-50	Alexander RV	B46/33F	1988	Arriva London, 2003
320	F130PHM	Volvo Citybus B10M-50	Alexander RV	B46/33F	1988	Arriva London, 2003
321	F131PHM	Volvo Citybus B10M-50	Alexander RV	B46/33F	1988	Arriva London, 2003
322	G147TYT	Volvo Citybus B10M-50	Alexander RV	B46/33F	1990	Arriva London, 2003
323	YJ04BKG	VDL Bus DB250	East Lancs Pyoneer	BC51/29F	2004	
324	G110FJW	MCW Metrobus DR102/70	MCW	B43/30F	1989	Travel West Midlands, 2004
325	F809TLV	MCW Metrobus DR132/17	MCW	B46/31F	1989	Arriva NW & Wales, 2004
326	F814TLV	MCW Metrobus DR132/17	MCW	B46/31F	1989	Arriva NW & Wales, 2004
327	F824TLV	MCW Metrobus DR132/16	MCW	B46/31F	1989	Arriva NW & Wales, 2004
328	F825TLV	MCW Metrobus DR132/16	MCW	B46/31F	1989	Arriva NW & Wales, 2004
329	G760VRT	Leyland Olympian ONCL10/1RZ	Alexander RL	B47/32F	1989	Mullany, Watford, 2005
330	G365YUR	Leyland Olympian ONCL10/1RZ	Alexander RL	B47/30F	1990	Mullany, Watford, 2005
331	L8YCL	Volvo Olympian	Alexander Royale	BC45/29F	1994	Harrogate & District, 2005
332	L9YCL	Volvo Olympian	Aiexander Royale	BC45/29F	1994	Harrogate & District, 2005
333	P476MBY	Volvo Olympian	Alexander RH	B43/32F	1996	Metroline, Harrow, 2006
334	P478MBY	Volvo Olympian	Alexander RH	B43/32F	1996	Metroline, Harrow, 2006
335	M85MYM	Volvo Olympian	Alexander RH	B45/24F	1995	Talisman, Great Bromley, 2006
336	M87MYM	Volvo Olympian	Alexander RH	B45/24F	1995	Talisman, Great Bromley, 2006

Lettered for school duties is Geldards' Scania 315, J816HMC, one of four in the fleet from Metroline. Only two saloons break the double-deck monopoly in the bus fleet. *Mark Bailey*

Also from Metroline are four Volvo Olympians with Northern Counties Palatine II bodies, including 340, L204SKD, seen here. Formed in 1991, Geldards now carry some 3000 pupils each day on their school services in North and West Yorkshire. *Mark Bailey*

337	G36HKY	Scania N113 DRB	Northern Counties Palatine	B47/33F	1990	Arriva Midlands, 2006
338	G711LKW	Scania N113 DRB	Northern Counties Palatine	B47/33F	1990	Arriva Midlands, 2006
339	G714LKW	Scania N113 DRB	Northern Counties Palatine	B47/33F	1990	Arriva Midlands, 2006
340	L204SKD	Volvo Olympian	Northern Counties Palatine II	B47/25D	1993	Metroline, Harrow, 2007
341	L205SKD	Volvo Olympian	Northern Counties Palatine II	B47/25D	1993	Metroline, Harrow, 2007
342	L207SKD	Volvo Olympian	Northern Counties Palatine II	B47/25D	1993	Metroline, Harrow, 2007
343	L215SKD	Volvo Olympian	Northern Counties Palatine II	B47/25D	1994	Metroline, Harrow, 2007
	YJ51ZVE	DAF DB250	Optare Spectra	N47/29F	2001	Ensignbus, Purfleet, 2008
	YJ51ZVH	DAF DB250	Optare Spectra	N47/29F	2001	Ensignbus, Purfleet, 2008
401	M763RCP	DAF SB220	Ikarus Citibus	B49F	1995	GM Buses, Manchester, 2007
908	YG52CGE	DAF SB3000	Van Hool T9 Alizée	C49FT	2002	
910	YJ04BKK	DAF SB4000	Van Hool T9 Alizée	C49FT	2004	
911	P867PWW	DAF SB3000	Van Hool Alizée HE	C51FT	1997	Brodyr Richards, Cardigan, 2005
912	R161GNW	DAF SB3000	Van Hool Alizée HE	C51FT	1998	Galloway, Mendlesham, 2005
914	T57AUA	DAF SB3000	Van Hool Alizée HE	C48FT	1998	Brodyr Richards, Cardigan, 2005
915	R178GNW	DAF SB3000	Van Hool Alizée HE	C53F	1998	Arriva Midlands, 2005
	YJ08DUA	DAF SB4000	Van Hool T9 Alizée	C50FT	2008	
	YJ08DUU	DAF SB4000	Van Hool T9 Alizée	C51FT	2008	

Previous registration:
R161GNW   R161GNW, 1440PP

Depot: Whitehall Road Industrial Estate, Leeds

GO WHIPPET

Whippet Coaches Ltd, Cambridge Road, Fenstanton, Huntingdon, PE28 9JB

KYN300X	Leyland Titan TNLXB2RR	Leyland	B44/32F	1981	Stagecoach Oxford, 1997
POG514Y	MCW Metrobus DR102/27	MCW	B43/30F	1982	Travel West Midlands, 2008
POG580Y	MCW Metrobus DR102/27	MCW	B43/30F	1983	Travel West Midlands, 2008
A911SYE	Leyland Titan TNLXB2RR	Leyland	B44/32F	1983	Westlink, 1996
B111WUV	Leyland Titan TNLXB2RR	Leyland	B44/32F	1984	London Central, 2000
E911DRD	Leyland Olympian ONLXCT/1RH	Optare	B42/26F	1988	Reading Buses, 2006
E912DRD	Leyland Olympian ONLXB/1RH	Optare	B42/26F	1988	Reading Buses, 2006
E441ADV	Volvo Citybus B10M-50	Alexander RV	BC47/35F	1988	Filer, Ilfracombe, 1991
E176OEW	Volvo Citybus B10M-50	Alexander RV	BC47/35F	1988	
F117PHM	Volvo Citybus B10M-50	Alexander RV	B46/29D	1988	Arriva London, 2003
G823UMU	Volvo Citybus B10M-50	Northern Counties	BC45/35F	1989	
G824UMU	Volvo Citybus B10M-50	Northern Counties	BC45/35F	1989	
G184JHG	Leyland Olympian ONLXB/2RZ	Alexander RL	BC47/31F	1989	Stagecoach, 2008
G340KKW	Leyland Olympian ONLXB/2RZ	Alexander RL	BC51/31F	1989	Stagecoach, 2008
G703TCD	Leyland Olympian ON2R56G13Z4	Alexander RL	BC51/31F	1990	Stagecoach, 2008
H348SWA	Leyland Olympian ON2R56G13Z4	Alexander RL	BC51/31F	1990	Stagecoach, 2008
H303CAV	Volvo Citybus B10M-50	Northern Counties	BC45/35F	1990	
J722KBC	Volvo B10M-60	Plaxton Paramount 3500 III	C53F	1992	
J723KBC	Volvo B10M-60	Plaxton Paramount 3500 III	C53F	1992	
J669LGA	Volvo B10M-60	Van Hool Alizée H	C53F	1992	Shearings, 1997
J670LGA	Volvo B10M-60	Van Hool Alizée H	C53F	1992	Skills, Nottingham, 1997
J687LGA	Volvo B10M-60	Van Hool Alizée H	C53F	1992	Priory, Gosport, 1997
J688LGA	Volvo B10M-60	Van Hool Alizée H	C53F	1992	Skills, Nottingham, 1997
J689LGA	Volvo B10M-60	Van Hool Alizée H	C53F	1992	Skills, Nottingham, 1997

Cambridge is the setting for this view of Go Whippet's S134EJE, one of two Volvo B10BLEs with Alexander ALX300 bodywork. Double-deck buses are progressively being replaced by single-decks. *Richard Godfrey*

Go Whippet was once known for its substantial fleet of former London MCW Metropolitans and later Leyland Titans. Its newest double-decks are three Olympians from East London Buses. These have East Lancs Pyoneer bodies with P345ROO seen here leaving the bus station in Cambridge. 2009 will see the celebration of Go Whippets 90th anniversary of providing passenger services. *Mark Doggett*

J808WFS	Leyland Olympian ON2R56G13Z4	Alexander RL	B47/32F	1992	Stagecoach, 2009
K713ASC	Leyland Olympian ON2R56G13Z4	Alexander RL	B47/32F	1992	Stagecoach, 2009
K699ERM	Volvo B10M-55	Alexander PS	B49F	1992	Stagecoach, 2009
K702DAO	Volvo B10M-55	Alexander PS	B49F	1992	Stagecoach, 2009
L51UNS	Volvo B10B	Northern Counties Paladin	B51F	1994	Whitelaws, Stonehouse, 2001
L56UNS	Volvo B10B	Northern Counties Paladin	B51F	1994	Whitelaws, Stonehouse, 2001
L455YAC	Volvo B6 9.9m	Alexander Dash	BC40F	1994	Stagecoach, 2004
L456YAC	Volvo B6 9.9m	Alexander Dash	BC40F	1994	Stagecoach, 2004
N630XBU	Scania L113CRL	Wright Access-ultralow	N43F	1996	City of Nottingham, 2007
N633XBU	Scania L113CRL	Wright Access-ultralow	N43F	1996	City of Nottingham, 2007
P343ROO	Volvo Olympian	East Lancs Pyoneer	B51/32F	1997	East Thames Buses, 2003
P345ROO	Volvo Olympian	East Lancs Pyoneer	B51/32F	1997	East Thames Buses, 2003
P348ROO	Volvo Olympian	East Lancs Pyoneer	B51/32F	1997	East Thames Buses, 2003
S134EJE	Volvo B10BLE	Alexander ALX300	N44F	1999	
V293UVY	Volvo B10BLE	Alexander ALX300	N46F	2000	Whitelaw, Stonehouse, 2005
W992BDP	Volvo B6BLE	East Lancs Spryte	N31F	2000	Thames Travel, Wallingford, 2005
FE51RAU	Volvo B7TL	East Lancs Vyking	N47/29F	2001	
FE51RBU	Volvo B6BLE	East Lancs Spryte	N41F	2001	
FE51RCU	Volvo B6BLE	East Lancs Spryte	N41F	2001	
FE51RDU	Volvo B6BLE	East Lancs Spryte	N41F	2001	
BD02HDG	Setra S315 GT	Setra	C48FT	2002	Anderson, Bermondsey, 2005
BD02HDJ	Setra S315 GT	Setra	C48FT	2002	Anderson, Bermondsey, 2005
AF52VMD	Scania L94UB	Wrightbus Solar	N43F	2002	

Previous registrations:

		J687LGA	J457HDS, LSK497
		J688LGA	J458HDS, LSK498
J669LGA	J459HDS, LSK499	J689LGA	J460HDS, LSK500
J670LGA	J456HDS, LSK496		

GREEN BUS SERVICE

Warstone Motors Ltd, The Garage, Jacobs Hall Lane, Great Wyrley, WS6 6AD

1	K136ARE	Mercedes-Benz 709D	Wright NimBus	B29F	1992	Arriva Midlands, 2003
2	M92JHB	Mercedes-Benz 709D	Wadham Stringer Wessex	B29F	1995	Stagecoach, 2008
3	NFR748T	Leyland Leopard PSU4E/2R	East Lancs	BC43F	1978	Stagecoach Ribble, 1997
4	GZ2248	Bedford OWB	Duple	B28F	1944,	preservation, 1996
5	J418PRW	Mercedes-Benz 811D	Wright Nimbus	B33F	1991	Stagecoach, 2003
6	NBZ1676	Leyland Leopard PSU4D/4R	East Lancs EL2000 (1995)	B47F	1976	Rider, York, 1992
7	M370LAX	Mercedes-Benz 709D	Alexander Sprint	B23F	1995	Stagecoach, 2007
8	L328CHB	Mercedes-Benz 811D	Marshall C16	B33F	1994	Stagecoach, 2003
9	L255NFA	Mercedes-Benz 709D	Wadham Stringer Wessex	B29F	1994	Arriva Midlands, 2004
10	L308YDU	Mercedes-Benz 709D	Alexander Sprint	B23F	1994	Stagecoach, 2007
11	N462VDD	Mercedes-Benz 709D	Alexander Sprint	B23F	1996	Stagecoach, 2007
14	K321YKG	Mercedes-Benz 709D	Alexander Sprint	B25F	1992	Stagecoach, 2005
15	GCA747	Bedford OB	Duple Vista	C29F	1950	Sargeant, Llanfaredd, 1973
17	N618VSS	Mercedes-Benz 709D	Alexander Sprint	B25F	1996	Stagecoach, 2005
18	K623UFR	Mercedes-Benz 709D	Alexander Sprint	B25F	1993	Stagecoach, 2005

Previous registration:

NBZ1676   RWT527R

Green Bus operates around south Staffordshire and into Wolverhampton from the north with a fleet now comprised mostly of midi and minibuses. 18, K623UFR is a Mercedes-Benz 709D new to Stagecoach Ribble.
Alan Blagburn

HALIFAX JOINT COMMITTEE

A R Blackman, Thrum Hall Industrial Park, Albert Road, Halifax, HX2 0DB

10	C424BUV	MCW Metrobus DR101/17	MCW	B43/28D	1985	Arriva London, 2000
14	K414MGN	Dennis Dart 9m	Plaxton Pointer	B35F	1993	Metroline, Harrow, 2003
17	BYX217V	MCW Metrobus DR101/12	MCW	B43/28D	1980	Go-Ahead Northern, 2003
24	KYO624X	MCW Metrobus DR101/14	MCW	B43/31F	1981	Arriva North East, 2002
30	KYV730X	MCW Metrobus DR101/14	MCW	B43/31F	1982	Arriva North East, 2002
34	F234YTJ	Leyland Olympian ONCL10/1RZ	Alexander RH	B45/30F	1989	Aintree Coachline, Bootle, 2005
38	H138MOB	Dennis Dart 8.5m	Carlyle Dartline	B28F	1991	Boyle, Longfield, 2003
39	H839PVW	Leyland Olympian ON2R50C13Z4	Alexander RH	B47/31F	1991	Dublin Bus, 2008
42	J642CEV	Leyland Olympian ON2R50C13Z4	Alexander RH	B47/31F	1991	Dublin Bus, 2008
43	KYV643X	MCW Metrobus DR101/14	MCW	B43/31F	1981	Arriva North East, 2002
47	G547VBB	Leyland Olympian ONCL10/1RZ	Northern Counties	B47/31F	1990	Arriva London, 2007
49	E749SKR	MCW Metrobus DR102/11	MCW	B46/31F	1988	Ensignbus, Purfleet, 2008
51	H751PVW	Leyland Olympian ON2R50C13Z4	Alexander RH	B47/31F	1991	Dublin Bus, 2008
53	G503VYE	Dennis Dart 8.5m	Duple Dartline	B28F	1990	Boyle, Longfield, 2003
64	H764PVW	Leyland Olympian ON2R50C13Z4	Alexander RH	B47/31F	1991	Dublin Bus, 2008
67	J607KCU	Dennis Dart 9.8m	Wright Handybus	B40F	1992	Ensignbus, Purfleet, 2008
82	J629CEV	Leyland Olympian ON2R50C13Z4	Alexander RH	B47/31F	1991	Dublin Bus, 2007
83	J630CEV	Leyland Olympian ON2R50C13Z4	Alexander RH	B47/31F	1991	Dublin Bus, 2007
85	K985CBO	Dennis Dart 8.5m	Wright Handybus	B29F	1993	Stagecoach, 2004
95	K995CBO	Dennis Dart 8.5m	Wright HandyBus	B29F	1993	Stagecoach, 2004
99	K993CBO	Dennis Dart 8.5m	Wright HandyBus	B29F	1993	Stagecoach, 2004
117	YJ07XND	BMC Condor 220	BMC	S57F	2007	*Operated for West Yorkshire PTE*
118	YJ57NFF	BMC Condor 220	BMC	S57F	2007	*Operated for West Yorkshire PTE*
119	YJ57WKC	BMC Condor 220	BMC	S57F	2007	*Operated for West Yorkshire PTE*
120	YJ57WKB	BMC Condor 220	BMC	S57F	2007	*Operated for West Yorkshire PTE*
162	B162WUL	MCW Metrobus DR101/17	MCW	B43/28D	1985	Arriva London, 2001
196	B196WUL	MCW Metrobus DR101/17	MCW	B43/28D	1985	London General, 2002
203	B203WUL	MCW Metrobus DR101/17	MCW	B43/32F	1985	London General, 2002
217	B217WUL	MCW Metrobus DR101/17	MCW	B43/28D	1985	Arriva London North, 2000
248	BYX248V	MCW Metrobus DR101/12	MCW	B43/31F	1980	Arriva North East, 2002

Special event vehicles:

FFY401	Leyland Titan PD2/3	Leyland	O30/26R	1947	Southport Corporation
FFY403	Leyland Titan PD2/3	Leyland	O30/26R	1947	Southport Corporation
BCP671	AEC Regent III 9612E	Park Royal	B33/26R	1950	Halifax JTC
LJX198	AEC Regent V 2D3RA	Weymann	B39/32F	1959	Hebble
PFN858	AEC Regent V 2LD3RA	Park Royal	B40/32F	1959	East Kent
214CLT	AEC Routemaster R2RH	Park Royal	B36/28R	1962	London Transport
3747RH	AEC Bridgemaster 3B3RA	Park Royal	B43/29F	1963	East Yorkshire, 2008
MDJ918E	AEC Regent V 2D3RA	MCW	B37/28R	1967	St Helens
DTG370V	MCW Metrobus DR101/15	MCW	O46/31F	1980	Newport

Previous registrations:

		J942CEV	92D133
H764PVW	91D1075	K985CBO	NDZ3140
J829CEV	91D10108	K993CBO	NDZ3149
J830CEV	91D10106	K995CBO	NDZ3154

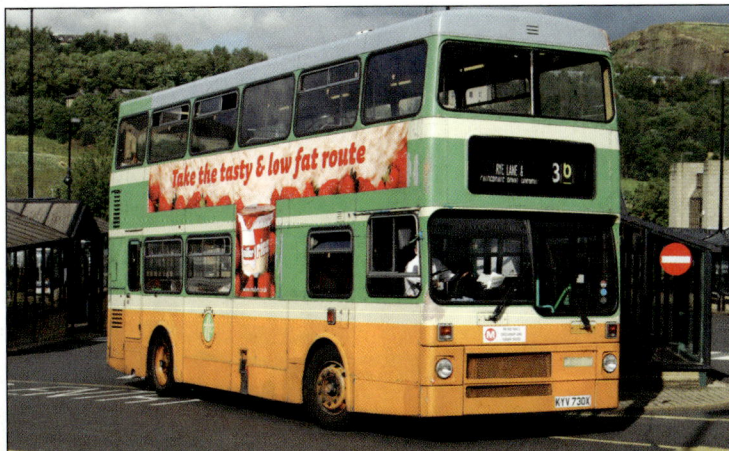

Continuing the traditional Halifax livery is the new Halifax Joint Committee. Metrobus 30, KYV730X is illustrated.
Alan Blagburn

HAM'S TRAVEL

DW PP & J Ham, The White House, London Road, Flimwell, Wadhurst, TN5 7PL

WYV64T	Leyland Titan TNLXB2RRSp	Park Royal	B44/32F	1979	Eastonways, Ramsgate, 1998
CUL162V	Leyland Titan TNLXB2RRSp	Park Royal	B(70)F	1980	Eastonways, Ramsgate, 1998
KYV312X	Leyland Titan TNLXB2RR	Leyland	B44/32F	1981	London Central, 1998
NHM465X	Leyland Titan TNLXB2RR	Leyland	B44/32F	1981	Driver Express, Horsham, 1999
KYV481X	Leyland Titan TNLXB2RR	Leyland	B44/32F	1982	Metroline, 1999
A305MKJ	Leyland Tiger TRCTL11/3R	Plaxton Paramount 3200	C57F	1983	
A933SYE	Leyland Titan TNLXB2RR	Leyland	B44/32F	1984	London Central, 1998
B522HAM	Volvo B10M-61	Plaxton Paramount 3200 III	C57F	1985	Crawley Luxury, 2002
B542HAM	Volvo B10M-61	Plaxton Paramount 3200 II	C53F	1986	Davis Coaches, Hawkhurst, 1997
C521DND	Volvo B10M-61	Plaxton Paramount 3200 II	C53F	1986	Davis Coaches, Hawkhurst, 1997
F865LCU	MCW MetroRider MF158	MCW	B31F	1988	Guide Friday, Stratford, 2002
F254HAM	Volvo B10M-60	Van Hool Alizée	C51FT	1989	Amport & District, Thruxton, 1993
H124THE	Dennis Dart 8.5m	Plaxton Pointer	B28F	1991	Arriva London, 2003
J229HGY	Dennis Dart 9m	Plaxton Pointer	B35F	1992	Metrobus, Crawley, 2004
B44HAM	Volvo B10M-60	Plaxton Première 350	C49FT	1992	Stagecoach Midland Red, 2000
L6HAM	Volvo B10M-62	Jonckheere Deauville 45	C53F	1994	Redwing, Camberwell, 1997
M20HAM	Ford Transit	Ford	M14	1995	Ford demonstrator, 1996
M84MYM	Volvo Olympian	Alexander Royale	B45/29F	1995	National Express, 2004
M91MYM	Volvo Olympian	Alexander Royale	B45/29F	1995	National Express, 2004
N60HAM	Volvo B10M-62	Plaxton Première 320	C57F	1995	Fowler, Holbeach Drove, 2001
N50HAM	Volvo B10M-62	Van Hool Alizée HE	C49FT	1995	Inland Travel, Flimwell, 2001
N1HAM	Volvo B10M-62	Van Hool Alizée HE	C51FT	1996	
P509NWU	Optare MetroRider MR13	Optare	B25F	1996	Metroline, Harrow, 2005
R16HAM	Volvo B10M-62	Berkhof Axial 50	C51FT	1998	Limebourne, Battersea, 1999
S3HAM	Volvo B10M-62	Van Hool T9 Alizée	C51FT	1998	
S4HAM	Volvo B10M-62	Berkhof Axial 50	C51FT	1999	

Now mostly confined to school duties, Titan KYV312X joined the Ham's Travel fleet ten years ago and was pictured in Tunbridge Wells during the 2008 summer. *Dave Heath*

T70HAM	Volvo B10M-62	Plaxton Excalibur	C53F	1999	Bus Eireann, 2003
T80HAM	Volvo B10M-62	Plaxton Excalibur	C53F	1999	Bus Eireann, 2003
T90HAM	Volvo B10M-62	Plaxton Excalibur	C53F	1999	Bus Eireann, 2003
T795TWC	Ford Transit	Ford	M8	1999	private owner, 2006
R17HAM	LDV Convoy	LDV	M16	2000	private owner, 2002
X8HAM	Scania N113DRB	East Lancs Cityzen	B47/31F	2000	
X7HAM	Volvo B12M	Berkhof Axial 50	C49FT	2001	
X9HAM	Volvo B10M-62	Berkhof Axial 50	C51FT	2001	
Y10HAM	LDV Convoy	LDV	M16	2001	
GU51UVE	Toyota Hiace	Autotrim	M8	2001	
MC02HAM	Volvo B12M	Jonckheere Mistral 50	C51FT	2002	Moroney, Bray, 2006
DW02HAM	Volvo B12B	Jonckheere Mistral 50	C49FT	2002	Callinan, Claregalway, 2006
DW52HAM	Volvo B7TL	East Lancs Vyking	NC46/32F	2003	
MC52HAM	Volvo B7TL	East Lancs Vyking	NC46/32F	2003	
PA52HAM	Scania K114IB4	Irizar InterCentury 12.32	C57F	2003	
EU03XOJ	LDV Convoy	LDV	M16	2003	
WR53VYW	Ford Transit	Ford	M8	2004	
KN04CSU	Ford Transit	Ford	M16	2004	National Vehicle Leasing, 2005
DW54HAM	Mercedes-Benz Vario O815	Sitcar Beluga	C33F	2004	
LN54VXL	Ford Transit	Ford	M8	2004	private owner, 2006
HG05URB	Ford Transit	Ford	M16	2005	private owner, 2006
DW05HAM	Volvo B7TL	East Lancs Vyking	NC49/31F	2005	
DW55HAM	Volvo B12B 12.7m	Van Hool Astrobel	C61/18DT	2006	
MC06HAM	TransBus Dart SLF	Neobus (2003)	N45F	2006	Maltese demonstrator
MC56HAM	ADL Dart 4	ADL Enviro 200	N39F	2006	
W5HAM	Volvo B12B	Van Hool Alicron	C57FT	2000	Moseley demonstrator, 2008
MC07HAM	Volvo B9TL	East Lancs Olympus	NC49/31F	2007	
BU57HAM	ADL E300	ADL Enviro 300	N45F	2007	
DW07HAM	Volvo B12B	Van Hool Alicron	C57F	2008	
-	Volvo B12B 12.7m	Van Hool Astrobel	C61/18DT	On order	

Previous registrations:

B44HAM	J446HDS	N50HAM	N750CYA
B142AKN	B420CMC, B20FUG, B44HAM	N60HAM	N54MDW, A10WHF
B522HAM	B270TLJ, 171CLC	NHM465X	KYV377X, 361CLT
B542HAM	C515DND	R16HAM	R927ULA
B552HAM	-	R17HAM	W187TNG
F254HAM	F263NUT	T70HAM	99D41332, T266UCH
DW02HAM	02G8794, WF52EUE	T80HAM	99D41317, T425UCH
L6HAM	L757YGE	T90HAM	99D41329, T308UCH
M20HAM	M437YTW	W5HAM	WA57JZP
MC02HAM	02WW1260, WJ51FKU	WYV64T	WYV64T, LIL2515

Depots: London Road, Flimwell and Cranbrook Road, Benenden.

The latest double-deck for Ham's Travel is MC07HAM, a Volvo B9TL with East Lancs Olympus bodywork. High-back seating is featured on the bus, seen at Wadhurst. *Dave Heath*

HEDINGHAM

Hedingham & District Omnibuses Ltd, Wethersfield Road, Sible Hedingham, Halstead, CO9 3LB

L84	HD	RGV284N	Leyland Leopard PSU3B/4R	Willowbrook	B55F	1974	
L124	SY	B124BOO	Leyland Tiger TRCTL11/2R	Plaxton Paramount 3200	C53F	1985	
L148	SY	WPH135Y	Leyland Tiger TRCTL11/2R	East Lancs EL2000 (1994)	B55F	1982	Kentish Bus, 1988
L150	TY	F150LTW	Leyland Lynx LX112L10ZR1	Leyland Lynx	B51F	1988	
L160	TY	H160HJN	Leyland Olympian ONCL10/1RZ	Alexander RL	B47/32F	1990	
L196	HD	H48NDU	Leyland Lynx LX2R11C15Z4R	Leyland Lynx 2	B51F	1990	Volvo demonstrator, 1992
L198	CN	K198EVW	Dennis Dart 9.8m	Alexander Dash	B43F	1992	
L200	CN	J295TWK	Leyland Lynx LX2R11C14Z4S	Leyland Lynx 2	B51F	1992	Volvo demonstrator, 1992
L206	HD	J724KBC	Leyland Lynx LX2R11V18Z4S	Leyland Lynx 2	B47F	1992	Westbus, Ashford, 1994
L207	TY	L207RNO	Volvo B6 9.9m	Alexander Dash	B40F	1994	
L208	SY	L208RNO	Volvo B6 9.9m	Alexander Dash	B40F	1994	
L210	CN	M210VEV	Dennis Dart 9m	Plaxton Pointer	B34F	1994	
L211	HD	M212WHJ	Dennis Lance SLF	Wright Pathfinder 320	N40F	1995	
L212	HD	M211WHJ	Dennis Lance SLF	Wright Pathfinder 320	N40F	1995	
L233	SY	F148SPV	Leyland Tiger TRCTL11/3ARZ	Plaxton Paramount 3200 III	C57F	1989	Partridge, Hadleigh, 1994
L241	KN	N241EWC	Dennis Dart 9m	Plaxton Pointer	B35F	1996	
L242	SY	J734CWT	Volvo B10M-60	Plaxton Excalibur	C53F	1992	Wallace Arnold, 1996
L243	TY	M262KWK	Volvo B6 9.9m	Alexander Dash	B40F	1995	Volvo demonstrator, 1996
L244	CN	M988NAA	Volvo B10M-62	Plaxton Première 320	C53F	1995	Excelsior, Bournemouth, 1996
L245	CN	SGR780V	Bristol VRT/SL3/6LXB	Eastern Coach Works	B43/31F	1979	NE Bus (Tees), 1996
L251	CN	MRJ8W	Bristol VRT/SL3/6LXB	Eastern Coach Works	BC41/29F	1980	Oxford Bus Company, 1996
L253	CN	PWY39W	Bristol VRT/SL3/6LXB	Eastern Coach Works	B43/31F	1980	Cambus (Viscount), 1996
L255	CN	STW30W	Bristol VRT/SL3/6LXC	Eastern Coach Works	B39/31F	1980	Cambus, 1996
L258	KN	M261KWK	Volvo B6 9.9m	Plaxton Pointer	B40F	1995	Volvo demonstrator, 1997
L281	KN	M107UWY	Volvo B10M-62	Plaxton Première 320	C53F	1995	Wallace Arnold, 1997

A notable difference between the Volvo B6 and the early Dart is seen in the window line of the Plaxton Pointer. Hedingham's L297, M832CVG, shows the B6 layout. *Mark Bailey*

Low-floor Dart L312, W312CJN, joined the fleet in 2000 and was the first of the type for Hedingham. It is seen arriving in Colchester. *Mark Bailey*

L282	CN	M571XKY	Volvo B10M-62	Plaxton Première 350	C53F	1995	A&R International, Bedfont, 97
L283	TY	L202HYE	Volvo B6 9.9m	Plaxton Pointer	BC36F	1994	Terminus, Crawley, 1997
L285	CN	M517NCG	Volvo B10M-62	Plaxton Première 350	C53F	1995	Excelsior, Bournemouth, 1997
L286	CN	P530CLJ	Volvo B10M-62	Plaxton Première 320	C53F	1996	Excelsior, Bournemouth, 1998
L287	CN	L833YDS	Volvo B6 9.9m	Alexander Dash	B40F	1994	Henderson, Hamilton, 1998
L290	CN	S290TVW	Volvo B10M-62	Plaxton Première 320	C53F	1998	
L291	SY	S291IVW	Volvo B10M-62	Plaxton Première 320	C53F	1998	
L295	CN	N664THO	Volvo B10M-62	Plaxton Première 320	C53F	1996	Excelsior, Bournemouth, 1998
L296	KN	M441CVG	Volvo B6 9.9m	Plaxton Pointer	B40F	1995	Lamberts, Beccles, 1999
L297	SY	M832CVG	Volvo B6 9.9m	Plaxton Pointer	B40F	1995	Lamberts, Beccles, 1999
L298	SY	N667THO	Volvo B10M-62	Plaxton Première 320	C53F	1996	Excelsior, Bournemouth, 1999
L299	CN	P395AAA	Volvo B10M-62	Plaxton Première 320	C53F	1997	Excelsior, Bournemouth, 1999
L300	KN	S300XHK	Volvo Olympian	Alexander RL	B51/36F	1999	
L301	HD	S376MVP	Volvo B10BLE	Alexander ALX300	N44F	1998	Volvo demonstrator, 1999
L303	TY	L67UNG	Volvo B6 9.9M	Alexander Dash	B38F	1994	Ambassador, Gt Yarmouth, '99
L304	CN	L69UNG	Volvo B6 9.9M	Alexander Dash	B38F	1994	Ambassador, Gt Yarmouth, '99
L305		WWL212X	Leyland Olympian ONLXB/1R	Eastern Coach Works	B47/32F	1982	Cityline, Oxford, 1999
L306	KN	BBW213Y	Leyland Olympian ONLXB/1R	Eastern Coach Works	B47/32F	1982	Cityline, Oxford, 1999
L307	CN	BBW218Y	Leyland Olympian ONLXB/1R	Eastern Coach Works	B47/28F	1982	Cityline, Oxford, 1999
L309	TY	R453FWT	Volvo B10M-62	Plaxton Première 350	C53F	1998	Wallace Arnold, 2000
L310	KN	VJO202X	Leyland Olympian ONLXB/1R	Eastern Coach Works	B47/32F	1982	The Oxford Bus Company, 2000
L311	KN	W311CJN	Volvo B10M-62	Plaxton Paragon	C53F	2000	
L312	HD	W312CJN	Dennis Dart SLF	Plaxton Pointer 2	N40F	2000	
L313	CN	VJO206X	Leyland Olympian ONLXB/1R	Eastern Coach Works	B47/32F	1982	The Oxford Bus Company, 2000
L314	CN	FWL779Y	Leyland Olympian ONLXB/1R	Eastern Coach Works	B45/32F	1983	The Oxford Bus Company, 2000
L315	TY	FWL780Y	Leyland Olympian ONLXB/1R	Eastern Coach Works	B45/32F	1983	The Oxford Bus Company, 2000
L316	TY	FWL778Y	Leyland Olympian ONLXB/1R	Eastern Coach Works	B45/32F	1983	Weavaway, Newbury, 2001
L317	TY	FWL781Y	Leyland Olympian ONLXB/1R	Eastern Coach Works	B45/32F	1983	Weavaway, Newbury, 2001
L319	KN	N326HUA	Volvo B6 9.9M	Alexander Dash	B40F	1995	Volvo demonstrator, 2001
L323	CN	EK51JBE	Volvo B10M-62	Plaxton Paragon	C53F	2002	
L324	CN	EK51JAU	Volvo B10M-62	Plaxton Paragon	C53F	2002	
L325	HD	EX02RYR	Dennis Dart SLF	Plaxton Pointer MPD	N29F	2002	
L326	CN	EJ02KYY	Dennis Dart SLF	Plaxton Pointer MPD	N29F	2002	
L328	HD	EU03BZK	TransBus Dart	TransBus Pointer	N37F	2003	
L329	TY	Y747HWT	Volvo B10M-62	Plaxton Paragon	C53F	2001	Wallace Arnold, 2003

L330	HD	EU53MVZ	Volvo B7TL	TransBus President	N47/28F	2003	
L331	CN	CUD223Y	Leyland Olympian ONLXB/1R	Eastern Coach Works	B47/28D	1983	Oxford Citybus, 2003
L332	CN	CUD224Y	Leyland Olympian ONLXB/1R	Eastern Coach Works	B47/28D	1983	Oxford Citybus, 2003
L333	SY	R259LGH	Volvo Olympian	Northern Counties Palatine	B47/31F	1997	London General, 2003
L334	SY	V250BNV	Dennis Dart SLF	Plaxton Pointer MPD	N29F	1999	Zak's, Birmingham, 2004
L335	CN	PJ02RHE	Dennis Dart SLF	Plaxton Pointer 2	N39F	2002	Pete's, West Bromwich, 2004
L336	CN	N540LHG	Volvo Olympian	Northern Counties Palatine	B47/33F	1996	London General, 2004
L337	HD	R279LHG	Volvo Olympian	Northern Counties Palatine	B47/31F	1998	London General, 2004
L338	CN	EU05AUT	ADL Dart 10.1m	ADL Pointer	N34F	2005	
L339	SY	EU05AUR	ADL Dart 10.1m	ADL Pointer	N34F	2005	
L340	HD	EU05CLJ	Volvo B7RLE	Wrightbus Eclipse Urban	N43F	2005	
L341	CN	R273LGH	Volvo Olympian	Northern Counties Palatine	B47/31F	1998	London General, 2004
L342	TY	L108HHV	Dennis Dart 9m	Northern Counties Paladin	B34F	1994	Trustline, Hunsdon, 2005
L343	TY	EU55BWC	ADL Dart 10.7m	ADL Pointer	N37F	2005	
L344	KN	P104OLX	Dennis Dart SLF 9.2m	Plaxton Pointer	N31F	1997	Metroline, Harrow, 2005
L345	CN	N427JBV	Volvo Olympian	Northern Counties Palatine 2	B43/30F	1995	London General, 2006
L346	CN	EU06KCX	ADL Dart 8.8m	ADL Mini Pointer	N29F	2006	
L347	CN	W649FUM	Volvo B10M-62	Plaxton Première 350	C48FT	2000	WA Shearings, 2006
L348	CN	EU56FLR	ADL Dart 10.7m	ADL Pointer	N37F	2006	
L349	TY	EU56FLP	ADL Dart 10.7m	ADL Pointer	N37F	2006	
L350	HD	EU56FLN	ADL Dart 10.7m	ADL Pointer	N37F	2006	
L351	KN	EU56FLM	ADL Dart 10.7m	ADL Pointer	N37F	2006	
L352	CN	P494MBY	Volvo Olympian	Alexander RH	B43/30F	1996	Metroline, Harrow, 2006
L353	CN	P489MBY	Volvo Olympian	Alexander RH	B43/30F	1996	Metroline, Harrow, 2006
L354	TY	M604TTV	Volvo B10B	Alexander Strider	B51F	1995	City of Nottingham, 2006
L355	HD	EU07GVY	ADL Trident 2	ADL Enviro 400	N47/33F	2007	
L356	CN	EY57FZE	ADL Dart 4 10.7m	ADL Enviro 200	N37F	2007	
L357	CN	J838TSC	Leyland Olympian ON2R56C13Z4	Alexander RH	B51/34F	1991	Lothian Buses, 2007
L358	CN	M276UKN	Volvo Olympian	Alexander RH	B47/31F	1995	Dublin Bus, 2007
L359	SY	M273UKN	Volvo Olympian	Alexander RH	B47/31F	1995	Dublin Bus, 2007
L360	SY	M282UKN	Volvo Olympian	Alexander RH	B47/31F	1995	Dublin Bus, 2007
L361	SY	M144UKN	Volvo Olympian	Alexander RH	B47/31F	1995	Dublin Bus, 2007
L362	SY	M146UKN	Volvo Olympian	Alexander RH	B47/31F	1995	Dublin Bus, 2007
L363	SY	M294UKN	Volvo Olympian	Alexander RH	B47/31F	1995	Dublin Bus, 2007
L364	KN	K107JWJ	Volvo Olympian	Northern Counties	B47/29F	1993	Stagecoach, 2008
L365	KN	L109LHL	Volvo Olympian	Northern Counties	B47/29F	1993	Stagecoach, 2008
L366	KN	S127RLE	Volvo Olympian	Alexander RH	B47/29F	1998	Metroline, Harrow, 2008
L367	KN	K357DWJ	Leyland Olympian ON2R50C13Z4	Northern Counties	B47/29F	1992	Stagecoach, 2008
L368	TY	K508ESS	Leyland Olympian ON2R50C13Z4	Alexander RL	BC43/27F	1992	
L369	TY	K518ESS	Leyland Olympian ON2R50C13Z4	Alexander RL	BC43/27F	1992	
L370	TY		ADL Dart 4 10.7m	ADL Enviro 200	N37F	2009	

Special event vehicle:

	HD	EX6566	Leyland Titan PD2/1	Leyland	B30/26R	1950	Great Yarmouth

Previous registrations:

M144UKN	95D202		M988NAA	A14XEL
M146UKN	95D203		N427JBV	N427JBV, WLT527
M273UKN	95D201		N664THO	XEL158
M276UKN	95D199		N667THO	XEL24
M282UKN	95D200		P395AAA	A12XEL
M294UKN	95D204		P530CLJ	A13XEL
M517NCG	XEL606			

Depots: Brunel Road, Clacton (CN); Stephenson Road, Clacton (CN); Wethersfield Road, Sible Hedingham (HD); High Street, Kelvedon (KN); Meekings Road, Sudbury (SY) and New Road, Tollesbury (TY).

HOPLEY'S COACHES

Hopleys Coaches Ltd, Gover Farm, Mount Hawke, Truro, TR4 8BH

IAZ2314	Leyland Olympian ONLXB/1R	Eastern Coach Works	B45/32F	1982	Arriva The Shires, 2002
508AHU	Volvo B10M-56	Plaxton Supreme V Express	C53F	1982	Tillingbourne, Cranleigh, 1994
BIG9704	Volvo B10M-61	Plaxton Paramount 3200 III	C53F	1987	WHM, Little Waltham, 2006
BIG9705	Volvo B10M-61	Plaxton Paramount 3500 III	C55F	1987	Oates, St Ives, 2006
AIG8286	Volvo B10M-61	Plaxton Paramount 3200 III	C53F	1988	WHM, Little Waltham, 2006
UOD541	Volvo B10M-61	Plaxton Paramount 3200 III	C59F	1988	WHM, Little Waltham, 2006
640UAF	Volvo B10M-60	Jonckheere Deauville P599	C51FT	1992	Vale of Llangollen, 2001
956CCE	Volvo B10M-62	Caetano Algarve 2	C55F	1994	Classic, Annfield Plain, 2006
P426SWC	Volvo Olympian	Alexander RH	B47/27D	1996	Dublin Bus, 2008
930FDV	Volvo B10M-62	Van Hool Alizée	S70F	1996	Pride of the Clyde, Glasgow, '06
YAF65	Volvo B10M-62	Jonckheere Deauville P599	C53F	1996	Clarke's of London, 2006
TSV302	Volvo B10M-62	Jonckheere Mistral 50	C51FT	1998	Stewarts, Grazeley Green, 2005
WK03BTE	Optare Solo M920	Optare	N31F	2003	
WK03BTF	Optare Solo M920	Optare	N31F	2003	
WK07AOJ	Mercedes-Benz Vario O814	Plaxton Beaver 2	B29F	2007	
WK57DNX	Volvo B12B	Plaxton Panther	C53F	2007	

Previous registrations:

508AHU	NUH262X		BIG9705	D756MRL, 831OCV
640UAF	J986GLG, VLT149, VLT293, VLT280		IAZ2314	MUH288X
930FDV	LSK835, N517PYS		P426SWC	96D308
956CCE	M1CLA		TSV302	S944WYB
AIG8286	E306OMG, WXC732		UOU541	E303OMG, HWY701
BIG9704	D263HFX, XCF447		YAF65	N539SJF

The Cornish Cathedral city of Truro is the location for this view of Hopley's Solo WK03BTF. The operator is base north of the A30 some thirteen kilometres from Truro. *Richard Godfrey*

HORNSBY TRAVEL

Hornsby Travel Services Ltd, 51 Ashby High Street, Scunthorpe, DN16 2NB

B44	8955RH	Dennis Javelin 12m	Plaxton Excalibur	C49FT	1998	
B48	T208KJV	Dennis Javelin 10m	Plaxton Première 320	C43F	1999	
B53	X94HTL	Dennis Dart SLF	Plaxton Pointer MPD	N29F	2000	
B54	W109MTL	Volvo B7R	Plaxton Prima	C57F	2000	
B58	Y39WVL	Dennis Dart SLF	Plaxton Pointer MPD	N37F	2001	
B62	W209CDN	DAF SB120	Wrightbus Cadet	N37F	2000	Arriva demonstrator, 200
B64	YG52CGZ	DAF SB120	Wrightbus Cadet	N39F	2003	
B65	YG52CGY	DAF SB120	Wrightbus Cadet	N39F	2003	
B67	W898RFA	Dennis Javelin 12m	Plaxton Première 320	C57F	2000	Bassett's, Tittensor, 2003
B68	2732RH	Dennis Javelin 12m	TransBus Profile	C57F	2003	
B69	Y466HUA	DAF SB120	Wrightbus Cadet	N39F	2001	Myall, Bassingbourne, 2003
B70	W966JNF	Mercedes-Benz Vario O814	Plaxton Beaver 2	B33F	2000	Brylaine, Boston, 2004
B71	7455RH	Volvo B12B	Plaxton Paragon	C49FT	2004	
B72	C334HWJ	Leyland Olympian ONLXB/1R	Eastern Coach Works	B45/32F	1986	Stagecoach, 2004
B74	FX54LLE	ADL Dart 8.8m	ADL Mini Pointer	N29F	2005	
B76	H479PVW	Leyland Olympian ON2R50C13Z4	Alexander RL	B47/31F	1990	Nu-Venture, Aylesford, 2006
B77	X674OBA	Dennis Dart SLF 10.7	Plaxton Pointer 2	N36F	2000	Jim Stones, Leigh, 2005
B78	SN03LGA	TransBus Dart 8.8m	TransBus Mini Pointer	N29F	2003	Waverley Travel, Edinburgh, 2005
B79	FX06JVF	Enterprise Plasma 7.9m	Plaxton Primo	N28F	2006	
B80	FX06JVG	Enterprise Plasma 7.9m	Plaxton Primo	N28F	2006	
B81	BX04DVV	BMC Falcon 1100	BMC	B40F	2004	Eirebus, Dublin, 2006
B	J845TSC	Leyland Olympian ON2R56C13Z4	Alexander RH	B51/31F	1991	Lothian Buses, 2008
B	WUK155	VDL Bus SB120	Plaxton Centro	N40F	2007	
B	FX08REU	Enterprise Plasma 7.9m	Plaxton Primo	N28F	2008	

Previous registrations:

8955RH	R220GFU	H479PVW	90D1018, DMN18R
BX04DVV	04D34633	X674OBA	H1JYM

Hornsby Travel's Plaxton-bodied VDL Bus SB120 has carried the Cherished index number WUK155 from new. The bus was pictured shortly after delivery. *Mark Bailey*

HULLEYS

Henry Hulley & Sons Ltd, Derwent Garage, Culver Road, Baslow, Bakewell, DE45 1RP

1	MX06ACV	Optare Solo M880 SL	Optare	N26F	2006	
2	P881PWW	Dennis Dart 9.8m	Plaxton Pointer	B40F	1997	Clarkson, South Elmsall, 2002
3	YN08JWY	MAN 12.240	Plaxton Centro	N37F	2008	
4	M437PUY	Dennis Javelin 12m	Plaxton Première 320	C53F	1995	Go-Whittle, Kidderminster, 2002
5	P273NRH	Optare Excel L1150	Optare	N45F	1996	East Yorkshire, 2002
6	R140XWF	Dennis Javelin	Plaxton Première 320	C53F	1998	
7	Y293HUA	DAF SB120	Wright Cadet	N39F	2001	K-Line, Honley, 2007
8	MX04AAE	Volvo B7R	Plaxton Prima	C53F	2004	Swan's, Chadderton, 2006
9	R280RAU	Mercedes-Benz Vario O810	Plaxton Beaver 2	B31F	1997	Trent Barton, 2004
10	MX56ACF	Optare Solo M880 SL	Optare	N26F	2006	
11	N360VRC	Dennis Lance 11m	Optare Sigma	B46F	1995	Trent Barton, 2004
12	N362VRC	Dennis Lance 11m	Optare Sigma	B46F	1995	Trent Barton, 2004
14	M803PRA	MAN 11.190	Optare Vecta	B40F	1994	Trent Barton, 2003
15	N361VRC	Dennis Lance 11m	Optare Sigma	B46F	1995	Trent Barton, 2004
16	AE05EUY	MAN 14.220	MCV Evolution	N40F	2005	
17	M802PRA	MAN 11.190	Optare Vecta	B40F	1994	Trent Barton, 2003
18	P659BUB	Dennis Lance 11m	Plaxton Verde	B49F	1996	Longstaff, Mirfield, 2004
19	P120GSR	Dennis Dart 10.1m	Plaxton Pointer	B40F	1996	Travel Dundee, 2003
20	AE55EHL	MAN 14.220	MCV Evolution	N40F	2005	

Previous registrations:

M437PUY	M437PUY, URH341	P659BUB	P2JJL

Web: www. hulleys-of-baslow.co.uk

Bakewell is the location for this view of Hulley's MAN 16, AE05EUY. It is one of two of this model added to the fleet in 2005, it carries a MCV Evolution body. *John Young*

JPT

City Nippy - JPT

P V Walsh, The Coach House, Joshua Lane, Middleton, Manchester, M24 2AZ

1	MX58VCN	ADL Dart 4 8.9m	ADL Enviro 200	N29F	2009	
2	MX08WCJ	Enterprise Plasma	Plaxton Primo	N28F	2008	
3	MX08WCK	Enterprise Plasma	Plaxton Primo	N28F	2008	
4	MX08WCT	Enterprise Plasma	Plaxton Primo	N28F	2008	
5	MX08PZH	MAN 12.240	Plaxton Centro	N37F	2008	
6	MX08TCU	MAN 12.240	Plaxton Centro	N37F	2008	
7	KX57OVU	ADL Dart 4 8.9m	ADL Enviro 200	N29F	2007	
8	KX57OVT	ADL Dart 4 8.9m	ADL Enviro 200	N29F	2007	
9	KX57OVR	MAN 12.240	Plaxton Centro	N37F	2007	
10	KX57OVY	MAN 12.240	Plaxton Centro	N37F	2007	
11	SF04HXR	Dennis Dart SLF 8.8m	Plaxton Pointer MPD	N29F	2004	Ashall, Clayton, 2008
13	V2JPT	Dennis Dart SLF 8.8m	Plaxton Pointer MPD	N29F	1999	Plaxton demonstrator, 2001
14w	V387HGG	Dennis Dart SLF 8.8m	Plaxton Pointer MPD	N28F	2000	Ashall, Clayton, 2002
15	V389HGG	Dennis Dart SLF 8.8m	Plaxton Pointer MPD	N29F	2000	Ashall, Clayton, 2002

JPT operates services to the north of Manchester and into the city centre. A recent arrival is Plaxton-bodied MAN, 9, KX57OVR, seen about to turn into Manchester's Piccadilly. *Alan Blagburn*

English Bus Handbook: Notable Independents

New to Pete's Travel, Dart 16, T430EBD, is seen arriving in Manchester on route 118. Recent changes to the fleet have seen Plaxton Primo buses displacing the older high-floor minibuses on the local services around Middleton.

16	T430EBD	Dennis Dart SLF 10.7m	Plaxton Pointer 2	N39F	1999	Pete's Travel, W Bromwich, 2004
17	T544HNH	Dennis Dart SLF 8.8m	Plaxton Pointer MPD	N29F	1999	Pete's Travel, W Bromwich, 2004
18	T547HNH	Dennis Dart CLF 8.8m	Plaxton Pointer 2	N29F	1999	Pete's Travel, W Bromwich, 2005
20	S3JPT	Dennis Dart SLF 10.2m	East Lancs Spryte	N35F	1998	Hackney Community, 2005
21	S737RNE	Mercedes-Benz Vario 0810	Plaxton Beaver 2	B27F	1998	Vale, Manchester, 2008
22	R277CBU	Mercedes-Benz Vario 0814	Plaxton Beaver 2	B31F	1998	Stagecoach, 2007
23	R785DUB	Mercedes-Benz Vario 0814	Plaxton Beaver 2	B27F	1998	Arriva Midlands, 2008
24	R899AVM	Mercedes-Benz Vario 0810	Plaxton Beaver 2	B27F	1997	Stagecoach, 2007
25	R261XDA	Mercedes-Benz Vario 0814	Alexander ALX100	B27F	1997	Travel West Midlands, 2007
26	P680RWU	Dennis Dart SLF 10m	Plaxton Pointer 2	N35F	1997	Norbus, Kirkby, 2005
27	P679RWU	Dennis Dart SLF 10m	Plaxton Pointer 2	N37F	1997	Norbus, Kirkby, 2005
28	P2JPT	Dennis Dart SLF	Plaxton Pointer 2	N37F	1997	Valet Parking, Luton Airport, 1999
29	P6JPT	Mercedes-Benz 811D	Plaxton Beaver	B31F	1997	
30	P123HCH	Mercedes-Benz Vario 0814	Alexander ALX100	B29F	1997	Bowers, Chapel-en-le-Frith, 2007
31	L149WAG	Dennis Dart 9m	Plaxton Pointer	B35F	1993	Arriva NW & Wales, 2007

Previous registrations:

S3JPT	S17GPS		V2JPT	V680FPB

JIM STONES

J B Stones, Derby Street West, Leigh, WN7 4PF

A499MHG	Leyland-DAB Tiger Cub	Leyland-DAB	B41F	1984	Leyland demonstrator, 1986
B10JYM	Leyland Tiger TRCTL11/3LZ	East Lancs Flyte (1999)	B53F	1985	MoD, 1996 (37KC41)
B11JYM	Leyland Tiger TRCTL11/3RZ	East Lancs Flyte (1999)	B53F	1986	MoD Police, 1998 (69KE46)
B16TYG	Leyland Tiger TRCTL11/3LZM	Plaxton Derwent	B54F	1990	MoD, 1996 (03KJ36)
B1BUS	Dennis Dart SLF 8.8m	Plaxton Mini Pointer	N29F	2003	
J5BUS	TransBus Dart SLF 8.8m	TransBus Mini Pointer	N29F	2004	
T1KET	TransBus Dart SLF 8.8m	TransBus Mini Pointer	N29F	2004	
BU52LEE	TransBus Dart SLF 8.8m	TransBus Mini Pointer	N29F	2004	
BUS1S	ADL Dart 10.1m	ADL Pointer	N34F	2004	
B1JYM	ADL Dart 10.1m	ADL Pointer	N34F	2004	
H1JYM	ADL Dart 10.7m	ADL Pointer	N37F	2005	
BUS51T	ADL Dart 4 10.7m	ADL Enviro 200	N38F	2006	
BUS1N	ADL Dart 4 8.9m	ADL Enviro 200	N29F	2007	
JB51BUS	ADL Dart 4 8.9m	ADL Enviro 200	N29F	2007	
BUS1T	ADL Dart 4 8.9m	ADL Enviro 200	N29F	2008	
M1BUS	ADL Dart 4 8.9m	ADL Enviro 200	N29F	2008	

Previous registrations:

A499MHG	A499MHG, BUS1T
B11JYM	C529GVU

Five Alexander Dennis Enviro 200 Darts are now in service with Jim Stones 'bus'n' around Leigh. BUS1N itself is seen arriving at the bus station. *Alan Blagburn*

KIME'S

R Kime & Co Ltd, 3 Sleaford Road, Folkingham, Sleaford, NG34 0SB

42	YAZ4142	DAF SB220	Northern Counties Paladin	B49F	1996	North Kent Express, Northfleet, '01
43	YAZ4143	DAF SB220	Northern Counties Paladin	B49F	1995	K-Line, Honley, 2001
46	PAZ9346	DAF SB220	Northern Counties Paladin	B49F	1995	Go Coastline, 2002
61	TAZ4061	Leyland Olympian ONCL10/1RZ	Alexander RH	B42/29F	1989	Go North East, 2004
62	TAZ4062	Leyland Olympian ON2R50C13Z4	Alexander RH	B47/31F	1990	Dublin Bus, 2004
63	TAZ4063	Leyland Olympian ON2R50C13Z4	Alexander RH	B47/31F	1990	Dublin Bus, 2004
64	TAZ4064	Leyland Olympian ON2R50C13Z4	Alexander RH	B47/34F	1990	Dublin Bus, 2002
73	YAZ8773	Leyland Olympian ON2R50C13Z4	Alexander RH	B47/33F	1990	Dublin Bus, 2003
74	YAZ8774	Leyland Olympian ON2R50C13Z4	Alexander RH	B47/33F	1990	Dublin Bus, 2002
78	WAZ8278	Leyland Olympian ONCL10/1RZ	Alexander RH	B42/29F	1989	Go North East, 2004
84	PAZ3184	Leyland Olympian ON2R50C13Z4	Alexander RH	B47/31F	1990	Dublin Bus, 2002
300	DAZ4300	Optare Excel L1150	Optare	N43F	1996	Trent Barton, 2005
301	DAZ4301	Optare Excel L1150	Optare	N43F	1996	Trent Barton, 2005
302	DAZ4302	Optare Excel L1150	Optare	N43F	1996	Trent Barton, 2005
D1	V929FMS	Dennis Trident	Alexander ALX400	N47/27F	1999	Portsmouth University, 2006
D2	PO51WNN	Dennis Trident	Plaxton President	N47/29F	2001	GHA, New Broughton, 2006
D3	T648KPU	Dennis Trident	Alexander ALX400	N51/26F	1999	Red Route, Northfleet, 2008
D4	GK04NZU	VDL Bus DB250	East Lancs Lowlander	N47/27F	2004	Dublin Bus, 2008
S1	YN51MJE	Scania L94 UB	Wrightbus Solar	N43F	2001	Fowlers, Holbeach Drove, 2003
S2	W178CDN	DAF SB220	Ikarus Polaris	N44F	2000	GM North, Manchester, 2008
S3	Y284HUA	DAF SB120	Wrightbus Cadet	N39F	2001	Stansted Transit, 2006
R4	FX04TJY	VDL Bus SB200	Wrightbus Commander	N44F	2004	
R5	YJ04BMV	VDL Bus SB200	Wrightbus Commander	N44F	2004	
R6	YJ55KZZ	VDL Bus SB200	Wrightbus Commander	N44F	2005	

Previous registrations:

GK04NZU	04D34571	TAZ4062	90D1039, H960PTW
DAZ4300	P151CTV	TAZ4063	90D1043, H961PTW
DAZ4301	P210BNR, WAZ8276	TAZ4064	90D1056, H664PTW
DAZ4302	V394KVY	WAZ8278	G675TCN
PAZ3184	90D1037, H690PTW	YAZ4142	P904TWW
PAZ9346	M848RCP	YAZ4143	M850RCP
PO51WNN	PO51WNN, DA51GHA	YAZ8773	90D1047, H950PTW
TAZ4061	G679TCN	YAZ8774	90D1053, H808PTW

Kimes operates three Dennis Tridents, the latest arrival being an Alexander-bodied example. D3, T648KPU was new to Stagecoach before passing on to the independent sector from East London Coaches.
David Longbottom

KONECTBUS

Konectbus Ltd, 7 John Goshawk Road, Rash's Green Ind Est, Dereham, NR19 1SY

100	J853TSC	Leyland Olympian ON2R56C13Z4	Alexander RH	B51/30D	1991	Lothian Buses, 2008
101	J855TSC	Leyland Olympian ON2R56C13Z4	Alexander RH	B51/30D	1991	Lothian Buses, 2008
109	TRN809V	Leyland National 10351A/2R		B44F	1979	Chase Coaches, Chasetown, 2007
200	R835FNG	Optare Excel L1150	Optare	N40F	1998	Bennett's, Gloucester, 2003
201	R836FNG	Optare Excel L1150	Optare	N40F	1998	Bennett's, Gloucester, 2003
202	R837FNG	Optare Excel L1150	Optare	N40F	1998	Bennett's, Gloucester, 2003
203	V203ENU	Optare Excel L1150	Optare	N45F	2000	Trent Buses, 2004
204	V204ENU	Optare Excel L1150	Optare	N45F	2000	Trent Buses, 2004
205	V205ENU	Optare Excel L1150	Optare	N45F	2000	Trent Buses, 2005
216	W216PRB	Optare Excel L1180	Optare	N45F	2000	Trent Buses, 2007
217	W217PRB	Optare Excel L1180	Optare	N45F	2000	Trent Buses, 2007
218	W218PRB	Optare Excel L1180	Optare	N45F	2000	Trent Buses, 2007
219	W219PRB	Optare Excel L1180	Optare	N42F	2000	Trent Buses, 2007
229	X229WRA	Optare Excel L1180	Optare	N44F	2000	Trent Buses, 2007
269	S169UAL	Optare Excel L1150	Optare	N43F	1998	Trent Buses, 2003
289	T789XVO	Optare Excel L1150	Optare	N45F	1999	Trent Buses, 2006
300	MX08UZT	Optare Versa V1100	Optare	N37F	2008	
301	MX08UZU	Optare Versa V1100	Optare	N37F	2008	

Konectbus covers the area around Dereham, Watton and Swaffham linking these towns to Norwich and Norwich University Hospital, as well as running the Park & Ride services. Recent arrivals include five VDL Bus DB250 double-decks with Wrightbus Pulsar Gemini bodywork. Seen heading for Thickthorn Park & Ride site is 504, YJ05PXE. *Mark Lyons*

English Bus Handbook: Notable Independents

The Optare Tempo model provides the main single-deck product for Konectbus, with two more on order for 2009. Heading for Norwich centre is 405, MX05ELJ, from the 2005 intake. *Richard Godfrey*

400	MX05EKW	Optare Tempo X1200	Optare	N42F	2005
401	MX05EKY	Optare Tempo X1200	Optare	N42F	2005
402	MX05EKZ	Optare Tempo X1200	Optare	N42F	2005
403	MX05ELC	Optare Tempo X1200	Optare	N42F	2005
404	MX05ELH	Optare Tempo X1200	Optare	N42F	2005
405	MX05ELJ	Optare Tempo X1200	Optare	N42F	2005
406	MX06YXU	Optare Tempo X1200	Optare	N42F	2006
407	YJ56WVA	Optare Tempo X1200	Optare	N42F	2007
408	YJ56WVB	Optare Tempo X1200	Optare	N42F	2007
409	YJ57EGY	Optare Tempo X1200	Optare	N43F	2007
410	YJ57EGX	Optare Tempo X1200	Optare	N43F	2007
411	MX58ABV	Optare Tempo X1200	Optare	N43F	2008
412		Optare Tempo X1200	Optare	N43F	2009
413		Optare Tempo X1200	Optare	N43F	2009
500	YJ05PXA	VDL Bus DB250	Wrightbus Pulsar Gemini	N41/24F	2005
501	YJ05PXB	VDL Bus DB250	Wrightbus Pulsar Gemini	N41/24F	2005
502	YJ05PXC	VDL Bus DB250	Wrightbus Pulsar Gemini	N41/24F	2005
503	YJ05PXD	VDL Bus DB250	Wrightbus Pulsar Gemini	N41/24F	2005
504	YJ05PXE	VDL Bus DB250	Wrightbus Pulsar Gemini	N41/24F	2005
900	VX51RHZ	Optare Solo M920	Optare	N33F	2002
901	VX51RJZ	Optare Solo M920	Optare	N33F	2002

Previous registrations:

R835FNG	R400BEN		R836FNG	R500BEN	
			R837FNG	R600BEN	

Depots: Rash's Green Industrial Estate, Dereham and The Drift, Fakenham. **Web:** www.konectbus.co.uk

LEVEN VALLEY

P W Thompson, 55 Marwood Drive, Great Ayton, Middlesbrough, TS9 6PD

8	T888EXG	Optare Solo M880 SL	Optare	N28F	2006
9	T9BHN	Optare Solo M880 SL	Optare	N28F	2007
10	T10BHN	Optare Solo M880 SL	Optare	N28F	2007
22	S222JDC	Optare Solo M880 SL	Optare	N27F	2008
23	S123RDC	Optare Solo M880 SL	Optare	N26F	2008
27	T27BXG	Optare Solo M880 SL	Optare	N28F	2005
28	T28BXG	Optare Solo M880 SL	Optare	N28F	2005
66	P766BJU	Mercedes-Benz 709D	Leicester Carriage Builders	BC29F	1996
76	P776BJU	Mercedes-Benz 709D	Leicester Carriage Builders	BC29F	1996
111	T111PDC	Optare Solo M710 SE	Optare	N23F	2007

Web: www.levenvalleybuses.com
Depot: Tilbury Road, South Bank, Middlesbrough

The Leven Valley fleet now comprises mostly Optare Solo buses of which 28, T28BXG, is seen heading for Middlesbrough. Registration marks are selected to have resonance with older vehicles that plied the streets of the town in the past. *David Longbottom*

English Bus Handbook: Notable Independents

MARSHALLS

Marshalls of Sutton-on-Trent Ltd, 11 Main Street, Sutton-on-Trent, Newark, NG23 6PF

VL42	H3YRR	Volvo B10M-60	Plaxton Paramount 3500 III	C53F	1991	Westerham Coaches, 1995
DS50	S3YRR	Dennis Javelin	Plaxton Première 320	C53F	1998	
DS51	S4YRR	Dennis Dart SLF	Plaxton Pointer 2	N39F	1998	
MN54	W3YRR	MAN 11.220	Berkhof Axial 35	C35F	2000	
MB56	X56LRY	Mercedes-Benz Vario 0814	Plaxton Beaver 2	BC33F	2001	
DD57	E925CDS	Leyland Lion LDTL11/1R	Alexander RH	BC49/37F	1987	City of Nottingham, 2001
DD58	A3YRR	Leyland Lion LDTL11/1R	Alexander RH	BC49/37F	1987	City of Nottingham, 2001
VL59	R3YRR	Volvo B10M-62	Plaxton Première 350	C49FT	1998	Biss Bros, Bishop's Stortford, 2002
OP60	YG02FWL	Optare Solo M920	Optare	N33F	2002	
DD63	P353ROO	Volvo Olympian	East Lancs	B51/35F	1997	East Thames Buses, 2003
DD64	R358XVX	Volvo Olympian	East Lancs	B51/35F	1997	East Thames Buses, 2003
VL66	YN04HJV	Volvo B7R	Transbus Profile	C53F	2004	
DD68	F127PHM	Volvo Citybus B10M-50	Alexander RV	B46/33F	1988	Localink, Stansted, 2004
DD69	H114ABV	Volvo Citybus B10M-50	Alexander RV	BC47/35F	1991	Blue Bus, Bolton, 2004
VL70	PM05JAM	Volvo B12B	Plaxton Panther	C49FT	2005	
IC71	YN05UGO	Irisbus EuroMidi CC80	Indcar Wing	C29F	2005	
OP72	YJ06VSP	Optare Tempo X1200	Optare	N42F	2006	
VL73	PM07JAM	Volvo B12B	Plaxton Panther	C53F	2007	
DD74	POG583Y	MVW Metrobus DR102/27	MCW	B43/30F	1982	Stagecoach, 2006
N75	OXK76	Neoplan Starliner N516 SHD	Neoplan	C38FT	2003	Ellison, St Helens, 2007
DD76	J837TSC	Leyland Olympian ON2R56C13Z4	Alexander RH	B51/34F	1991	Lothian Buses, 2007
VL77	PM04JAM	Volvo B12M	Plaxton Paragon	C53F	2004	McGeehan, Flintown, 2007
MB78	BN08OOW	Mercedes-Benz Tourismo	Mercedes-Benz	C36F	2008	
OP79	V269DRC	Optare Solo M920	Optare	N33F	1999	City of Nottingham, 2008
OP80	V270DRC	Optare Solo M920	Optare	N33F	1999	City of Nottingham, 2008
OP81	MX54WMK	Optare Solo M920	Optare	N33F	2005	Ace, Aintree, 2008
OP82	YJ57EHD	Optare Solo M950	Optare	N31F	2008	Optare demonstrator, 2008

Special event vehicles:

B30	RAU624R	Bedford YLQ	Plaxton Supreme III	C45F	1976	Gash, Newark, 1988
1409	FW5696	Leyland Tiger TS7	Burlingham(1950)	B35F	1935	Lincolnshire Road Car
DD7	LNN353	Daimler CVD6	Duple	L27/26RD	1950	Gash of Newark

Previous registrations:

A3YRR	E163YGB, WLT364, E938CDS	OXK76	YN03AXA
E925CDS	E161YGB, FSU661	PM04JAM	04D59533
H3YRR	H836AHS	R3YRR	R919HTW

Depots: Main Street, Sutton-on-Trent and Vicarage Lane, North Muskham.

Newark bus station is the location for this view of Marshalls' only Dart DS51, S4YRR. *Richard Godfrey*

MIDLAND RIDER

Midland Rider Ltd, 6B CMT Trading Estate, Bradwell Road, Oldbury, B69 4BQ

1005	J636KCU	Dennis Dart 9.8m	Wright Handybus	B40F	1992	Go North East, 2001
1009	JDZ2349	Dennis Dart 8.5m	Wright Handybus	B31F	1990	Camden, West Kingsdown, 2002
1012	K860PCN	Dennis Dart 9.8m	Wright Handybus	B40F	1992	MacEwan, Amisfield Town, 2004
1014	J377AWT	Leyland Lynx LX2R11C15Z4S	Leyland	B49F	1991	Arriva Yorkshire, 2005
1015	CAZ6617	Dennis Dart 9.8m	Wright Handybus	B39F	1994	Ulsterbus, 2005
1016	CAZ6641	Dennis Dart 9.8m	Wright Handybus	B39F	1994	Ulsterbus, 2005
1017	KDZ5805	Dennis Dart 9m	Wright Handybus	B34F	1991	Huyton Travel, 2006
1019	K859PCN	Dennis Dart 9.8m	Wright Handybus	B40F	1992	People's Express, W Bromwich, 06
1021	G738RTY	DAF SB220	Optare Delta	B49F	1989	Acorn, Tamworth, 2007
1022	CAZ6602	Dennis Dart 9.8m	Wright Handybus	B39F	1994	Chalkwell, Sittingbourne, 2007
1023	CAZ6619	Dennis Dart 9.8m	Wright Handybus	B39F	1994	Chalkwell, Sittingbourne, 2007
1024	R694WAW	Dennis Dart SLF 10.9m	East Lancs Spryte	N38F	1998	Go Ahead Meteor, 2008
1025	R695WAW	Dennis Dart SLF 10.9m	East Lancs Spryte	N38F	1998	Go Ahead Meteor, 2008
1026	N450DWJ	Volvo B6LE	Wright Crusader	N32F	1996	First, 2008
	N451DWJ	Volvo B6LE	Wright Crusader	N32F	1996	First, 2008
	K250CBA	DAF SB220	Ikarus Citibus	B48F	1993	Arriva NW & Wales, 2009

Previous registrations:

K250CBA	K133TCP		R695WAW	R286OFV
R694WAW	R287OFV			

Midland Rider operates into Birmingham centre from the Oldbury area using a fleet of single-deck buses. The fleet is represented by Wright Handybus-bodied Dart 1012, K860PCN. *Richard Godfrey*

MINSTERLEY MOTORS

Minsterley Motors Ltd, Stiperstones, Minsterley, SY5 0LZ

SNM71R	Bedford YMT	Plaxton Supreme III	C40F	1977	
KTA356V	Bedford YMT	Duple Dominant II Express	C53F	1980	Thomas, Relubbus, 1995
BGR684W	Bedford YMT	Duple Dominant	B53F	1980	Jolly, South Hylton, 1995
T329KDM	Volvo B10M-62	Plaxton Première 350	C49FT	1999	Bakerbus, Biddulph, 2007
T330KDM	Volvo B10M-62	Plaxton Première 350	C49FT	1999	Bakerbus, Biddulph, 2007
T581JTD	Dennis Dart SLF	Plaxton Pointer MPD	N28F	1999	Jim Stones, Leigh, 2003
DY52DZF	Volvo B12B	Sunsundegui Sideral	C49FT	2003	
YN53EJL	Mercedes-Benz Vario O814	TransBus Cheetah	C33F	2003	
FJ54ZCL	Volvo B7R	Jonckheere Modulo	C53FT	2004	
FJ54ZCO	Volvo B7R	Jonckheere Modulo	C53FT	2004	
YN05UVB	Mercedes-Benz Vario O814	Plaxton Cheetah	C29F	2005	
YN05GZV	Scania K114 EB4	Berkhof Axial 50	C53FT	2005	
YN06NZH	Scania L94UB	Wrightbus Solar	N42F	2006	
YN06NZR	Scania K114 EB4	Berkhof Axial 50	C53FT	2006	
YN06NZS	Scania K114 EB4	Berkhof Axial 50	C53FT	2006	
YN06NZV	Scania K114 EB4	Berkhof Axial 50	C53FT	2006	
YN06NZW	Scania K114 EB4	Berkhof Axial 50	C53FT	2006	
DX57JXS	Volvo B7RLE	Plaxton Centro	N44F	2007	
DX57REU	Volvo B7RLE	Plaxton Centro	N44F	2007	
DX57TVW	Volvo B7RLE	Plaxton Centro	N45F	2007	Lloyds, Machynlleth, 2008
YN08NKH	Volvo B7RLE	Plaxton Centro	N44F	2008	
YN08NLG	Volvo B7RLE	Plaxton Centro	N44F	2008	
YN08NLJ	Volvo B7RLE	Plaxton Centro	N44F	2008	

Previous registrations:

510DMY	-	T329KDM	9500RU
DX57TVW	LC57WYN	T330KDM	5658RU
KTA356V	FUJ941V, 3504CD	T581JTD	B1BUS

Web: www.minsterleymotors.co.uk

Six Plaxton Centro-bodied Volvo buses are now the principal vehicles on Minsterley's service that links Shrewsbury with Bishop's Castle. YN08NKH is seen arriving in Shrewsbury. *Tony Wilson*

NIBS

WH Nelson Coaches (Wickford) Ltd, The Coach Station, Bruce Grove, Wickford, SS11 8BZ

C51CHM	Leyland Olympian ONLXB/1RH	Eastern Coach Works	B42/30F	1986	Isle of Man Transport, 2002
C73CHM	Leyland Olympian ONLXB/1RH	Eastern Coach Works	B42/30F	1986	Isle of Man Transport, 2002
G601KTX	Scania N113DRB	Alexander RH	B47/33F	1990	Cardiff Bus, 2000
G603KTX	Scania N113DRB	Alexander RH	B47/33F	1990	Cardiff Bus, 2000
G604KTX	Scania N113DRB	Alexander RH	B47/33F	1990	Cardiff Bus, 2000
G605KTX	Scania N113DRB	Alexander RH	B47/33F	1990	Cardiff Bus, 2000
G607KTX	Scania N113DRB	Alexander RH	B47/33F	1990	Cardiff Bus, 2000
BIL4710	Scania N113DRB	Alexander RH	B45/31F	1990	Travel West Midlands, 2002
H228LOM	Scania N113DRB	Alexander RH	B45/31F	1990	Travel West Midlands, 2002
H229LOM	Scania N113DRB	Alexander RH	B45/31F	1990	Travel West Midlands, 2002
H237LOM	Scania N113DRB	Alexander RH	B45/31F	1990	Town & Country, 2006
H241LOM	Scania N113DRB	Alexander RH	B45/31F	1990	Travel West Midlands, 2002
H243LOM	Scania N113DRB	Alexander RH	B45/31F	1990	Travel West Midlands, 2002
H245LOM	Scania N113DRB	Alexander RH	B45/31F	1990	Travel West Midlands, 2002
H247LOM	Scania N113DRB	Alexander RH	B45/31F	1990	Town & Country, 2006
H258MFX	Dennis Dominator DDA1033	East Lancs	B47/33F	1990	Transdev Yellow Buses, 2006
H264MFX	Dennis Dominator DDA1033	East Lancs	B47/33F	1990	Transdev Yellow Buses, 2006
J266SPR	Dennis Dominator DDA1033	East Lancs	B47/33F	1990	Transdev Yellow Buses, 2006
J269SPR	Dennis Dominator DDA1033	East Lancs	B47/33F	1990	Transdev Yellow Buses, 2006
BIL9406	Dennis Dart SLF 10.7m	Plaxton Pointer 2	N36F	1998	Tellings-Golden Miller, 2008
R507SJM	Dennis Dart SLF 10.7m	Plaxton Pointer 2	N36F	1998	Tellings-Golden Miller, 2008
R510SJM	Dennis Dart SLF 10.7m	Plaxton Pointer 2	N36F	1998	Tellings-Golden Miller, 2008
BIL4419	Dennis Trident	East Lancs Lolyne	N45/28F	1999	Metrobus, Crawley, 2007
BIL4539	Dennis Trident	East Lancs Lolyne	N45/28F	1999	Metrobus, Crawley, 2007
BIL6538	Dennis Trident	East Lancs Lolyne	N45/28F	1999	Metrobus, Crawley, 2007
EU05ECV	Scania OmniDekka N94UD	East Lancs	N51/32F	2005	
MX08UZK	Optare Solo M850 SL	Optare	N28F	2008	
YN08OCR	Scania OmniCity N270 UD	East Lancs Olympus	N51/32F	2008	
EU58BLJ	Scania OmniCity N270 UD	East Lancs Olympus	N51/32F	2008	

Previous registrations:

BIL4419	T401SMV		BIL9406	R506SJM
BIL4539	T410SMV		C51CHM	C51CHM, EMN204U
BIL4710	H226LOM		C73CHM	C73CHM, EMN213U
BIL6538	T404SMV			

Web: www.nibsbuses.com

NIBS' Trident BIL4419 was pictured at Billericay while undertaking rail replacement work.
Richard Godfrey

NORFOLK GREEN

Go West Travel Ltd, Hamlin Way, King's Lynn, PE30 4NG

101	YJ56WUG	Optare Tempo X1200	Optare	N42F	2007	
102	YJ56WUH	Optare Tempo X1200	Optare	N42F	2007	
103	YJ57YCC	Optare Tempo X1200	Optare	N42F	2008	
104	YJ57YCD	Optare Tempo X1200	Optare	N42F	2008	
105	YK57FHM	Optare Tempo X1200	Optare	N43F	2008	Optare demonstrator, 2008
106	YN58FXH	Optare Tempo X1200	Optare	N43F	2008	Optare demonstrator, 2008
201	S170UAL	Optare Excel L1150	Optare	N43F	1998	Konnectbus, Dereham, 2008
202	S171UAL	Optare Excel L1150	Optare	N43F	1998	Konnectbus, Dereham, 2008
203	S172UAL	Optare Excel L1150	Optare	N43F	1998	Konnectbus, Dereham, 2008
204	R186DDX	Optare Excel L1150	Optare	N37F	1997	Ipswich Buses, 2008
205	P190SGV	Optare Excel L1150	Optare	N38F	1997	Ipswich Buses, 2008
206	P195SGV	Optare Excel L1150	Optare	N38F	1997	Ipswich Buses, 2008
207	S156UAL	Optare Excel L1150	Optare	N42F	1998	Trent Barton, 2008
208	J508GCD	Dennis Dart 9.8m	Alexander Dash	B40F	1992	Stagecoach, 2006
230	J530GCD	Dennis Dart 9.8m	Alexander Dash	B40F	1992	Stagecoach, 2005
235	J535GCD	Dennis Dart 9.8m	Alexander Dash	B40F	1992	Stagecoach, 2005
241	J541GCD	Dennis Dart 9.8m	Alexander Dash	B40F	1992	Stagecoach, 2005
278	K578NHC	Dennis Dart 9.8m	Alexander Dash	B40F	1992	Stagecoach, 2006
301	YN07EGE	Optare Solo M950 SL	Optare	N32F	2007	
302	YN07EGF	Optare Solo M950 SL	Optare	N32F	2007	
303	YN07EHO	Optare Solo M950 SL	Optare	N32F	2007	
304	YK08EUC	Optare Solo M950 SL	Optare	N32F	2007	
316w	L686CDD	Mercedes-Benz 709D	Alexander Sprint	B25F	1994	Stagecoach, 2003
319w	K879ODY	Mercedes-Benz 709D	Alexander Sprint	B25F	1993	Stagecoach South, 2002
400	V909SEG	Mercedes-Benz Vario 0814	Plaxton Beaver 2	B31F	2000	
402	N967ENA	Mercedes-Benz 709D	Plaxton Beaver	B27F	1996	Jim Stones, Leigh, 2000
404	S294UAL	Mercedes-Benz Vario 0814	Plaxton Beaver 2	B31F	1999	Trent Buses, 2006
405	P685HND	Mercedes-Benz Vario 0814	Plaxton Beaver 2	B27F	1996	Dawsonrentals, 2001
406	V993DNB	Mercedes-Benz Vario 0814	Plaxton Beaver 2	B31F	1999	HAD, Shotts, 2003
408	S292UAL	Mercedes-Benz Vario 0814	Plaxton Beaver 2	B31F	1999	Trent Buses, 2006
409	T619VEW	Mercedes-Benz Vario 0814	Plaxton Beaver 2	B31F	2000	
410	P570APJ	Mercedes-Benz 709D	Plaxton Beaver	B27F	1996	Epsom Buses, 2002

Seen about to depart from King's Lynn bus station for Spalding is Optare Tempo 102, YJ56WUH, one of six Optare Tempo buses now used. *Mark Doggett*

Coasthopper **operates along the north Norfolk coast linking Sheringham, Wells and Hunstanton every half an hour during high summer. Pictured in the dedicated livery for this service is Optare Solo 304, YK08EUC.**
David Longbottom

411	V561JFL	Mercedes-Benz Vario O814	Plaxton Beaver 2	B31F	1999	
413	S313DLG	Mercedes-Benz Vario O814	Plaxton Beaver 2	B31F	1998	Legg, Byfleet, 2002
414	S309DLG	Mercedes-Benz Vario O814	Plaxton Beaver 2	B31F	1998	Mobus, Leven, 2002
415	S154NNH	Mercedes-Benz Vario O814	Plaxton Beaver 2	B31F	1998	Mancini, Faversham, 2003
600	S41FWY	Optare Solo M850	Optare	N27F	1999	AIA Travel, Birkenhead, 2003
601	Y54HBT	Optare Solo M920	Optare	N34F	2001	
602	Y56HBT	Optare Solo M920	Optare	N34F	2001	
603	Y57HBT	Optare Solo M920	Optare	N34F	2001	
604	Y58HBT	Optare Solo M920	Optare	N34F	2001	
605	AP02XOO	Optare Solo M920	Optare	N33F	2002	
606	YN03NCV	Optare Solo M920	Optare	N33F	2003	
607	YN03NCX	Optare Solo M920	Optare	N33F	2003	
608	YN03NCY	Optare Solo M920	Optare	N33F	2003	
609	X384VVY	Optare Solo M850	Optare	N30F	2001	Centra, Heathrow, 2004
610	YG52DHL	Optare Solo M850	Optare	N27F	2002	Pink Elephant, Stansted, 2003
611	X249VWR	Optare Solo M850	Optare	N27F	2000	Centra, Heathrow, 2004
612	W119UCF	Optare Solo M920	Optare	N34F	2000	Goodwin, Witheridge, 2004
613	MX55WCU	Optare Solo M920	Optare	N33F	2005	
614	MX06BSZ	Optare Solo M920	Optare	N33F	2006	Alton Towers, 2007
615	MX03YDB	Optare Solo M920	Optare	N33F	2003	Bakerbus, Biddulph, 2006
616	W5JPT	Optare Solo M920 SL	Optare	N29F	2007	Tanner, Thatto Heath, 2006
617	MX55WCV	Optare Solo M920	Optare	N31F	2005	Ace, Aintree, 2008
618	MX54WMJ	Optare Solo M920	Optare	N33F	2005	Ace, Aintree, 2008
619	X49VVY	Optare Solo M850	Optare	N27F	2001	Thames Travel, Wallingford, 2007
620	AU53KSO	Optare Solo M920	Optare	N33F	2004	Konectbus, Dereham, 2008

Ancillary vehicle:

666	L159LBW	Volvo B10M-62	Jonckheere Deauville P45	TV	1994	Stagecoach, 2007

Previous registrations:

N967ENA	M1BUS, N967ENA, 55BUS	S154NNH	S8STM, S154NNH, J101747(GBJ)

Depots: Hall Road, Cromer; Fleet Hargate, Spalding; Tattersett, East Rudham; Highfields, Fakenham; Hamlin Way, King's Lynn; Bexwell, Downham Market; Station Road, Reepham; Old Station Yard, Swaffham; Pious Drive, Upwell and the car park, Walsingham. **Web:** www.norfolkgreen.co.uk

English Bus Handbook: Notable Independents

NU-VENTURE

Nu-Venture Coaches Ltd, F Deacon Trading Estate, Forstal Road,
Aylesford, Maidstone, ME20 7SP
Invictabus Ltd, 86 Mill Hall, Aylesford, ME20 7JN

T347	KYV347X	Leyland Titan TNLXB2RR	Leyland	B39/30F	1981	Stagecoach, 2005
T348	KYV348X	Leyland Titan TNLXB2RR	Leyland	B44/24F	1981	Stagecoach, 2007
T415	KYV415X	Leyland Titan TNLXB2RR	Leyland	B44/32F	1982	Metroline, 1996
T455	KYV455X	Leyland Titan TNLXB2RR	Leyland	B44/30F	1982	Stagecoach Fife Scottish, 1999
T552	NUW552Y	Leyland Titan TNLXB2RR	Leyland	B44/31F	1982	Stagecoach London, 2001
T592	NUW592Y	Leyland Titan TNLXB2RR	Leyland	B44/27F	1982	Stagecoach, 2006
T652	NUW652Y	Leyland Titan TNLXB2RR	Leyland	B44/32F	1982	Stagecoach London, 2001
T857	A857SUL	Leyland Titan TNLXB2RR	Leyland	B44/32F	1983	Kent CC, 2003
T867	A867SUL	Leyland Titan TNLXB2RR	Leyland	B44/32F	1983	Stagecoach, 2004
T873	A873SUL	Leyland Titan TNLXB2RR	Leyland	B44/32F	1983	Stagecoach, 2004
T905	A905SUL	Leyland Titan TNLXB2RR	Leyland	B44/29F	1983	Stagecoach, 2006
T984	A984SYE	Leyland Titan TNLXB2RR	Leyland	B44/29F	1984	Stagecoach, 2005
101	N101CKN	Mercedes-Benz 711D	UVG CitiStar	B25FL	1996	Wealden PSV, Tonbridge, 1997
102	N102CKN	Mercedes-Benz 711D	UVG CitiStar	B25FL	1996	Wealden PSV, Tonbridge, 1997
164	G614CEF	Leyland Lynx LX2R11C154ZR	Leyland	B49F	1990	Stagecoach, 2006
165	G615CEF	Leyland Lynx LX2R11C154ZR	Leyland	B49F	1990	Stagecoach, 2008
166	F166SMT	Leyland Lynx LX112L10ZR1	Leyland	B51F	1989	Metrobus, Crawley, 2003
167	F167SMT	Leyland Lynx LX112L10ZR1	Leyland	B51F	1989	Kent CC, 2008
168	F168SMT	Leyland Lynx LX112L10ZR1	Leyland	B51F	1989	Kent CC, 2008
383	X383VVY	Optare Solo M850	Optare	N30F	2001	Thames Travel, Wallingford, 2008
700	RIB7002	Dennis Dart 9m	Carlyle Dartline	B36F	1990	RML Travel, Burslem, 2008
702	N2FPK	Dennis Dart 9.8m	UVG UrbanStar	B38F	1996	Flightparks, Horley, 2000
703	N3FPK	Dennis Dart 9.8m	UVG UrbanStar	B38F	1996	Flightparks, Horley, 2000
704	M74VJO	Dennis Dart 9.8m	Plaxton Pointer	B40F	1995	Stagecoach, 2008
718	R718BNF	Dennis Dart SLF	Marshall Capital	N41F	1998	White Rose, Thorpe, 2002
719	GN53YUF	TransBus Dart 8.8m	TransBus Mini Pointer	N28F	2003	
720	R720BNF	Dennis Dart SLF	Marshall Capital	N41F	1998	White Rose, Thorpe, 2002
	MHS5P	Dennis Javelin 12m	Plaxton Première 320	C49FT	1997	Sullivan Bus, Potters Bar, 2005

**One of Nu-Venture's
many Titans is
A984SYE, which
continues to carry its
London Buses
identity T984.**
Richard Godfrey

Leyland Lynx 164, G614CEF, was new to Cleveland Transit and is one of a pair from that source currently with Nu-Venture. It is seen in its home town of Tunbridge Wells. *Dave Heath*

721	SN53LWM	TransBus Dart 8.8m	TransBus Mini Pointer	N28F	2003	*Operated for Kent CC*
722	SN53LWO	TransBus Dart 8.8m	TransBus Mini Pointer	N28F	2003	*Operated for Kent CC*
723	SN53LWP	TransBus Dart 8.8m	TransBus Mini Pointer	N28F	2003	*Operated for Kent CC*
746	T446HRV	Dennis Dart SLF	Caetano Compass	N44F	1999	Alexcars, Cirencester, 2005
770	V270BNV	Dennis Dart 8.8m	Plaxton Pointer MPD	N29F	1999	Pegasus.com, Guiseley, 2008
785	X185BNH	Dennis Dart 8.8m	Plaxton Pointer MPD	N29F	2000	Waverley Travel, Edinburgh, 2008
811	LK55ABU	BMC Condor 220	BMC	S55F	2005	*Operated for First Student*
812	LK55ABV	BMC Condor 220	BMC	S55F	2005	*Operated for First Student*
813	LK55ABX	BMC Condor 220	BMC	S55F	2005	*Operated for First Student*
905	N905HWY	Optare MetroRider MR13	Optare	B26F	1996	Metrobus, Oprington, 2003
2138	IUI2138	Leyland Olympian ONLXB/1RV	Alexander RL	B43/27F	1986	Stagecoach, 2004
	8421RU	Scania K92CRB	Van Hool Alizée H	C53F	1988	Swallow, Rainham, 1996
	3558RU	MAN 18.310	Marcopolo Continental 340	C51FT	2001	

Special event vehicles:

2563	BXI2563	Bristol RELL6G	Alexander Q	B52F	1983	Citybus, Belfast
V2	A102SUU	Volvo-Ailsa B55-10	Alexander RV	B47/31D	1984	London Buses
625	B625DWF	Leyland Tiger TRCTL11/2RH	Alexander P	B52F	1985	East Midland (NBC)

Previous registrations:

3558RU	Y2NVC		KYV415X	KYV415X, IUI2140
8421RU	F100CWG		MHS5P	P215TGP
IUI2138	C465SSO, MHS5P		RIB7002	CMN12A, H403HOY
IUI2139	GUW500W		YDZ3458	A226VWD, AKG213A

Depots: Forstal Road, Aylesford and Vicarage Road, Hoo Marina.
Web: www.nu-venture.co.uk

OLYMPUS

Olympus Bus & Coach Ltd; Roadrunner Coaches Ltd, Templebank,
Riverway, Harlow, CM20 2DY
SM Coaches Ltd, 9 Burnt Mill Ind Est, Elizabeth Way, Harlow, CM20 2HT

FIG2843	Dennis Javelin 12m	Duple 320	S70F	1988	Reynolds, Watford, 2007
MUI7843	Bova Futura FHD12.290	Bova	C53F	1989	Southern Star, Harlow, 2007
HUI4165	Bova Futura FHD12.290	Bova	C53F	1990	Moor-Dale, 2002
G99NBD	Mercedes-Benz 709D	Dormobile Routemaker	B29F	1990	Northern Blue, Burnley, 2003
K408FHJ	Dennis Dart 9.8m	Plaxton Pointer	B38F	1993	Arriva The Shires, 2008
L192MAU	Mercedes-Benz 811D	Dormobile Routemaker	BC33F	1993	City of Nottingham, 2002
DAZ1558	Volvo B10M-62	Plaxton Première 320	S70F	1994	Ulsterbus, 2006
DAZ1559	Volvo B10M-62	Plaxton Première 320	C53F	1994	Ulsterbus, 2006
EAZ2576	Volvo B10M-62	Plaxton Première 320	C53F	1994	Ulsterbus, 2006
EAZ2588	Volvo B10M-62	Plaxton Première 320	C53F	1994	Ulsterbus, 2006
KKZ7562	Bova Futura FHD 12.340	Bova	C53F	1994	Holmeswood Coaches, 2008
M733AOO	Iveco TurboDaily 59.12	Marshall DailyBus	B25F	1995	Arriva The Shires, 2008
M736AOO	Iveco TurboDaily 59.12	Marshall DailyBus	B25F	1995	Arriva The Shires, 2003
M153RBH	Iveco TurboDaily 59.12	Marshall DailyBus	B25F	1995	Arriva The Shires, 2003
M530VWT	Iveco TurboDaily 59.12	Mellor	B13F	1995	WG Services, Harlow, 2004
N139GMJ	Iveco TurboDaily 59.12	Marshall DailyBus	BC31F	1996	WG Services, Harlow, 2004
P498GKJ	Iveco Daily 49.10	-	B25F	1996	Kent CC, 2002
P677RWU	Dennis Dart 10.1m	Plaxton Pointer	B35F	1997	Stansted Transit, 2007
P678RWU	Dennis Dart 10.1m	Plaxton Pointer	B35F	1997	Stansted Transit, 2007
P681RWU	Dennis Dart 10.1m	Plaxton Pointer	B35F	1997	Stansted Transit, 2007
P684RWU	Dennis Dart 10.1m	Plaxton Pointer	B35F	1997	Stansted Transit, 2007
P685RWU	Dennis Dart 10.1m	Plaxton Pointer	B35F	1997	Stansted Transit, 2007
P831BUD	Dennis Dart SLF 10.7m	Marshall Capital	N38F	1997	Kent CC, 2008
S342SET	Scania L94 IB4	Irizar Century 12.35	C49FT	1998	Shamrock, Pontypridd, 2004
S343SET	Scania L94 IB4	Irizar Century 12.35	C49FT	1998	Shamrock, Pontypridd, 2004
S6APH	Mercedes-Benz Vario O814	Plaxton Cheetah	C25F	1999	OFJ Connections, Heathrow, 2006

Olympian Coaches operates local services around Harlow along with commuter services into London from Harlow and Bishop's Stortford. Seen in London's Victoria is EAZ2588, a Volvo B10M with Plaxton Première 320 bodywork. *Colin Lloyd*

Parliament Square is the location for this view of Olympian's Scania coach, L100SCS. Featuring the Irizar Century coachwork which is Scania's partner coachbuilder, it is seen in one of the contract liveries. *Colin Lloyd*

S234HGU	Iveco EuroRider 391E.12.35	Beulas Stergo E	C49FT	1999	WG Services, Harlow, 2004
T936HTR	Iveco EuroRider 391E.12.35	Beulas Stergo E	C52F	1999	WG Services, Harlow, 2004
T289CGU	Iveco EuroRider 391E.12.35	Beulas Stergo E	C53F	1999	Holmeswood Coaches, 2008
V282SBW	Volvo B10M-62	Plaxton Excalibur	C45FT	1999	Oxford Bus, 2007
V283SBW	Volvo B10M-62	Plaxton Excalibur	C45FT	1999	Oxford Bus, 2007
LV02LKD	Irisbus EuroRider 391E.12.35A	Beulas Stergo E	C53F	2002	Redwing, Herne Hill, 2005
FN02VBK	Irisbus EuroMidi CC80E18M/P	Indcar Maxim 2	C29F	2002	WG Services, Harlow, 2004
LK03PZW	Irisbus EuroMidi CC80E18M/P	Indcar Maxim 2	C29F	2003	WG Services, Harlow, 2004
MV03AJY	LDV Convoy	Concept	M16	2003	Brew, Ponders End, 2006
FD54EOO	Irisbus Scolabus	Vehixel	S68F	2005	
FD54EOP	Irisbus Scolabus	Vehixel	S68F	2005	
EU06KHK	Irisbus Agoraline	Irisbus	N44F	2006	
EU07XGK	Irisbus LoGo 65C18	Unvi Compa	BC24F	2007	
EU07XGL	Irisbus LoGo 65C18	Unvi Compa	BC24F	2007	
EU07XGM	Irisbus LoGo 65C18	Unvi Compa	BC24F	2007	
EU08FHB	ADL Dart 4 8.9m	ADL Enviro 200	N28F	2008	
EU08FHC	ADL Dart 4 8.9m	ADL Enviro 200	N28F	2008	
EU08FHD	ADL Dart 4 8.9m	ADL Enviro 200	N28F	2008	

Previous registrations:

FIG2843	F683CYC, NIL7253	S342SET	98D54637
HUI4165	G119HVK, WSV573, NMS100, G918NUP, HKR11, G561XRG		
KKZ7562	L385RYC, 2439RU, L2GHW	S343SET	98D53532
MUI7843	F148NVE, VWH836	T936HTR	MJI2378
MV03AJY	MV03AJY, H3BOE	V282SBW	V15OXF
N139GMJ	N139GMJ, N1TRS	V283SBW	V14OXF

English Bus Handbook: Notable Independents

PENNINE

ND, JD & MB Simpson, Broughton Road, Skipton, BD23 1TE

	Reg	Chassis	Body	Layout	Year	
	JIL2426	Leyland Leopard PSU3E/4R	Plaxton Supreme IV	C53F	1981	Wing, Sleaford, 1984
D1	J615KCU	Dennis Dart 9.8m	Wright HandyBus	B40F	1991	Avon, Prenton, 2003
D3	J954MFT	Dennis Dart 9.8m	Wright HandyBus	B40F	1992	Bakerbus, Biddulph, 2003
D4	K986SCU	Dennis Dart 9.8m	Wright HandyBus	B40F	1993	Go-Ahead Northern, 2003
D5	K984SCU	Dennis Dart 9.8m	Wright HandyBus	B40F	1993	Go-Ahead Northern, 2003
D6	K989SCU	Dennis Dart 9.8m	Wright HandyBus	B40F	1993	Go-Ahead Northern, 2003
D7	M828RCP	Dennis Dart 9.8m	Northen Counties Paladin	B40F	1993	Countryman, Ibstock, 2004
D8	N3BLU	Dennis Dart 9.8m	Plaxton Pointer	B40F	1995	Countryman, Ibstock, 2004
D9	P696HND	Dennis Dart 9.8m	Plaxton Pointer	B40F	1996	Countryman, Ibstock, 2004
D10	N260PRJ	Dennis Dart 9.8m	Plaxton Pointer	B40F	1995	Metrobus, Crawley, 2007
D11	N134XND	Dennis Dart 9.8m	Plaxton Pointer	B40F	1995	Ensignbus, Rainham, 2007
D12	N53KBW	Dennis Dart 9.8m	Plaxton Pointer	B40F	1996	Stagecoach, 2008
D13	R703YWC	Dennis Dart 9.8m	Plaxton Pointer	B40F	1997	Stagecoach, 2008
D14	R706YWC	Dennis Dart 9.8m	Plaxton Pointer	B40F	1997	Stagecoach, 2008
D15	R717YWC	Dennis Dart 9.8m	Plaxton Pointer	B40F	1997	Stagecoach, 2008
LN5	JIL2795	Leyland National 1051/1R/0501 (Volvo)		B44F	1973	Executive, Greasby, 1994
LN15	JIL7417	Leyland National 10351A/2R (Volvo)		B44F	1979	Arriva North West, 1999
LN17	JIL2428	Leyland National 10351A/2R (Volvo)		B44F	1979	Arriva North West, 1999
LN18	JIL7416	Leyland National 10351A/2R (Volvo)		B44F	1979	Arriva North West, 2000
LN19	BYW432V	Leyland National 10351A/2R (Volvo)		B44F	1979	Arriva North West, 2001
LN20	BYW430V	Leyland National 10351A/2R (Volvo)		B44F	1979	Arriva North West, 2001

Special event vehicles:

	Reg	Chassis	Body	Layout	Year	
	MTD235	Leyland Royal Tiger PSU1/15	Leyland	C41C	1951	Leyland demonstrator
	OWY197K	Leyland Leopard PSU3B/4R	Plaxton Panorama Elite III	C53F	1972	Laycock, Barnoldswick

Ancillary vehicle:

	Reg	Chassis	Body	Layout	Year	
	HWR449T	Leyland Leopard PSU3E/4R	Plaxton Supreme III Express	RV	1979	Fishwick, Leyland, 1979

Previous registrations:

HWR449T	DCK452T, JIL4613	JIL2795	XDL800L
J954MFT	J954MFT, 9423RU	JIL7416	BYW413V
JIL2426	BTL485X	JIL7417	BYW437V
JIL2428	BYW412V	P696HND	N609WND

Depots: West Close Garage, Barnoldswick; New Road Garage, Ingleton; White Friars Garage, Settle and Albion Square, Skipton

Pennine's Dart D12, N53KBW, stands in Skipton bus station prior to running service 212 to Carleton. *Dave Heath*

PREMIERE

Premiere Travel Ltd, Trent Wharf, Meadow Lane, Nottingham, NG2 3HR

Reg	Chassis	Body		Seating	Year	History
THX209S	Leyland National 10351A/2R			B42F	1978	Arriva Midlands, 2007
TRN808V	Leyland National 10351B/1R			B42F	1979	Arriva Midlands, 2007
C59HOM	Volvo B10M-55	Alexander P		B50F	1986	Rapsons, Inverness, 2007
F660RTL	Volvo B10M-60	Plaxton Paramount 3200 III		C53F	1989	Henshaw, Moreton, 2006
G611CEF	Leyland Lynx LX2R11C15Z4S	Leyland		B49F	1989	Stagecoach, 2007
G620CEF	Leyland Lynx LX2R11C15Z4S	Leyland		B49F	1989	Stagecoach, 2007
G112ENV	Volvo B10M-55	Duple 300		BC49F	1989	Invincible, Tamworth, 2008
G113ENV	Volvo B10M-55	Duple 300		BC49F	1989	Invincible, Tamworth, 2008
G114ENV	Volvo B10M-55	Duple 300		BC49F	1989	Invincible, Tamworth, 2008
LUI7665	Dennis Dart 9.8m	Plaxton Pointer		B39F	1994	Birmingham Coach Co, 2007
LUI6233	Dennis Lance 11m	Plaxton Verde		B46F	1994	Diamond, Birmingham, 2008
LUI7627	Dennis Lance 11m	Plaxton Verde		B46F	1994	Birmingham Coach Co, 2007
LUI7662	Dennis Lance 11m	Plaxton Verde		B46F	1994	Birmingham Coach Co, 2007
LUI7672	Dennis Lance 11m	Plaxton Verde		B46F	1994	Diamond, Birmingham, 2008
L211YAG	Dennis Lance 11m	Plaxton Verde		B46F	1994	Green, Birmingham, 2008
M101BLE	Dennis Dart 9.8m	Plaxton Pointer		B39F	1994	Diamond, Birmingham, 2008
M890GBB	Dennis Dart 9.8m	Plaxton Pointer		B40F	1994	Diamond, Birmingham, 2008
B15PTL	Volvo B10M-60	Jonckheere Deauville 45		C51FT	1994	MM Coachlines, Walderslade, '07
LUI8478	Volvo B10M-62	Caetano Enigma		C53FT	1995	Walkers, Nottingham, 2003
LUI7668	Volvo B10M-62	Plaxton Excalibur		C49FT	1995	Walkers, Nottingham, 2003
M453LLJ	Dennis Dart 9.8m	East Lancs		B40F	1995	Transdev Yellow Buses, 2007
NIL9017	Dennis Dart 9.8m	Marshall C37		B40F	1995	Tally Ho!, Kingsbridge, 2003
N777ELK	Dennis Lance 11m	Plaxton Verde		B49F	1996	Ellen Travel, Shenstone, 2008
P509OUG	Optare Excel L1070	Optare		N35F	1996	Diamond Bus, Birmingham, 2008
P510OUG	Optare Excel L1070	Optare		N35F	1996	Diamond Bus, Birmingham, 2008
P683RWU	Dennis Dart SLF	Plaxton Pointer 2		N35F	1997	Stansted Transit, 2008
P689RWU	Dennis Dart SLF	Plaxton Pointer 2		N35F	1997	Stansted Transit, 2008
R207DKG	Optare Excel L1150	Optare		N42F	1997	Central Connect, 2008
R212DKG	Optare Excel L1150	Optare		N42F	1997	Rotala, 2008
R215DKG	Optare Excel L1150	Optare		N42F	1997	Rotala, 2008
V264DRC	Optare Solo M920	Optare		N33F	1999	Nottingham City Transit, 2008
V275DRC	Optare Solo M920	Optare		N33F	1999	Nottingham City Transit, 2008
V278DRC	Optare Solo M920	Optare		N33F	1999	Nottingham City Transit, 2008
V279DRC	Optare Solo M920	Optare		N33F	1999	Nottingham City Transit, 2008
V281DRC	Optare Solo M920	Optare		N33F	1999	Nottingham City Transit, 2008
B10PTL	Volvo B10M-62	Caetano Enigma		C49FT	2002	Classic, Annfield Plain, 2007
B12PTL	Volvo B10M-62	Caetano Enigma		C49FT	2002	Tellings-Golden Miller, 2007
BD52HKN	LDV Convoy	LDV		M16	2002	Arriva Hire, 2007
BD03YFW	LDV Convoy	LDV		M16	2003	Arriva Hire, 2007
KX03HYW	Mercedes-Benz Vario 0814	Plaxton Beaver		B33F	2003	LB Tower Hamlets, 2007
KV03ZGL	TransBus E300	TransBus Enviro 300		N44F	2003	Reading Buses, 2007
KV03ZGM	TransBus E300	TransBus Enviro 300		N44F	2003	Reading Buses, 2007
KV03ZGN	TransBus E300	TransBus Enviro 300		N44F	2003	Reading Buses, 2007
KX53SBZ	Optare Solo M920	Optare		N32F	2004	Courtney, Bracknell, 2008
MX04VLV	Optare Solo M920	Optare		N33F	2004	Little, Ilkeston, 2008
B17PTL	Volvo B12M	VDL Jonckheere Mistral 50		C51FT	2004	Murphy, Chalfont St Peter, 2008
FSV428	Dennis R	Caetano Enigma		C49FT	2004	
FN04FSG	Dennis R	Caetano Enigma		C49FT	2004	
YN05VRR	Dennis R	Plaxton Panther		C53F	2005	
SF06VYT	Volvo B12M	VDL Jonckheere Mistral 50		C53F	2006	Park's of Hamilton, 2008
CL07PTL	Volvo B12B	Caetano Enigma		C53F	2007	
KX57BWE	ADL Dart 4	ADL Enviro 200		N37F	2008	
KX08DNC	Mercedes-Benz Vario 0816	Plaxton Cheetah		C29F	2008	

Previous registrations:

B10PTL	FN02RXA	LUI7665	M812HCU
B12PTL	5877MW, RG02DNN	LUI7668	M665WCK, KBZ1410
B15PTL	L4VLT, VLT149, VLT935, L536YCC, L44MMC	LUI7672	L202YAG
B17PTL	LSK879, SJ04KBF	LUI8478	N620XBU
F660RTL	F287OFE, 5517RH	N777ELK	N903PFC
FSV428	FN04FSP, N30TTS	NIL9017	M583WLV
LUI6233	L204YAG	SF06VYT	HSK657
LUI7627	L212YAG, WLT461, L942RJN	TRN808V	TRN808V, FSV428
LUI7662	L201YAG		

QUANTOCK MOTOR SERVICES

Quantock Motor Services Ltd, Rosebank, Langley Marsh, Wiveliscombe, TA4 2UJ

Reg	Chassis	Body	Seating	Year	Operator
DUF179	Leyland Tiger TS8	Harrington	C32R	1937	Southdown
JG9938	Leyland Tiger TS8	Park Royal	C32R	1937	East Kent
AJA132	Bristol L5G	Burlingham (1950)	B35R	1938	North Western
GNU750	Daimler COG5	Willowbrook	C35F	1939	Tailby & George, Willington
CCX777	Daimler CWA6	Duple	L27/28R	1945	Halifax
ACH441	AEC Regal III O682	Windover	C32F	1948	Trent
JTE546	AEC Regent III 6811E	Park Royal	B33/26R	1948	Morecambe & Heysham
KTF594	AEC Regent III 9621E	Park Royal	O33/26R	1949	Morecambe & Heysham
EVD406	Crossley DD42/7	Roe (1955)	B31/25R	1949	Wood, Mirfield
LJH665	Dennis Lancet III	Duple	C35F	1949	Lee, Barnet
JFJ875	Daimler CVD6 SD	Weymann	B35F	1950	Exeter
CDB206	Bristol L5G	Weymann	B35R	1950	North Western
KFM767	Bristol L5G	Eastern Coach Works	B35R	1950	Crosville
LFM302	Leyland Tiger PS1/1	Weymann	B35F	1950	Crosville
DCK219	Leyland Titan PD2/3	East Lancashire	CL27/22RD	1951	Ribble
CHG541	Leyland Tiger PS2/14	East Lancashire	B39F	1954	Burnley, Colne & Nelson
BAS563	Bristol Lodekka LD6G	Eastern Coach Works	O33/27R	1956	Southern Vectis
VDV752	Bristol Lodekka LDL6G	Eastern Coach Works	O37/33RD	1957	Devon General
WAL782	Leyland Tiger PS1/1	Willowbrook	L31/30RD	1957	Barton
890ADV	AEC Reliance 2MU3RV	Willowbrook	C41F	1959	Devon General
805EVT	AEC Reliance 2MU3RV	Weymann	B41F	1960	PMT
851FNN	AEC Regent V 2D3RA	Northern Counties	L37/33F	1960	Barton
DPV65D	AEC Regent V	Neepsend	B37/28R	1966	Ipswich

Quantock Motors Serivces specialise in restoring vintage buses and operating them on special services. Pictured at Winkleigh is DCK219, one of the original Ribble 'White Lady' coach-seated buses that operated the express services that mostly originated from Manchester and Liverpool. *Dave Godley*

Now located many miles from Buxton at the operator's base in Norton Fitzwarren, CDB206 has recently been restored by Quantock. It is a Bristol L5G with Weymann bodywork dating from the middle of the last century.
Dave Godley

XTF98D	Leyland Titan PD3/4	East Lancs	B41/32F	1966	Haslingden
HJA965E	Leyland Titan PD2/40	Neepsend	B37/26R	1967	Stockport
GNH258F	Daimler CVG6 DD	Roe	B33/26R	1967	Northampton
A24OVL	Leyland Tiger TRCTL11/2R	Duple Dominant	B62F	1983	Embling, Guyhirn, 2007
TYR95	Van Hool Alizée T815	Van Hool	C53F	1987	Nottinghamshire CC, 2005
E50TYG	Leyland Royal Tiger RTC	Leyland Doyen	C53F	1988	Wilkins, Cymmer, 2007
MIB767	Leyland Royal Tiger RTC	Leyland Doyen	C53F	1988	Wilkins, Cymmer, 2007
TCZ6123	Leyland Tiger TRCL10/ARZM	Plaxton Paramount 3200 III	C53F	1989	Mitcham Belle, 2005
TCZ6124	Leyland Tiger TRCL10/ARZA	Plaxton Paramount 3200 III	C53F	1989	Edinburgh Tours, 2004
TIL8148	Volvo B10M-61	Jonckheere Deauville P599	C53F	1990	Taunton Coaches, 2008
G995VWV	Leyland Lynx LX112L10ZR1R	Leyland	B47F	1990	Hedingham & District, 2006
TCZ6121	Leyland Tiger TRCL10/3ARZA	Plaxton Paramount 3500 III	C53F	1991	Edinburgh Tours, 2004
TCZ6122	Leyland Tiger TRCL10/3ARZA	Plaxton Paramount 3500 III	C53F	1991	Edinburgh Tours, 2004
UIL2724	Van Hool Alizée T815	Van Hool	C53F	1991	Cheney, Banbury, 2008
XIL8438	Van Hool Alizée T815	Van Hool	C53F	1991	Joplin, Tittleshall, 2007
K33GOW	Dennis Dart 9.8m	Northern Counties Paladin	B42F	1992	Whittle, Kidderminster, 2006
VYD333	DAF SB3000	Van Hool Alizée	C51FT	1993	Eavesway, Wigan, 2004
HKO169	Volvo Olympian	Northern Counties	B47/30F	1995	Arriva Southern Counties, 2005
M26XEH	Dennis Dart 9.8m	Northern Counties Paladin	B39F	1995	Travel London, 2007
N209ONL	Mercedes-Benz 609D	TBP	BC16F	1995	Sunset, Sancreed, 2007
FNZ1052	Scania K113 CRB	Van Hool Alizée HE	C49FT	1996	D&H Travel, Deepdale, 2007
LUI9951	Bova Futura FHD 12.340	Bova	C51FT	1997	Crown, Benfleet, 2007
157TYB	DAF SB3000	Van Hool T9 Alizée	C48FT	2001	Eavesway, Wigan, 2004
YN55RDV	Scania N94 UD	East Lancs Pyoneer	PO43/26F	2005	

Previous registrations:

157TYB	Y298HUA	TCZ6123	F635UBL, MIB767
BAS563	MDL952	TCZ6124	G68DFS
E50TYG	E50TYG, OIJ864	TIL8148	G382RNH, VLT288, G563JJC, 3432RE, G742MEP
FNZ1052	N136YMS	TYR95	D731WCH
HKO169	M924PKN	UIL2724	H203DVM
MIB767	E51TYG, 551OVW, E269ODE	VYD333	K105TCP
TCZ6121	H71NFS	XIL8438	H202DVM, 848KMX
TCZ6122	H72NHS		

Note: A significant number of other vehicles are owned, both restored or awaiting restoration, but they are not licenced for use.
Depot: Taunton Trading Estate, Norton Fitzwarren

RED ROSE

Red Rose Travel Ltd, 2 Brook End, Weston Turville, HP22 5PF

F49ENF	Leyland Tiger TRBL10/3ARZA	Alexander N	B53F	1987	Timeline, 1998
H620ACK	Volvo B10M-55	Alexander PS	B51F	1990	Stagecoach, 2007
M79CYJ	Dennis Dart 9.8m	Plaxton Pointer	B40F	1995	Brighton & Hove, 2008
N802GRV	Mercedes-Benz 709D	UVG CitiStar	B29F	1996	
N803GRV	Mercedes-Benz 709D	UVG CitiStar	B29F	1996	
N784JBM	Mercedes-Benz 711D	UVG CitiStar	B29F	1996	
P507NWU	Optare MetroRider MR17	Optare	B25F	1996	Optare demonstrator, 1997
P263NRH	Dennis Dart SLF 10.7m	Plaxton Pointer 2	N41F	1996	East Yorkshire, 2007
P264NRH	Dennis Dart SLF 10.7m	Plaxton Pointer 2	N41F	1996	East Yorkshire, 2007
R843FWW	Optare MetroRider MR17	Optare	B31F	1997	
R502YWC	Mercedes-Benz Vario 0814	Plaxton Beaver 2	B29F	1997	Lewis Meridian, Greenwich, 2008
R796GSF	Mercedes-Benz Vario 0814	Plaxton Beaver 2	B29F	1997	
T341FWR	Optare MetroRider MR17	Optare	B31F	1998	
T458JRH	Mercedes-Benz Vario 0814	Plaxton Beaver 2	B29F	1999	East Yorkshire, 2008
V108LVH	Optare MetroRider MR17	Optare	B31F	1999	
Y358LCK	Dennis Dart SLF 10.8m	East Lancs Spryte	N41F	2001	
Y359LCK	Dennis Dart SLF 10.8m	East Lancs Spryte	N41F	2001	
VU02TPZ	Dennis Dart 8.8m	Plaxton Pointer	N29F	2002	Norbus, Kirkby, 2006
RR03BUS	Optare Solo M920	Optare	N33F	2003	
FJ06URR	ADL Dart SLF 9m	Caetano Nimbus	N31F	2006	
OU07FKH	ADL Dart 4 8.9m	ADL Enviro 200	N29F	2007	
OU07FKJ	ADL Dart 4 8.9m	ADL Enviro 200	N29F	2007	
OU08EHO	ADL Dart 4 8.9m	ADL Enviro 200	N29F	2008	
OU08EHP	ADL Dart 4 8.9m	ADL Enviro 200	N29F	2008	

Web: www.redrosetravel.com
Depot: Oxford Road, Dinton

Watford is the location for this image of Red Rose R796GSF, one of three Beaver 2-bodied Vario buses in use. The operator is based at Dinton between Aylesbury and Thame. *Richard Godfrey*

REGAL BUSWAYS

Regal Busways Ltd, 21 Havengore Close, Great Wakering, SS3 0PH

201	YJ06YSK	Optare Solo M710 SL	Optare	N23F	2006	
202	YJ56AOT	Optare Solo M710 SL	Optare	N23F	2006	
203	YJ56AOU	Optare Solo M710 SL	Optare	N23F	2006	
204	YJ58CDE	Optare Solo M710 SL	Optare	N23F	2008	
205	YJ58CDF	Optare Solo M710 SL	Optare	N23F	2008	
501	K716PCN	Dennis Dart 9.8m	Alexander Dash	B40F	1992	Solent Blue Line, 2007
503	J605XHL	Dennis Dart 9m	Plaxton Pointer	B34F	1991	Birmingham Coach Co, 2007
504	L4YTB	Dennis Dart 9.8m	Plaxton Pointer	B40F	1993	Stagecoach, 2007
505	J105DUV	Dennis Dart 8.5m	Plaxton Pointer	B28F	1992	London United, 2002
506	K955PBG	Dennis Dart 9.8m	Plaxton Pointer	B36F	1993	Arriva NW & Wales, 2007
507	K877UDB	Dennis Dart 9.8m	Plaxton Pointer	B40F	1992	Arriva NW & Wales, 2007
508	K414FHJ	Dennis Dart 9.8m	Plaxton Pointer	B40F	1992	Arriva the Shires, 2008
509	M69CYJ	Dennis Dart 9.8m	Plaxton Pointer	B40F	1992	Brighton & Hove, 2008
510	L110HHV	Dennis Dart 9m	Northern Counties Paladin	B34F	1994	NIBS, Wickford, 2008
540	J140DUV	Dennis Dart 8.5m	Plaxton Pointer	B24F	1992	London United, 2003
601	YN04PZY	ADL Dart SLF 8.8m	ADL Mini Pointer	N29F	2004	
602	YN04PZZ	ADL Dart SLF 8.8m	ADL Mini Pointer	N29F	2004	
603	W772URP	Dennis Dart SLF 8.8m	Plaxton Pointer MPD	N22F	2000	Airparks, Birmingham, 2006
611	X211ONH	Dennis Dart SLF 8.8m	Plaxton Pointer MPD	N29F	2001	Centra, Heathrow, 2007
701	AE55EHM	ADL Dart SLF 10.7m	MCV Evolution	N37F	2005	
1201	OUI6274	Leyland Olympian ONLXB/1R	Eastern Coach Works	B45/32F	1982	Linburg, Darnall, 2007
1203	B203DTU	Leyland Olympian ONLXB/1R	Eastern Coach Works	B45/32F	1985	Arriva Midlands, 2004
1204	C264XEF	Leyland Olympian ONLXB/1R	Eastern Coach Works	BC42/28F	1986	Arriva North East, 2005
P01	YJ56WVX	Optare Tempo X1200	Optare	NC36F	2006	
P02	YJ56WVC	Optare Tempo X1200	Optare	NC36F	2007	
P03	YJ56WVY	Optare Tempo X1200	Optare	NC36F	2007	
	7345FM	Volvo B10M-62	Plaxton Panther	C49FT	2001	Plan-it Travel, Chelsfield, 2008

Previous registrations:

7345FM	FE51RCO, M2WMT		
C80CHM	C80CHM, EMN203U	OUI6274	JTY401X
K716PCN	K716PCN, NFX667	W772URP	W222APS

Depot: Ongar Road West, Cooksmill Green; **web:** www.regalbusways.com

One of three Tempo buses that are fitted with luxury leather high-back seats is P02, YJ56WVC, which carries *Essex Pullman* livery. *Richard Godfrey*

RELIANCE

JH & M Duff, Reliance Garage, York Road, Sutton-on-the-Forest, York, YO61 1ES

K100BLU	DAF SB3000	Van Hool Alizée	C55F	1993	Blue Bus, Bolton, 2001
R362DJN	Volvo Olympian	East Lancs Pyoneer	N51/32F	1998	East Thames Buses, 2005
R363DJN	Volvo Olympian	East Lancs Pyoneer	N51/32F	1998	East Thames Buses, 2005
W475RKS	Mercedes-Benz Atego 1190L	Ferqui/Optare Solera	C35F	2000	Radical Travel, Edinburgh, 2006
X381XON	Volvo B7L	Wrightbus Eclipse	N42F	2001	Mortons, Dublin, 2008
X618VWR	Volvo B10BLE	Wrightbus Renown	NC46F	2001	
YJ51NXK	Volvo B6BLE	Wrightbus Crusader	N42F	2001	
YJ52HEA	Volvo B7TL	Wrightbus Eclipse Gemini	N45/29F	2002	Volvo demonstrator, 2005
YJ05UKR	Volvo B7RLE	Wrightbus Eclipse Urban	N43F	2005	
BV55UAZ	Volvo B7RLE	Wrightbus Eclipse Urban	N44F	2005	Volvo demonstrator, 2006

Previous registration:
X381XON   GCZ9023, X381XON, 01D93727

Web: www.reliancemotorservices.co.uk

New to East Thames Buses, R362DJN is one of two from that source now operating with Reliance. It carries an East Lancs Pyoneer body. *Mark Bailey*

RENOWN

Renown - Cavendish

Renown Coaches Ltd, 13 Sea Road, Bexhill-on-Sea, TN40 1EE

1	V931VUB	Optare Solo M850	Optare	N27F	1999	*Operated for East Sussex CC*
2	YV03UTY	Optare Solo M850	Optare	N27F	2003	*Operated for East Sussex CC*
3	YV03UTX	Optare Solo M850	Optare	N27F	2003	*Operated for East Sussex CC*
4	W84NDW	Optare Solo M850	Optare	N29F	2000	North West, Liverpool, 2007
5	W283EYG	Optare Solo M850	Optare	N28F	2000	Arriva Scotland, 2007
7	YG52DHD	Optare Solo M850	Optare	N25F	2002	*Operated for East Sussex CC*
8	YJ05WCA	Optare Solo M850 SL	Optare	N25F	2005	*Operated for East Sussex CC*
9	V932VUB	Optare Solo M850	Optare	N27F	1999	*Operated for East Sussex CC*
10	V649EEF	Optare Solo M920	Optare	N31F	1999	Stockton DC, 2007
11	YJ05WCD	Optare Solo M850 SL	Optare	N25F	2005	*Operated for East Sussex CC*
12	W936JNF	Dennis Dart SLF 8.5m	Plaxton Pointer MPD	N29F	2002	Norbus, Kirkby, 2005
13	L881YVK	Dennis Dart SLF 10.1m	Plaxton Pointer	B40F	1994	Boomerang Bus, Tewkesbury, '05
14	SN04EFL	TransBus Dart 10.1m	TransBus Pointer	NC36F	2004	
15	YJ05WCE	Optare Solo M850 SL	Optare	N25F	2005	*Operated for East Sussex CC*
15	H115THE	Dennis Dart SLF 8.5m	Reeve Burgess Pointer	B32F	1991	Metroline, Harrow, 2000
16	P690RUU	Dennis Dart SLF 10.1m	Plaxton Pointer 2	N35F	1997	First Stop, Renfrew, 2004
17	P696RUU	Dennis Dart SLF 10.1m	Plaxton Pointer 2	N35F	1997	Gibson Direct, Renfrew, 2004
18	P699RUU	Dennis Dart SLF 10.1m	Plaxton Pointer 2	N35F	1997	Dickson, Erskine, 2004
19	FN54FLC	TransBus Dennis Dart	Caetano Nimbus	N28F	2004	
20	X212ONH	Dennis Dart SLF 10.1m	Plaxton Pointer 2	N37F	1997	Flights Hallmark, Hounslow, 2008
21	P341OEW	Dennis Dart SLF 10.1m	Marshall Capital	N37F	1997	Chesterbus, 2008
22	P342OEW	Dennis Dart SLF 10.1m	Marshall Capital	N37F	1997	Chesterbus, 2008
23	P343OEW	Dennis Dart SLF 10.1m	Marshall Capital	N37F	1997	Chesterbus, 2008
24	P344OEW	Dennis Dart SLF 10.1m	Marshall Capital	N37F	1997	Chesterbus, 2008

Shown passing through Newhaven, Renown's Dart 68, M68CYJ, illustrates the early Pointer body style. As we go to press it was announced that the Cavendish element has been sold to Stagecoach along with twenty-eight buses. *Alan Blagburn*

Arriva London was the source of several Leyland Olympians that are fitted with Northern Counties bodywork. Number 48, G518VBB, is seen in Eastbourne during August 2008. *Richard Godfrey*

25	C25CHM	Leyland Olympian ONLXB/1R	Eastern Coach Works	B42/26D	1986	Arriva London, 2005
27	C329HWJ	Leyland Olympian ONLXB/1R	Eastern Coach Works	BC40/32F	1985	Stagecoach, 2007
28	D168FYM	Leyland Olympian ONLXB/1RH	Eastern Coach Works	B42/26F	1986	Arriva The Shires, 2005
30	D180FYM	Leyland Olympian ONLXB/1RH	Eastern Coach Works	B42/26D	1986	Arriva London, 2005
32	D162FYM	Leyland Olympian ONLXB/1RH	Eastern Coach Works	B42/26D	1986	Arriva London, 2005
34	D214FYM	Leyland Olympian ONLXB/1RH	Eastern Coach Works	B42/26D	1986	Arriva London, 2005
39	D169FYM	Leyland Olympian ONLXB/1RH	Eastern Coach Works	B42/26F	1986	Arriva The Shires, 2005
41	D211FYM	Leyland Olympian ONLXB/1RH	Eastern Coach Works	B42/26D	1986	Arriva The Shires, 2005
42	D167FYM	Leyland Olympian ONLXB/1RH	Eastern Coach Works	B42/26F	1986	Arriva The Shires, 2005
43	R112RLY	Dennis Dart SLF 10.1m	Plaxton Pointer 2	N36F	1997	Metroline, Harrow, 2006
44	G994VWV	Leyland Lynx LX112L10ZR1	Leyland	B47F	1990	Brighton Buses, 2005
45	R175VI A	Dennis Dart SLF 10.1m	Plaxton Pointer 2	N35F	1998	Metroline, Harrow, 2006
46	P746HND	Dennis Dart SLF 10.1m	Plaxton Pointer 2	N39F	1997	Trustline, Hunsdon, 2006
47	R117RLY	Dennis Dart SLF 10.1m	Plaxton Pointer 2	N36F	1997	Metroline, Harrow, 2006
48	G518VBB	Leyland Olympian ONCL10/1RZ	Northern Counties	B47/31F	1990	Arriva London, 2006
49	G537VBB	Leyland Olympian ONCL10/1RZ	Northern Counties	B47/31F	1990	Arriva London, 2007
50	G540VBB	Leyland Olympian ONCL10/1RZ	Northern Counties	B47/31F	1990	Arriva London, 2006
51	G541VBB	Leyland Olympian ONCL10/1RZ	Northern Counties	B47/31F	1990	Arriva London, 2006
52	G542VBB	Leyland Olympian ONCL10/1RZ	Northern Counties	B47/31F	1990	Arriva London, 2006
53	G543VBB	Leyland Olympian ONCL10/1RZ	Northern Counties	B47/31F	1990	Arriva London, 2006
54	S794XUG	Optare Solo M850	Optare	N25F	1998	East Sussex CC
55	G525VBB	Leyland Olympian ONCL10/1RZ	Northern Counties	B47/31F	1990	Arriva London, 2006
56	G556VBB	Leyland Olympian ONCL10/1RZ	Northern Counties	B47/31F	1990	Arriva London, 2006
57	P697RWU	Dennis Dart SLF 10.1m	Plaxton Pointer 2	N35F	1997	Wilson, Greenock, 2006
58	P738RYL	Dennis Dart SLF 10.1m	Plaxton Pointer 2	N36F	1996	Blue Triangle, Rainham, 2006
59	ANZ8799	Dennis Dart 8.8m	Carlyle Dartline	B28F	1991	Ensignbus, Purfleet, 2007
60	ANZ8798	Dennis Dart 8.8m	Carlyle Dartline	B28F	1991	Ensignbus, Purfleet, 2007
61	P691RWU	Dennis Dart SLF 10.1m	Plaxton Pointer 2	N35F	1997	Wilson, Greenock, 2006
62	L720OMV	Dennis Dart 9.8m	Plaxton Pointer	B35F	1994	Ensignbus, Purfleet, 2007
63	T63KLD	Dennis Dart SLF 10.1m	Marshall Capital	N31F	1999	Metroline, Harrow, 2008
64	T64KLD	Dennis Dart SLF 10.1m	Marshall Capital	N31F	1999	Metroline, Harrow, 2008
65	M65CYJ	Dennis Dart 9.8m	Plaxton Pointer	B40F	1995	Brighton & Hove, 2006
67	M97WBW	Dennis Dart 9.8m	Plaxton Pointer	B43F	1995	Stagecoach, 2006
68	M68CYJ	Dennis Dart 9.8m	Plaxton Pointer	B40F	1995	Brighton & Hove, 2006
70	K710PCN	Dennis Dart SLF 9.8m	Alexander Dash	B40F	1992	Stagecoach, 2007
72	J702KCU	Dennis Dart SLF 9.8m	Plaxton Pointer	B40F	1992	Speldhurst, Shefford, 2007
73	K703PCN	Dennis Dart SLF 9.8m	Alexander Dash	B40F	1992	Stagecoach, 2007

75	T65KLD	Dennis Dart SLF	Marshall Capital	N31F	1999	Metroline, Harrow, 2008
78	T458HNH	Dennis Dart SLF 10.1m	Plaxton Pointer 2	N37F	1999	Simonds, Botesdale, 2008
79	T459HNH	Dennis Dart SLF 10.1m	Plaxton Pointer 2	N37F	1999	Simonds, Botesdale, 2008
81	T461HNH	Dennis Dart SLF 10.1m	Plaxton Pointer 2	N37F	1999	NCP, Birmingham, 2008
103	P943EBB	Volvo B10M-62	Plaxton Première 350	C53F	1997	Fleetwing, Aldershot, 2006
105	N805NHS	Volvo B10M-62	Jonckheere Mistral 45	C53F	1996	Oakfield, Broxbourne, 2004
113	XIL1273	Volvo B10M-62	Van Hool Alizée HE	C49FT	1995	Shearings, 2002
114	XIL1274	Volvo B10M-62	Van Hool Alizée HE	C49FT	1995	Shearings, 2002
-	WX53OXZ	Mercedes-Benz Sprinter 413cdi	UV Modular	B16F	2003	Operated for East Sussex CC
-	R396XDA	DAF SB220	Northern Counties Paladin	N42F	1990	UK North, Manchester, 2008
-	R397XDA	DAF SB220	Northern Counties Paladin	N42F	1990	UK North, Manchester, 2008
-	T164AUA	DAF SB220	Ikarus Citibus	N43F	1990	UK North, Manchester, 2008
-	W176CDN	DAF SB220	Ikarus Citibus	N44F	1990	UK North, Manchester, 2008
-	T439EBD	Dennis Dart SLF 10.7m	Plaxton Pointer 2	N39F	1999	Waters, Corringham, 2008
-	T440EBD	Dennis Dart SLF 10.7m	Plaxton Pointer 2	N39F	1999	TM Travel, Sheffield, 2008
-	T417MNH	Dennis Dart SLF 10.7m	Plaxton Pointer 2	N39F	1999	Preston Bus, 2008
-	T342PNV	Dennis Dart SLF 10.7m	Plaxton Pointer 2	N37F	1999	Weavaway, Newbury, 2008
-	T553HNH	Dennis Dart SLF 10.7m	Plaxton Pointer 2	N41F	1999	Menzies, Heathrow, 2008
-	T927PNV	Dennis Dart SLF 10.7m	Plaxton Pointer 2	N39F	1999	Menzies, Heathrow, 2008
-	W558JVV	Dennis Dart SLF 10.7m	Plaxton Pointer 2	N39F	2000	Velvet, Eastleigh, 2008
-	W566JVV	Dennis Dart SLF 10.7m	Plaxton Pointer 2	N39F	2000	Veolia Cymru, 2008
-	W378SVV	Dennis Dart SLF 10.7m	Plaxton Pointer 2	N39F	2000	Kent CC, 2008
-	W671TNV	Dennis Dart SLF 10.7m	Plaxton Pointer 2	N39F	2000	Menzies, Heathrow, 2008
-	X215ONH	Dennis Dart SLF 10.7m	Plaxton Pointer 2	N37F	2001	Flights Hallmark, Hounslow, 2008
-	YJ51ZVK	DAF DB250	Optare Spectra	N47/27F	2001	Ensignbus, Purfleet, 2008
-	YJ51ZVM	DAF DB250	Optare Spectra	N47/27F	2001	Ensignbus, Purfleet, 2008
-	KP51UFL	Dennis Dart SLF 10.7m	Plaxton Pointer 2	N37F	2001	Thames Travel, Wallingford, 2008
-	KU52RXL	Dennis Dart SLF 10.7m	Plaxton Pointer 2	N41F	2002	Veolia England, 2008
RML2324	CUV324C	AEC Routemaster R2RH1	Park Royal	B40/32R	1965	Arriva London, 2005
RML2736	SMK736F	AEC Routemaster R2RH1	Park Royal	B40/32R	1965	Arriva London, 2005

Previous registrations:

ANZ8798	H139MOB	T439EBD	T439EBD, 99D81367
ANZ8799	H167NON	T461HNH	T461HNH, 99D81366
D180FYM	D180FYM, 480CLT	T927PNV	T5BUS
L881YVK	L881YVK, RIL9772	W378SVV	Q588RRP
P943EBB	P3CLA	XIL1273	M627KVU
T342PNV	99D80588	XIL1274	M644KVU

Depots: Beeching Road, Bexhill-on-Sea; East Quay; Newhaven and Dittons Road, Polegate.

As we go to press it is reported that the Cavendish element of this fleet will pass to Stagecoach along with the twenty-eight buses that are marked in blue.

RICHARDSON TRAVEL

Richardson Travel Ltd, Russell House, Bepton Road, Midhurst, GU29 9NZ

120	F120PHM	Volvo Citybus B10M-50	Alexander RV	B46/37F	1988	
139	F139PHM	Volvo Citybus B10M-50	Alexander RV	B46/37F	1988	Blue Triangle, Rainham, 2003
140	F140PHM	Volvo Citybus B10M-50	Alexander RV	B46/29D	1988	Arriva London, 2003
144	F144PHM	Volvo Citybus B10M-50	Alexander RV	B46/37F	1988	Blue Triangle, Rainham, 2003
148	G148TYT	Volvo Citybus B10M-50	Alexander RV	B46/29D	1990	Arriva London, 2003
291	YN57PYJ	Mercedes-Benz Vario 0816	Plaxton Cheetah	C29F	2008	
471	P514UUG	Volvo B10B	Wright Endeavour	BC47F	1997	Stagecoach, 2008
494	YN54WWR	Volvo B12B	Plaxton Panther	C49FT	2005	
495	YN06MXK	Volvo B12B	Plaxton Panther	C49FT	2006	
531	YN53VCD	Volvo B7R	Plaxton Profile	C53F	2004	
533	YN07NUF	Volvo B12M	Plaxton Panther	C53FT	2007	
574	YR52MDY	Volvo B7R	Plaxton Prima	C57F	2002	
575	YS02YYD	Volvo B7R	Plaxton Profile	C57F	2002	Gordon, Rotherham, 2006
576	YN57BWE	Volvo B12M	Plaxton Paragon	C57F	2007	
791	PN05SYF	Volvo B7TL	East Lancs Vyking	NC47/32F	2005	
801	PN52XBP	Volvo B7TL	East Lancs Vyking	NC47/33F	2002	

Depot: Pitsham Lane, Midhurst

ROSS TRAVEL

P&M Ross, The Garage, Allison Street, Featherstone, WF7 5BC

1	YJ58PGK	Optare Solo M950	Optare	N32F	2008	
2	YK04ENN	Mercedes-Benz Vario 0814	TransBus Beaver 2	B31F	2004	
3	Y753HVY	Mercedes-Benz Vario 0814	Plaxton Beaver 2	B31F	2001	
4	YJ05WCR	Optare Solo M950	Optare	N33F	2005	
5	V967RCX	Mercedes-Benz Vario 0814	Plaxton Beaver 2	B31F	1999	
7	MX07NTK	Optare Solo M950	Optare	N32F	2007	
8	Y751HVY	Mercedes-Benz Vario 0814	Plaxton Beaver 2	B31F	2001	
9	YJ05WCP	Optare Solo M950	Optare	N33F	2005	
10	YJ54XGE	Mercedes-Benz Vario 0814	Plaxton Beaver 2	B31F	2004	
11	L289ETG	Scania N113 DRB	Alexander Strider	B50F	2004	
	HIL7644	Volvo B10M-61	Van Hool Alizée	C53F	1988	Longstaff, Mirfield, 2003
	OIJ721	Bova Futura FHD12.340	Bova	C49FT	1999	
	S257JUG	Mercedes-Benz 614D	UVG	B14F	1998	Leeds City Council, 2005
	S262JUG	Mercedes-Benz 614D	UVG	B14F	1998	Leeds City Council, 2005
	T9RTG	Scania K124 IB4	Van Hool T9 Alizée	C49FT	2001	Reay's, Wigton, 2004
	YJ53CFF	Mercedes-Benz Vario 0815	Sitcar Beluga	C33F	2003	Hilton, Newton-le-Willows, 2006
	YJ07DWM	Volvo B12B	Van Hool Alicron	C49FT	2007	
	YJ08VPR	VDL Bova Futura FHD127.	VDL Bova	C FT	2008	

Cardiff Bus, 2004 *(for row 11 L289ETG)*

Previous registrations:

		T9RTG	Y739WSM
HIL7644	E447LCX, JJL945	OIJ721	T868FWW

Web: www.rosstravelgroup.co.uk

Carrying *Featherstone Rover* branding on its Ross Travel livery is 9, YJ05WCP, an Optare Solo M950 seen passing through Castleton en route for Wakefield. *Alan Blagburn*

SAFEGUARD

Safeguard Coaches - Farnham Coaches

Safeguard Coaches Ltd, Ridgemount, Guildford Park Road, Guildford, GU2 7TH

F	247FCG	Volvo B10M-60	Plaxton Paramount 3200 III	C53F	1990	Excelsior, Bournemouth, 1990
	277FCG	Volvo B10M-60	Plaxton Paramount 3200 III	C55F	1990	Excelsior, Bournemouth, 1991
	H577MOC	Dennis Dart 8.5m	Carlyle Dartline	B28F	1990	Boomerang Bus, Tewkesbury, '00
	N611WND	Dennis Dart 9.8m	Northern Counties Paladin	B39F	1995	Pink Elephant, Heathrow, 1998
F	N561UPF	Volvo B10M-62	Plaxton Première 350	C53F	1996	
	DSK559	Volvo B10M-62	Van Hool Alizée	C49FT	1997	
	DSK558	Volvo B10M-62	Van Hool Alizée	C49FT	1998	Park's of Hamilton, 2002
	XHY378	Volvo B10M-62	Van Hool Alizée	C53F	1998	Park's of Hamilton, 2002
	R433FWT	Volvo B10M-62	Plaxton Première 350	C53F	1998	Wallace Arnold, 2001
	S503UAK	Dennis Javelin	Plaxton Première 320	C57F	1998	
F	DSK560	Volvo B10M-62	Plaxton Première 350	C49FT	1999	Wallace Arnold, 2003
	WPF926	Volvo B10M-62	Van Hool T9 Alizée	C49FT	1999	Clyde Coast, Ardrossan, 2003
	515FCG	Mercedes-Benz O404-15R	Hispano Vita	C49FT	1999	Austin, Earlston, 2001
	W203YAP	Mercedes-Benz Vario O814	Plaxton Cheetah	C29F	2000	Airlinks, West Drayton, 2001
F	538FCG	Setra S250	Setra Special	C48FT	2000	
F	159FCG	Setra S315 GT-HD	Setra	C53F	2000	Redwing, Herne Hill, 2003
	X307CBT	Optare Excel L1150	Optare	N39F	2000	Tillingbourne, Cranleigh, 2001
	X308CBT	Optare Excel L1150	Optare	N39F	2000	Tillingbourne, Cranleigh, 2001

Optare Tempo YJ06FXM joined Safeguard in 2006 and is seen in Woodbridge Road, Guildford, while operating local service number 5. *Mark Lyons*

Safeguard operates a special event vehicle in the form of AEC Reliance 200APB. Showing its Burlingham body it is seen in all its splendour. *Dave Heath*

	Y748HWT	Volvo B10M-62	Plaxton Paragon	C49FT	2001	Wallace Arnold, 2003
	Y758HWT	Volvo B10M-62	Plaxton Paragon	C49FT	2001	Wallace Arnold, 2004
F	196FCG	Volvo B7R	Plaxton Prima	C57F	2001	Bebb, Llantwit Fardre, 2002
	VU02TTJ	Dennis Dart SLF 8.5m	Plaxton Pointer MPD	N29F	2002	
	RX03XKH	Dennis Javelin 12m	Plaxton Profile	S70F	2003	
	FY03WZN	Mercedes-Benz Atego 1223L	Ferqui/Optare Solera	C39F	2003	Chivers, Elstead, 2007
	YJ03UMM	Optare Excel L1180	Optare	N41F	2003	
F	YN04WTL	Volvo B12M	Plaxton Paragon	C49FT	2004	Logan, Dunloy, 2005
	YN04WTM	Volvo B12M	Plaxton Paragon	C49FT	2004	Logan, Dunloy, 2005
	YN05HUY	Volvo B12M	Plaxton Paragon	C53F	2005	
	YJ06FXM	Optare Tempo X1150	Optare	N37F	2006	
	BX56VIM	Mercedes-Benz Tourino OC510	Mercedes-Benz	C36F	2006	
F	WA07CXX	Volvo B12B	Van Hool T9 Alicron	C53F	2007	
F	WA57CYY	Volvo B12B	Van Hool T9 Alicron	C57F	2007	
	YN57BWW	Volvo B12M	Plaxton Paragon	C57F	2007	
	MX58ABF	Optare Versa V1110	Optare	N38F	2008	
	WA58EOO	Volvo B12B	Van Hool T9 Alicron	C53F	2008	
F	YN58NCC	Volvo B12M	Plaxton Panther	C53FT	2009	
	YN58NDD	Volvo B12M	Plaxton Panther	C53FT	2009	

Special event vehicle:

	200APB	AEC Reliance MU3RV	Burlingham	B44F	1956	

Previous registrations:

159FCG	W257UGX	DSK558	LSK501, R410EOS
196FCG	CN51XNO	DSK559	P46GPG
247FCG	G514EFX	DSK560	T530EUB
277FCG	G520EFX, G518EFX	H577MOC	H577MOC, WLT339, RIL9774
515FCG	W417HOB	WPF926	S132PGB
531FCG	-	N561UPF	N561UPF, 531FCG
538FCG	W354EOL	XHY378	LSK504, R398EOS

Web: www.safeguardcoaches.co.uk; www.farnhamcoaches.co.uk
Depots: Guildford Park Road, Guildford and Odiham Road, Farnham (F = Farnham Coaches)

SANDERS

Sanders Coaches Ltd, Heath Drive, Hempstead Road Ind Est, Holt, NR25 6JU

PIL4682	Bedford Venturer YNV	Plaxton Paramount 3200 II	C57F	1985	Spratts, Wreningham, 2003
SJI1632	DAF MB200	Plaxton Paramount 3200	C53F	1985	Mayne, Buckie, 1995
SJI1626	Bedford Venturer YNV	Plaxton Paramount 3200 II	C57F	1987	East Midland, 1992
HIL7391	Bedford Venturer YNV	Plaxton Paramount 3200 II	C57F	1987	Spratts, Wreningham, 2002
SJI1621	DAF MB230	Plaxton Paramount 3500 III	C55F	1987	Grey-Green, 1995
RDV903	DAF MB230	Van Hool Alizée H	C51F	1988	Robinsons, Great Harwood, 1995
LIL2493	DAF SB2305	Van Hool Alizée	C53F	1990	McCutcheon, Lisbellaw, 1997
SJI1622	Volvo B10M-60	Plaxton Paramount 3500 III	C57F	1990	Birmingham Coach, Tividale, 2000
SJI1617	Volvo B10M-60	Plaxton Paramount 3500 III	C49FT	1990	Ruffle's, Castle Hedingham, 2000
WOA521	DAF MB230	Van Hool Alizée H	C51FT	1990	Robinsons, Great Harwood, 1999
WSV503	DAF MB230	Van Hool Alizée H	C51FT	1990	Robinsons, Great Harwood, 1999
259VYC	DAF MB230	Van Hool Alizée H	C51F	1990	Robinsons, Great Harwood, 1999
354TRT	DAF MB230	Van Hool Alizée H	C51FT	1990	Robinsons, Great Harwood, 1999
SGF965	DAF MB230	Van Hool Alizée H	C51FT	1990	Robinsons, Great Harwood, 1999
J201BVO	DAF SB220	Optare Delta	B49F	1991	Konectbus, Dereham, 2007
J810KHD	DAF SB220	Ikarus CitiBus	B49F	1992	Capital, West Drayton, 2000
K640FAU	DAF SB220	Optare Delta	B47F	1992	Konectbus, Dereham, 2007
RJI8604	DAF MB230	Van Hool Alizée H	C49F	1993	Swiftsure, Burton, 2004
L517EHD	DAF SB220	Ikarus CitiBus	B48F	1993	Jowitt, Royston, 2003
M612RCP	DAF SB220	Ikarus CitiBus	B49F	1994	Hallmark, Coleshill, 2002
6546FN	DAF DB250	Northern Counties Palatine 2	B47/30F	1995	East Thames Buses, 2002
M764RCP	DAF SB220	Ikarus CitiBus	B49F	1995	First Bristol, 2000
M806RCP	DAF SB220	Ikarus CitiBus	B48F	1994	Capital, West Drayton, 2000
M832RCP	DAF SB220	Ikarus CitiBus	B47F	1995	London Luton Airport, 2001
M834RCP	DAF SB220	Ikarus CitiBus	B47F	1995	London Luton Airport, 2001
M836RCP	DAF SB220	Ikarus CitiBus	B47F	1995	London Luton Airport, 2001
N593DWY	DAF SB220	Ikarus CitiBus	B51F	1995	London Luton Airport, 2001
N594DWY	DAF SB220	Ikarus CitiBus	B51F	1995	London Luton Airport, 2002
N597DWY	DAF SB220	Ikarus CitiBus	B48F	1995	Arriva Fox County, 2002
SJI1615	DAF SB3000	Van Hool Alizée HE	C51FT	1995	First Lowland, 1998
N67FWU	DAF SB220	Ikarus CitiBus	B51F	1996	London Luton Airport, 2002
N988FWT	DAF SB220	Ikarus CitiBus	B49F	1996	Simonds, Botesdale, 2001

One of the few Optare Delta buses remaining with Sanders is J201BVO, seen here in Norwich. The SB120 model, and its sister products have proved popular with Sanders.
David Longbottom

Five more recent DAF SB220 buses, to the new low-floor specification, joined Sanders from the airport service provider Centra. Initially dual-doored these have been refurbished as single-door buses with W262CDN seen preparing to depart for Holt. *Mark Doggett*

N63MDW	Mercedes-Benz 814D	Autobus Classique Nouvelle	B33F	1996	Bebb, Llantwit Fardre, 1998
N68MDW	Mercedes-Benz 814D	Autobus Classique Nouvelle	B33F	1996	Bebb, Llantwit Fardre, 1998
N33SCS	Mercedes-Benz 811D	Plaxton Beaver	BC31F	1996	
N44SCS	Mercedes-Benz 811D	Plaxton Beaver	BC31F	1996	
P77SCS	Mercedes-Benz 811D	Plaxton Beaver	BC31F	1996	
P767PCL	Dennis Javelin 12m	Plaxton Première 320	C57F	1996	
P768PCL	Dennis Javelin 12m	Plaxton Première 320	C57F	1996	
P34KWA	Mercedes-Benz 814D	Plaxton Beaver	BC32F	1997	
P35KWA	Mercedes-Benz 811D	Plaxton Beaver	BC31F	1997	
P542SCL	Mercedes-Benz 814D	Plaxton Beaver	BC31F	1997	
P543SCL	Mercedes-Benz 814D	Plaxton Beaver	BC31F	1997	
P33TCC	Dennis Javelin GX	Plaxton Première 350	C53F	1997	Stort Valley, Stansted, 2001
PVF377	Dennis Javelin GX	Plaxton Première 350	C53F	1997	Stort Valley, Stansted, 2001
YVJ677	Dennis Javelin GX	Plaxton Première 350	C53F	1997	Stort Valley, Stansted, 2001
6542FN	DAF DB250	Northern Counties Palatine 2	B43/29F	1997	East Thames Buses, 2003
537FN	DAF SB3000	Van Hool T9 Alizée	C49FT	1998	Caroline Seagull, Gt Yarmouth, '08
R31GNW	DAF SB220	Optare Delta	B53F	1998	Speedlink, 1998
S156JUA	DAF SB220	Optare Delta	B53F	1998	
S157JUA	DAF SB220	Optare Delta	B53F	1998	
T270EWW	DAF SB220	Optare Delta	B49F	1999	
T793TWX	DAF SB220	Ikarus Polaris	N47F	1999	Arriva Bus & Coach, 2008
536FN	DAF SB3000	Van Hool T9 Alizée	C49FT	2000	Galloway, Mendlesham, 2007
538FN	DAF SB3000	Van Hool T9 Alizée	C49FT	2000	Galloway, Mendlesham, 2007
W261CDN	DAF SB220	East Lancs Myllennium	N43F	2000	Centra, Edinburgh Airport, 2006
W262CDN	DAF SB220	East Lancs Myllennium	N43F	2000	Centra, Luton Airport, 2006
W263CDN	DAF SB220	East Lancs Myllennium	N43F	2000	NCP, Heathrow, 2008
W265CDN	DAF SB220	East Lancs Myllennium	N43F	2000	NCP, Heathrow, 2008
W266CDN	DAF SB220	East Lancs Myllennium	N43F	2000	NCP, Heathrow, 2008
W80HOD	Setra S315 GT-HD	Setra	C49FT	2000	Hodson, Navenby, 2005
OGR647	Setra S315 GT-HD	Setra	C49FT	2000	Redwing, Herne Hill, 2003
3990ME	Setra S315 GT-HD	Setra	C49FT	2000	Redwing, Herne Hill, 2003
SIB7515	VDL Bova Futura FHD12.370	VDL Bova	C49FT	2001	Barnes, Aldbourne, 2007
YJ03PKX	DAF SB120	Wrightbus Cadet	N38F	2003	Veolia Cymru, 2008
YJ03PKY	DAF SB120	Wrightbus Cadet	N38F	2003	Veolia Cymru, 2008
YJ05PZA	VDL Bus SB4000	Van Hool T9 Alizée	C51FT	2005	Landtourers, Brighton Hill, 2006

In 2007, YJ07JPV, a Dutch-built DB250 bus with East Lancs Lowlander bodywork joined the fleet. It is seen in Norwich. *Mark Bailey*

YJ05PZE	VDL Bus DB250	East Lancs Lowlander	NC51/29F	2005
YJ55WPA	VDL Bus DB250	East Lancs Lowlander	NC51/29F	2005
YN07DTO	Mercedes-Benz Vario 0814	Plaxton Cheetah	C29F	2007
YN07DVF	Mercedes-Benz Vario 0814	Plaxton Cheetah	C29F	2007
MX07NTL	Optare Solo M880	Optare	N28F	2007
MX07NTM	Optare Solo M880	Optare	N28F	2007
YJ07JNO	VDL Bus SB4000	Van Hool T9 Alizée	C51FT	2007
YJ07JPV	VDL Bus DB250	East Lancs Lowlander	NC51/29F	2007
PL08YLZ	Volvo B9TL	East Lancs Olympus	NC51/29F	2008
PL08YMA	Volvo B9TL	East Lancs Olympus	NC51/29F	2008

Previous registrations:

259VYC	H236AFV	SGF965	G230NCW
354TRT	F224YHG	SIB7515	Y221NYA
3990ME	W213UGX	SJI1615	M640RCP, 3990ME
536FN	W186CDN, 3379PP	SJI1617	G371REG
537FN	R37GNW	SJI1621	E160TVR, OGR647
538FN	V215EGV, 6399PP	SJI1622	H556WTS
HIL7391	D422KVF	SJI1626	D342SWB
LIL2493	F614HGO	SJI1627	C115AFX
OGR647	W217UGX	SJI1632	C301CFP, CXI7390, C57USS, C134USS
PIL4682	C357FBO	T793TWX	T115AUA, 99D36983
PVF377	P44TCC	WOA521	G231NCW
RDV903	E221KFV	WSV503	G228NCW
RJI8604	L513EHD	YVJ677	P55TCC

Depots: Heath Drive, Holt; Cornish Way, North Walsham and Claypit Lane, Fakenham.
Web: www.sanderscoaches.com

SARGEANTS

Sargeant Brothers Ltd, Mill Street, Kington, HR5 3AL

	SBZ5810	Leyland Tiger TRCTL11/3R	Jonckheere Jubilee P50	C53F	1983	Teme Valley Motors, 1997
w	C900JGA	Bedford YNT (Volvo)	Plaxton Paramount 3200 II	C53F	1986	Evans, Tregaron, 1996
	L54CNY	Volvo B10M-60	Plaxton Première 320	C57F	1993	Mid Wales, Penrhyncoch, 2004
w	L91WBX	Renault Master	Cymric	M16L	1993	
	M960VWY	Mercedes-Benz O405	Optare Prisma	B49F	1995	
	M311KRY	Volvo B10M-62	Jonckheere Mistral	C53F	1995	Dunn-Line, Nottingham, 2001
	YIL1720	Volvo B10M-62	Van Hool Alizée	S70F	1996	Focus, Much Hoole, 2008
	P509PUM	Ford Transit VE6	Ford	M14	1997	Bradford MBC, 2002
	R399EOS	Volvo B10M-62	Van Hool T9 Alizée	C53F	1998	KW Beard, Cinderford, 2008
	S285MGB	Renault Master	Renault	M8	1998	
	S361AHC	Renault Master	Crystals	M8	1998	Dundee City C, 2002
	S84XCJ	Mercedes-Benz Vario O814	Plaxton Beaver 2	B31F	1998	
	T74JBO	Mercedes-Benz Vario O814	Autobus Nouvelle 2	BC33F	1999	Village Green, Shobdon, 2007
	W408UCJ	Optare Excel L1150	Optare	N43F	2000	
	Y876PWT	Optare Solo M920	Optare	N31F	2001	
	Y877PWT	Optare Solo M920	Optare	N31F	2001	
	VA51SAR	DAF DB250	Optare Spectra	N47/27F	2001	
	YR52UNB	LDV Convoy	Jaycas	M16	2002	Tandy, Bayston Hill, 2006
	YE52FHA	Optare Solo M920	Optare	N31F	2002	
	YN53ELC	Optare Solo M850	Optare	N27F	2003	
	ML53YKU	Ford Transit	Ford	M16	2003	Cowie, Saughill, 2005
	VX04ULT	TransBus Dart 8.8m	TransBus Mini Pointer	N29F	2004	
	YJ05WCW	Optare Tempo X1200	Optare	N42F	2005	
	YX05FFS	Volkswagen T030-174	Bluebird	M13	2005	

Previous registrations:

L54CNY	L54CNY, CNZ1524	SBZ5810	BRN701Y, 8177VT, 9975VT, RJU94Y, 991FOT, NTP95Y, TVM263
MIL7610	B560DSE	T74JBO	T74JBO, 8056UA
ML53YKU	ML53YKU, BU53MCA	YIL1720	N713CYC
R399EOS	KSK950, R399EOS, WSV573, R7DCC		

Sargeants' only double-deck bus is Optare Spectra-bodied DAF DB250 VA51SAR, one of the first low-floor versions of the Spectra. It operates the link from Hereford to Kington which requires high capacity during school term.
David Longbottom

SHAMROCK BUSES

Shamrock Buses Limited, 50a Holton Road, Holton Heath Trading Park, Poole, BH16 6LT

100	G645BPH	Volvo Citybus B10M-50	Northern Counties	B45/35F	1989	Red Arrow, Huddersfield, 2008
101	B863XYR	Volvo Citybus B10M-50	East Lancs (1992)	B44/30D	1985	2 Travel, Swansea, 2005
102	B865XYR	Volvo Citybus B10M-50	East Lancs (1992)	B44/30D	1985	2 Travel, Swansea, 2005
103	B866XYR	Volvo Citybus B10M-50	East Lancs (1992)	B44/30D	1985	2 Travel, Swansea, 2005
111	DLJ111L	Daimler Fleetline CRG6LX	Alexander AL	O43/31F	1973	Bournemouth PTA, 2007
200	KUI2269	Leyland AT68M/2RFT	Northern Counties Paladin (1994) BC42F		1971	MacEwan, Amisfield, 2002
201	UIB3987	Leyland AT68M/2RFT	Northern Counties Paladin (1994) BC42F		1971	MacEwan, Amisfield, 2002
210	R685MEW	Dennis Dart SLF	Marshall Capital	N39F	1998	Click, Skelmersdale, 2008
211	R689MEW	Dennis Dart SLF	Marshall Capital	N39F	1998	Click, Skelmersdale, 2008
212	X242ABU	Dennis Dart SLF	Alexander ALX200	N27F	1995	Southdown PSV, Copthorne, 2008
248	E848DPN	Leyland Olympian ONCL10/1RZ	Northern Counties	B43/34F	1988	East Yorkshire, 2007
249	E849DPN	Leyland Olympian ONCL10/1RZ	Northern Counties	B43/34F	1988	East Yorkshire, 2007
250	E850DPN	Leyland Olympian ONCL10/1RZ	Northern Counties	B43/34F	1988	East Yorkshire, 2007
253	E853DPN	Leyland Olympian ONCL10/1RZ	Northern Counties	B43/34F	1988	East Yorkshire, 2007
254	E854DPN	Leyland Olympian ONCL10/1RZ	Northern Counties	B43/34F	1988	East Yorkshire, 2007
255	E855DPN	Leyland Olympian ONCL10/1RZ	Northern Counties	B43/34F	1988	East Yorkshire, 2007
257	E857DPN	Leyland Olympian ONCL10/1RZ	Northern Counties	B43/34F	1988	Eastbourne, 2006
258	E858DPN	Leyland Olympian ONCL10/1RZ	Northern Counties	B43/34F	1988	Eastbourne, 2006
282	D82UTF	Leyland Olympian ONLXCT/1RH	Eastern Coach Works	B43/27F	1986	Reading Buses, 2006
283	D83UTF	Leyland Olympian ONLXCT/1RH	Eastern Coach Works	B43/27F	1986	Reading Buses, 2006
284	D84UTF	Leyland Olympian ONLXCT/1RH	Eastern Coach Works	B43/23F	1986	Reading Buses, 2005

One of many Northern Counties Olympians that carry the Greater Manchester-style body is 248, E848DPN which was new to Eastbourne, arriving via East Yorkshire. It is seen at Boscombe Pier in July 2008. *Richard Godfrey*

Lettered for the Boscombe Beach Park & Ride is open-top Fleetline 111, DLJ111L. The Alexander AL body was built in 1973 for Bournemouth later undertaking an open-top conversion. *Mark Doggett*

285	F85MJH	Leyland Olympian ONLXCT/1RH	Optare	B43/27F	1988	Reading Buses, 2005
286	F86MJH	Leyland Olympian ONLXCT/1RH	Optare	B43/27F	1988	Reading Buses, 2005
287	F87MJH	Leyland Olympian ONLXCT/1RH	Optare	B43/27F	1988	Reading Buses, 2005
290	E921KYR	Leyland Olympian ONLXB/1RH	Northern Counties	B43/30F	1987	Focus, Much Hoole, 2006
291	E927KYR	Leyland Olympian ONLXB/1RH	Northern Counties	B43/30F	1987	Focus, Much Hoole, 2006
292	B135WNB	Leyland Olympian ONLXB/1R	Northern Counties	B43/30F	1985	Focus, Much Hoole, 2006
294	B49PJA	Leyland Olympian ONLXB/1R	Northern Counties	B43/30F	1984	Focus, Much Hoole, 2006
301	JFR11W	Leyland Olympian ONLXB/1R	Eastern Coach Works	B45/32F	1981	Lancashire United, 2004
302	DBV132Y	Leyland Olympian ONLXB/1R	Eastern Coach Works	B45/32F	1983	Burnley & Pendle, 2005
304	OFV14X	Leyland Olympian ONLXB/1R	Eastern Coach Works	B45/32F	1981	Lancashire United, 2003
306	OFV23X	Leyland Olympian ONLXB/1R	Eastern Coach Works	B45/32F	1981	Burnley & Pendle, 2004
307	JFR10W	Leyland Olympian ONLXB/1R	Eastern Coach Works	B45/32F	1981	Lancashire United, 2003
308	C178ECK	Leyland Olympian ONLXB/1R	Eastern Coach Works	BC42/32F	1984	Lancashire United, 2004
311	C481YWY	Leyland Olympian ONLXB/1R	Eastern Coach Works	BC42/29F	1985	Burnley & Pendle, 2004
313	JFR13W	Leyland Olympian ONLXB/1R	Eastern Coach Works	B45/32F	1981	Burnley & Pendle, 2004
804	G804GSX	Leyland Olympian ON2R56C13Z4	Alexander RH	B51/30D	1990	Lothian Buses, 2006
809	G809GSX	Leyland Olympian ON2R56C13Z4	Alexander RH	B51/30D	1990	Lothian Buses, 2006
810	G810GSX	Leyland Olympian ON2R56C13Z4	Alexander RH	B51/30D	1990	Lothian Buses, 2006
817	G817GSX	Leyland Olympian ON2R56C13Z4	Alexander RH	B51/30D	1990	Lothian Buses, 2007
828	G828GSX	Leyland Olympian ON2R56C13Z4	Alexander RH	B51/30D	1990	Lothian Buses, 2007
830	G830GSX	Leyland Olympian ON2R56C13Z4	Alexander RH	B51/30D	1990	Lothian Buses, 2007
832	G832GSX	Leyland Olympian ON2R56C13Z4	Alexander RH	B51/30D	1990	Lothian Buses, 2007

Special event vehicle:

H174	MFX174W	Leyland Fleetline FE30AGR	Alexander AL	B43/31F	1981	Bournemouth

Previous registrations:

B135WNB	B135WNB, YIL4415		E927KYR	E927KYR, YIL4028
E921KYR	E921KYR, YIL4029		JFR10W	B152TRN

Web: www.shamrockbuses.co.uk

SIMONDS

Simonds of Botesdale Ltd, Roswald House, Oak Drive, Diss, IP22 4GX

SIA488	Volvo B10M-61	Plaxton Paramount 3200	C57F	1983	Berkeley, Paulton, 1993
JUI4236	Volvo B10M-61	Van Hool Alizée H	C57F	1984	Riches, Stradbrooke, 2006
7236PW	Volvo B10M-61	Plaxton Paramount 3500 II	C53F	1985	Worthing Coaches, 1988
TCF496	Volvo B10M-61	Plaxton Paramount 3200 II	C57F	1986	Soames, Otley, 2001
VRY841	Volvo B10M-61	Plaxton Paramount 3500 III	C53F	1987	Berkeley, Paulton, 1992
538ELX	Volvo B10M-61	Plaxton Paramount 3200 III	C53F	1988	Joplin, Tittleshall, 2004
9983PW	Volvo B10M-61	Van Hool Alizée H	C53F	1988	Excelsior, Bournemouth, 1991
DSK648	Volvo B10M-61	Van Hool Alizée H	C53F	1989	Kenzies, Shepreth, 1994
WVE284	Volvo B10M-60	Plaxton Paramount 3500 III	C50F	1989	Soames, Otley, 2002
9383MX	Volvo B10M-60	Van Hool Alizée H	C53F	1990	Kenzies, Shepreth, 1995
2091PW	Volvo B10M-62	Van Hool T9 Alizée	C49FT	1993	Kenzies, Shepreth, 2002
SLK886	Volvo B10M-60	Van Hool Alizée	C53F	1993	Epsom Coaches, 1999
224ENG	Volvo B6	Alexander Dash	B40F	1994	Mulleys, Ixworth, 2007
HIB644	Volvo B10M-62	Van Hool Alizée HE	C53F	1994	Kenzies, Shepreth, 1997
256JPA	Volvo B10M-62	Van Hool Alizée HE	C53F	1996	Wickson, Walsall Wood, 2003
TVG397	Volvo B10M-62	Van Hool Alizée HE	C57F	1996	Wickson, Walsall Wood, 2004
4940VF	Volvo B10M-62	Van Hool Alizée HE	C49FT	1998	Kenzies, Shepreth, 2005
KIA891	Volvo B10M-62	Van Hool T9 Alizée	C49FT	1999	Hayton, Burnage, 2001
V378SVV	Mercedes-Benz Vario O814D	Plaxton Beaver 2	B31F	1999	Ali, Worcester, 2004
X181BNH	Mercedes-Benz Vario O814D	Plaxton Beaver 2	B31F	1999	Thames Travel, Wallingford, 2004
X182BNH	Mercedes-Benz Vario O814D	Plaxton Beaver 2	B31F	1999	Thames Travel, Wallingford, 2004
4512UR	Volvo B7R	Plaxton Prima	C53F	2000	
YVF158	Volvo B10M-62	Van Hool T9 Alizée	C49FT	2000	Mayne's, Buckie, 2003
KDX108	Volvo B10M-62	Van Hool T9 Alizée	C49FT	2002	Wickson, Walsall Wood, 2006
378BNG	Volvo B12M	Van Hool T9 Alizée	C49FT	2002	Reay's, Wigton, 2006
166UMB	Volvo B12M	Van Hool T9 Alizée	C51F	2002	Southern, Barrhead, 2007
KF02ZXX	Mercedes-Benz Vario O814	Plaxton Beaver 2	B29F	2002	Ferguson, Cleland, 2006
AO52NXD	Mercedes-Benz Vario O814	Plaxton Beaver 2	BC33F	2002	

Simonds operates many coaches, the Van Hool-bodied Volvo being the preferred choice. Seen outside Stratford Leisure Centre is 378BNG, a Volvo B12M with the later Van Hool T9 Alizée body. When pictured the coach had its original index mark, YD02PXN. *Mark Doggett*

Recent arrivals with Simonds include four MAN 14.220s with MCV Evolution bodywork. AE07DZH is seen arriving in Norwich. *Mark Bailey*

AU03FSA	Volvo B12B	Plaxton Panther	C49FT	2003	
KV03ZFJ	TransBus Enviro 300	TransBus	N44F	2003	
KV03ZFK	TransBus Enviro 300	TransBus	N44F	2003	
XJO46	Mercedes-Benz Sprinter 413cdi	Ferqui/ Optare Soroco	C16F	2003	Lawton, Stickney, 2007
VX53AVJ	Mercedes-Benz Vario 0814	Plaxton Beaver 2	BC33F	2004	Hatts Coaches, Foxham, 2008
EG04ZHT	Ford Transit	Ford	M8	2004	private owner, 2005
AU54JRV	Mercedes-Benz Vario 0814	Plaxton Beaver 2	B33F	2005	
YJ05XWV	Volvo B12B	Van Hool T9 Alicron	C49FT	2005	
DM05GSM	Volvo B12B	Van Hool T9 Alicron	C49FT	2005	Mayne's, Buckie, 2008
WJ55TRV	Volvo B12B	Van Hool T9 Alizée	C53F	2005	Moroney, Bray, 2008
AE06XSA	MAN 14.220	MCV Evolution	N43F	2006	
AE06XSB	MAN 14.220	MCV Evolution	N43F	2006	
AE07DZG	MAN 14.220	MCV Evolution	N43F	2007	
AE07DZH	MAN 14.220	MCV Evolution	N43F	2007	

Previous registrations:

98TNO	-	JGV929	-
166UMB	SB02SOU	JUI4236	A767UHT, 791WHT
224ENG	L71UNG	LAZ6577	-
256JPA	KSK950, KSK953, N414PYS, WT2899	KDX108	WT02WCT, AR02HCY
378BNG	YD02PXN, TKU717	KIA891	T99HAY
538ELX	E460PEE, 388XYC, E899HFW,	SIA488	OOU855Y
	5517RH, E883HTL	SIJ82	-
2091PW	K53TER	SLK886	K288GDT, K289GDT, A9HRR
4512UR	X779XEX	TCF496	636VHX, C373CTP
4940VF	R81NAV	TVG397	N996HWJ, WJ8355
7236PW	C196WJT	VRY841	D807SGB
8333UR	-	WJ55TRV	05WW4200
9383MX	G41SAV	WVE284	F428DUG
9983PW	E306OPR, XEL941, E407SEL	XJO46	FX53JWE
DSK648	F38DAV	YVF158	X777GSM
HIB644	L54REW		

Depots: Oak Drive, Diss and Pulham Road, Starston.
Web: www.simonds.co.uk

SOMERBUS

Somerbus Ltd, 64 Brookside, Paulton, BS39 7YR

J248LLK	Van Hool A308	Van Hool	N25F	1991	British Airways, 2006
J129GMP	Dennis Dart 9m	Plaxton Pointer	BC35F	1991	Andybus, Dauntsey, 2008
N357VRC	Dennis Lance 11m	Optare Sigma	B46F	1995	TM Travel, Staveley, 2005
NHG541	Optare Solo M850 SL	Optare	N27F	2004	
MX54KXZ	Optare Solo M880 SL	Optare	N28F	2005	Henderson, Hamilton, 2008
ER05BUS	Optare Solo M850 SL	Optare	NC27F	2005	
	Optare Solo M880 SL	Optare	NC27F	2009	

Depot: Wick Lane, Stanton Wick

SOUTH LANCS TRAVEL

Green Triangle Buses Ltd; B&D Coaches Ltd, 22/23 Chanters Ind Est, Arley Way, Atherton, Wigan, M46 9BP

1	T11SLT	Dennis Dart SLF 8.8m	Plaxton Pointer MPD	N28F	1999	
2	V22SLT	Dennis Dart SLF 8.8m	Plaxton Pointer MPD	N28F	1999	
3	V33SLT	Dennis Dart SLF 8.8m	Plaxton Pointer MPD	N28F	1999	
4	W44SLT	Dennis Dart SLF 8.8m	Plaxton Pointer MPD	N28F	2000	
5	W55SLT	Dennis Dart SLF 8.8m	Plaxton Pointer MPD	N28F	2000	
6	Y66SLT	Dennis Dart SLF 8.8m	Plaxton Pointer MPD	N28F	2001	
7	Y77SLT	Dennis Dart SLF 8.8m	Plaxton Pointer MPD	N28F	2001	
12	W187YBN	Dennis Dart SLF 8.8m	Plaxton Pointer MPD	N29F	2000	Blue Bus, Bolton, 2004
13	T973TBA	Dennis Dart SLF 8.8m	Plaxton Pointer MPD	N28F	1999	Blue Bus, Bolton, 2004
14	T974TBA	Dennis Dart SLF 8.8m	Plaxton Pointer MPD	N28F	1999	Blue Bus, Bolton, 2004
16	K16SLT	Optare Solo M950 SL	Optare	N32F	2007	
17	K17SLT	Optare Solo M950 SL	Optare	N32F	2007	
18	K18SLT	Optare Solo M950 SL	Optare	N32F	2007	
19	W19SLT	Mercedes-Benz Vario O814	Plaxton Beaver 2	B27F	2000	
20	W20SLT	Mercedes-Benz Vario O814	Plaxton Beaver 2	B27F	2000	
22	K2SLT	Optare Solo M950	Optare	N33F	2004	
23	K3SLT	Optare Solo M950	Optare	N33F	2004	
24	S24SLT	Scania L94UB	Wrightbus Solar	N44F	2005	
25	S25SLT	Scania L94UB	Wrightbus Solar	N44F	2005	
26	S26SLT	Scania L94UB	Wrightbus Solar	N44F	2005	
27	YN51MGU	Scania L94UB	Wrightbus Solar	N43F	2001	Veolia England, 2007
28	V928FMS	Dennis Dart SLF 10.7m	Alexander ALX200	N40F	1999	Blue Bus, Bolton, 2004
30	R104GNW	Dennis Dart SLF 10.7m	UVG	N38F	1997	TPT, Buntingford, 1999
31	T731DGD	Dennis Dart SLF 10.7m	Marshall Capital	N43F	1999	Arnott, Erskine, 2003
40	K40SLT	Optare Solo M920	Optare	N33F	2004	
50	K50SLT	Optare Solo M920	Optare	N33F	2004	
57	L657MYG	Mercedes-Benz 711D	Plaxton Beaver	B27F	1993	Gibson Direct, Renfrew, 2001
58	N258DUR	Mercedes-Benz 709D	Plaxton Beaver	B27F	1995	Pete's Travel, West Bromwich, '99
60	K60SLT	Optare Solo M950	Optare	N33F	2004	Beeston, Hadleigh, 2006
67	M167LNC	Mercedes-Benz 709D	Alexander AM	B23F	1994	Arriva North West, 2002
68	M166LNC	Mercedes-Benz 709D	Alexander AM	B23F	1994	Arriva North West, 2002
69	N819RFP	Mercedes-Benz 709D	Plaxton Beaver	B29F	1995	Trent, Derby, 2005
70	K70SLT	Optare Solo M950	Optare	N33F	2004	Henderson, Hamilton, 2007
80	YJ54UWN	Optare Solo M950	Optare	N33F	2004	Henderson, Hamilton, 2007

The South Lancs Travel fleet includes three Dennis Darts with the less common Caetano Compass bodywork. New to Amos of Daventry, 96, X196FOR, is seen at Leigh bus station about to depart for Wigan. *Richard Godfrey*

An interesting repaint of Olympian 146, S858DGX has seen the re-appearance of the former Leigh Corporation colours. The bodywork is to the East Lancashire Pyoneer design. *Alan Blagburn*

82	M88SLT	Dennis Dart SLF 8.8m	Plaxton Pointer MPD	N29F	2000	Blue Bus, Bolton, 2004
83	M99SLT	Dennis Dart SLF 8.8m	Plaxton Pointer MPD	N29F	2000	Blue Bus, Bolton, 2004
84	T84JBA	Dennis Dart SLF 11.3m	Plaxton Pointer SPD	N44F	1999	Anglian Coaches, Ellough, 2002
85	T85JBA	Dennis Dart SLF 11.3m	Plaxton Pointer SPD	N44F	1999	Anglian Coaches, Ellough, 2002
86	T86JBA	Dennis Dart SLF 11.3m	Plaxton Pointer SPD	N44F	1999	Anglian Coaches, Ellough, 2002
90	YJ54UWO	Optare Solo M950	Optare	N33F	2004	Henderson, Hamilton, 2007
95	X195FOR	Dennis Dart SLF 10.7m	Caetano Compass	N42F	2000	Burton's, Haverhill, 2006
96	X196FOR	Dennis Dart SLF 10.7m	Caetano Compass	N42F	2000	Burton's, Haverhill, 2006
98	X198FOR	Dennis Dart SLF 10.7m	Caetano Compass	N42F	2000	Burton's, Haverhill, 2006
145	X645RDA	Volvo B7TL	East Lancashire Vyking	N47/29F	2000	Kenneallys, Waterford, 2007
146	S846DGX	Volvo Olympian	East Lancashire Pyoneer	B51/29F	1998	Metrobus, Crawley, 2004
147w	P347ROO	Volvo Olympian	East Lancashire Pyoneer	B51/28D	1997	East Thames, Hackney, 2003
148	S858DGX	Volvo Olympian	East Lancashire Pyoneer	B51/29F	1998	Metrobus, Crawley, 2004
154	M54PRA	Volvo B10M-60	Alexander Q	BC51F	1994	Trent, Derby, 2004
180	X80SLT	Dennis Trident	East Lancs Lolyne	N53/37F	2002	
200	W396PRC	Scania L94UB	Wrightbus Solar	N43F	2000	Dunn-Line, Nottingham, 2006
300	W397PRC	Scania L94UB	Wrightbus Solar	N43F	2000	Dunn-Line, Nottingham, 2006
400	W398PRC	Scania L94UB	Wrightbus Solar	N43F	2000	Dunn-Line, Nottingham, 2006

Special event vehicles:

	MMY991C	AEC Reliance 2U3RA	Harrington Grenadier	C51F	1965	Valiant, London
466	AFB597V	Bristol LH6L	Eastern Coach Works	B43F	1980	Bristol Omnibus
2573	BXI2573	Bristol RELL6G	Alexander Q	B52F	1984	Citybus, Belfast

Ancillary vehicle:

	KHU323P	Bristol LH6L	Eastern Coach Works	Tow car	1976	preservation, 2004

Previous registrations:

K60SLT	YN53YGY		T731BGD	T500CBC
K70SLT	YN04LWS		T973TBA	T10BLU
MM51XVB	M99SLT		T974TBA	T11BLU
M88SLT	X14BLU, X882OBA		W187YBN	W12BLU
M99SLT	X13BLU, X883OBA		X645RDA	X645RDA, 01D17885, 00KE11

Depots: Arley Way, Atherton and Mill Lane, Appley Bridge

English Bus Handbook: Notable Independents

SOUTHDOWN PSV

Southdown PSV Ltd, 3 Silverwood, Snow Hill, Copthorne, RH10 3EN

T132AUA	DAF DB250	Plaxton President	B45/19D	1999	Buzzlines, Hythe, 2006
V380SVV	Dennis Dart SLF 10.7m	Plaxton Pointer 2	N36F	1999	Park & Fly, Manchester, 2006
V392SVV	Dennis Dart SLF 10.7m	Plaxton Pointer 2	N39F	1999	Stansted Bus, 2006
W554JVV	Dennis Dart SLF 10.7m	Plaxton Pointer 2	N38F	2000	Centra, Heathrow, 2006
W569JVV	Dennis Dart SLF 10.7m	Plaxton Pointer 2	N39F	2000	First, 2006
W107RNC	Dennis Dart SLF 10.7m	Plaxton Pointer 2	N40F	2000	Bluebird, Middleton, 2008
X188BNH	Dennis Dart SLF 10.7m	Plaxton Pointer 2	N37F	2001	Solent Blue Line, 2006
SN53LWL	TransBus Dart SLF 10.7m	TransBus Pointer 2	N42F	2004	Cushing & Littlewood, Acle, '08
GX06AOE	Volvo B7TL	East Lancs Vyking	N45/27F	2006	
GX57BXG	ADL Dart 4 10.7m	ADL Enviro 200	N37F	2008	
GX57BXH	ADL Dart 4 10.7m	ADL Enviro 200	N37F	2008	

Previous registration:

W107RNC   W107RNC, W4BLU

Newest of the buses with Southdown PSV is GX57BXH, an Alexander Dennis Enviro 200. As can be seen in the picture, as it passes through Charlwood, it is lettered for the Crawley-Horley circular routes 526 and 527. Southdown PSV also continues as a dealer in both new and pre-used stock, principally buses. *Dave Heath*

SPEEDWELL

Speedwellbus Ltd, Raglan Street, Hyde, SK14 2DX

0001	M115XLV	Mercedes-Benz 709D	Carlyle Citybus 2	B27F	1995	Stanley Taxis, 2002
0002	N645HSX	Mercedes-Benz 811D	Leicester Carriage Eagle	B33F	1995	Leven Valley, South Bank, 2005
0003	R944AMB	Mercedes-Benz Vario 0814	Plaxton Beaver 2	B31F	1997	Collinsdale, Low Fell, 2002
	T156AUA	Mercedes-Benz Vario 0814	Alexander ALX100	B27F	1999	MacPherson, Donisthorpe, 2007
0006	T445WWT	Mercedes-Benz Vario 0814	Plaxton Beaver 2	B29F	1999	Stringer, Pontefract, 2005
0007	T446WWT	Mercedes-Benz Vario 0814	Plaxton Beaver 2	B29F	1999	Stringer, Pontefract, 2005
0008	V983DNB	Mercedes-Benz Vario 0814	Plaxton Beaver 2	B27F	1999	Norfolk Green, King's Lynn, 2003
0009	V994DNB	Mercedes-Benz Vario 0814	Plaxton Beaver 2	B27F	1999	Konectbus, Dereham, 2003
0010	R269XDA	Mercedes-Benz 709D	Alexander Sprint	B27F	1995	M&M, Accrington, 2007
0011	R501YWC	Mercedes-Benz Vario 0814	Plaxton Beaver 2	B27F	1999	Stagecoach, 2007
0012	R504YWC	Mercedes-Benz Vario 0814	Plaxton Beaver 2	B27F	1999	Stagecoach, 2007
	R505YWC	Mercedes-Benz Vario 0814	Plaxton Beaver 2	B27F	1999	Stagecoach, 2007
0013	R506YWC	Mercedes-Benz Vario 0814	Plaxton Beaver 2	B27F	1999	Stagecoach, 2007
	R508YWC	Mercedes-Benz Vario 0814	Plaxton Beaver 2	B27F	1999	Stagecoach, 2008
0014	R510YWC	Mercedes-Benz Vario 0814	Plaxton Beaver 2	B27F	1999	Stagecoach, 2007
0015	R511YWC	Mercedes-Benz Vario 0814	Plaxton Beaver 2	B27F	1999	Stagecoach, 2007
0016	R512YWC	Mercedes-Benz Vario 0814	Plaxton Beaver 2	B27F	1999	Stagecoach, 2007
0017	R513YWC	Mercedes-Benz Vario 0814	Plaxton Beaver 2	B27F	1999	Stagecoach, 2007
0018	R514YWC	Mercedes-Benz Vario 0814	Plaxton Beaver 2	B27F	1999	Evans, Bromborough, 2007
1001	MX55WCW	Optare Solo M950	Optare	N33F	2006	
1003	MX04VLY	Optare Solo M920	Optare	N33F	2004	
1004	MX04VLZ	Optare Solo M920	Optare	N33F	2004	
1005	MX54KYH	Optare Solo M780	Optare	N24F	2004	
1006	MX56NLD	Optare Solo M950	Optare	N32F	2007	
1007	MX56NLU	Optare Solo M950	Optare	N32F	2007	

Many of Speedwell's Plaxton Beaver 2s originated with Stagecoach, and 0014, R510YWC, is seen working from Stockport. *John Young*

Marple is the location of this view of Speedwell's Dart 3005, VU02TTZ. Speedwell's services operate in Derbyshire's High Peak, Glossopdale and extend as far as Tameside, Stockport and Oldham. *John Young*

1009	MX06ABO	Optare Solo M780	Optare	N24F	2006	
1010	MX06ABU	Optare Solo M780	Optare	N24F	2006	
1012	MX06BOV	Optare Solo M920	Optare	N33F	2006	
1013	MX06BPU	Optare Solo M920	Optare	N33F	2006	
2001	OHV806Y	Leyland Titan TNLXB2RR	Leyland	B44/.26D	1983	TM Travel, Staveley, 2005
2003	A984SYF	MCW Metrobus DR101/17	MCW	B43/28D	1984	TM Travel, Staveley, 2005
2004	B192WUL	MCW Metrobus DR101/17	MCW	B43/28D	1985	TM Travel, Staveley, 2005
2005	OJD822Y	MCW Metrobus DR101/16	MCW	B43/28D	1983	TM Travel, Staveley, 2005
	G501SFT	Leyland Olympian ONCL10/1RZA	Northern Counties	B47/30F	1989	Arriva NW & Wales, 2008
	G521WJF	Leyland Olympian ONLXB/1RZ	Alexander RH	B45/30F	1989	Thamesdown, 2008
	G522WJF	Leyland Olympian ONLXB/1RZ	Alexander RH	B45/30F	1989	Thamesdown, 2008
	G523WJF	Leyland Olympian ONLXB/1RZ	Alexander RH	B45/30F	1989	Thamesdown, 2008
3002	K825NKH	Dennis Dart 9m	Plaxton Pointer	B34F	1992	Fair Rider, Huddersfield, 2007
3003	K827NKH	Dennis Dart 9m	Plaxton Pointer	B34F	1992	Fair Rider, Huddersfield, 2007
3004	K432OKH	Dennis Dart 9m	Plaxton Pointer	B34F	1992	Fair Rider, Huddersfield, 2007
3005	VU02TTZ	Dennis Dart SLF 10.7m	Plaxton Pointer 2	N37F	2002	UniLink, Southampton, 2003

STANSTED BUS

Stansted Transit Ltd, 7/500 The Hub, Avenue West, Great Notley, CM77 7AA

Reg	Chassis	Body	Type	Year	History
NML616E	AEC Routemaster R2RH1	Park Royal	B42/32R	1967	Stagecoach, 2004
D194FYM	Leyland Olympian ONLXB/1R	Eastern Coach Works	B42/26D	1986	Arriva London, 2005
D219FYM	Leyland Olympian ONLXB/1R	Eastern Coach Works	B42/26D	1986	Arriva London, 2005
D245FYM	Leyland Olympian ONLXB/1R	Eastern Coach Works	B42/26D	1986	Arriva London, 2005
F259YTJ	Leyland Olympian ONCL10/2RZ	Northern Counties Palatine	B47/30F	1990	Cherry, Bootle, 2007
G957KJX	DAF SB220	Optare Delta	B47F	1990	BAA, Stansted, 2006
J102DUV	Dennis Dart 8.5m	Plaxton Pointer	B24F	1992	London United, 2002
J202VHN	DAF SB220	Optare Delta	B51F	1992	Go-North East, 2006
K710KGU	Dennis Dart 9m	Plaxton Pointer	B35F	1992	Metrobus, Crawley, 2004
N5BLU	Dennis Dart 9.8m	Alexander Dash	B40F	1995	Blue Bus, Bolton, 2000
R739TMO	Dennis Dart SLF 11.3m	Plaxton Pointer SPD	N41F	1997	Pete's Travel, West Bromwich, '03
S45UBO	Volvo B10M-62	Plaxton Première 350	C49FT	1998	Dawsonrentals, 2007
T13VCC	Dennis Javelin 12m	Plaxton Première 350	C49FT	1999	Burton's, Haverhill, 2006
T13VVC	Dennis Javelin 12m	Plaxton Première 350	C49FT	1999	Burton's, Haverhill, 2006
T223SAS	Dennis Dart SLF 11.3m	Plaxton Pointer SPD	N43F	1999	Rapson, 2003
T402LGP	Dennis Dart SLF 10.7m	Caetano Compass	N31D	1999	Centra, Heathrow, 2005
T414LGP	Dennis Dart SLF 10.7m	Caetano Compass	N31D	1999	Centra, Heathrow, 2005
T74WWV	Dennis Dart SLF	Plaxton Pointer	N33D	1999	National Express, W Drayton, '06
T75WWV	Dennis Dart SLF	Plaxton Pointer	N33D	1999	National Express, W Drayton, '06
V682FPO	Dennis Dart SLF	Caetano Compass	N44F	1999	Appleby, Conisholme, 2001
V237LWU	Optare MetroRider MR15	Optare	B31F	1999	First, 2005
W483OUF	Dennis Dart SLF	Caetano Compass	N44F	2000	London Transit, 2003
KM51BFZ	Dennis Dart SLF 10.7m	Plaxton Pointer 2	N37F	2002	Kent CC, 2007
KP51SXU	Dennis Dart SLF 8.8m	Plaxton Pointer MPD	N29F	2001	Puma, Renfrew, 2007
KP51SXZ	Dennis Dart SLF 8.8m	Plaxton Pointer MPD	N29F	2001	Supertravel, Speke, 2007
KV51KZC	Dennis Dart SLF 8.8m	Plaxton Pointer MPD	N29F	2001	Nippy Bus, Yeovil, 2007
YT51EAG	Dennis Dart SLF 8.8m	Plaxton Pointer MPD	N29F	2002	Birmingham Motor Traction, 2007
YT51EAK	Dennis Dart SLF 8.8m	Plaxton Pointer MPD	N29F	2002	Classic, Annfield Plain, 2007
YT51EAM	Dennis Dart SLF 8.8m	Plaxton Pointer MPD	N29F	2001	Birmingham Motor Traction, 207

Six Enviro 200 Darts joined Stansted Bus during 2007. This 8.9 metre model is the replacement for the Mini Pointer Dart with KX57FMA showing off the styling at Bishop's Stortford. *Richard Godfrey*

The Mini Pointer Dart is well established at Stansted Bus, the type being represented by 1007, KE53VDP, seen in Saffron Walden having arrived on route 301 from Bishop's Stortford. *Alan Blagburn*

KV03ZGP	TransBus Dart 8.8m	TransBus Mini Pointer	N29F	2003	
KV03ZGR	TransBus Dart 8.8m	TransBus Mini Pointer	N29F	2003	
KE53VDP	TransBus Dart 8.8m	TransBus Mini Pointer	N29F	2003	
KE53VDY	TransBus Dart 8.8m	TransBus Mini Pointer	N29F	2003	
YC53NZZ	LDV Convoy	LDV	M16	2004	private owner, 2005
YK53JPU	LDV Convoy	LDV	M16	2004	Arrow Self Drive, 2005
EU06KPA	ADL Dart 8.8m	ADL Mini Pointer	N29F	2006	
EU06KOW	ADL Dart 8.8m	ADL Mini Pointer	N29F	2006	
EU06KOX	ADL Dart 8.8m	ADL Mini Pointer	N29F	2006	
KX07HDG	Volvo B12B	Plaxton Panther	C49FT	2007	
KX07HDH	Volvo B12B	Plaxton Panther	C49FT	2007	
KX07HDJ	Volvo B12B	Plaxton Panther	C49FT	2007	
KX57FMA	ADL Dart 4 8.9m	ADL Enviro 200	N29F	2007	
KX57FMC	ADL Dart 4 8.9m	ADL Enviro 200	N29F	2007	
KX57FMD	ADL Dart 4 8.9m	ADL Enviro 200	N29F	2007	
KX57FME	ADL Dart 4 8.9m	ADL Enviro 200	N29F	2007	
KX57FMF	ADL Dart 4 8.9m	ADL Enviro 200	N29F	2007	
KX57FMG	ADL Dart 4 8.9m	ADL Enviro 200	N29F	2007	
AE08DLF	ADL Dart 10.7m	MCV Evolution	N37F	2008	
KX08HLU	Volvo B12B	Plaxton Panther	C49FT	2008	
KX08HLR	Volvo B12B	Plaxton Panther	C49FT	2008	

Previous registrations:

D219FYM	D219FYM, 519CLT	T13VVC	T400BCL, T162SBJ
T13VCC	T13VCC, T50BCL		

Depots: Great Notley; Bridge Street, Saffron Walden and Ninth Avenue, Stansted Airport
Web: www.stanstedtransit.co.uk

STEPHENSONS OF ESSEX

Stephensons of Essex Ltd, Riverside Ind Est, South Street, Rochford, SS4 1BS

114	N70SLK	Volvo B10M-62	Plaxton Expressliner 2	C49FT	1996	Hursts, Wigan, 2005
117	A17SOE	Volvo B10M-62	Plaxton Première 350	C49FT	1996	Skills, Nottingham, 2002
133	A14SOE	Volvo B10M-61	Van Hool Alizée	C57F	1984	APT Travel, Rayleigh, 2007
233	M233TBV	Volvo B10M-55	Alexander PS	BC48F	1995	Burnley & Pendle, 2005
235	M453VCW	Volvo B10M-55	Alexander PS	BC48F	1995	Burnley & Pendle, 2005
321	EU07FVL	Optare Solo M780	Optare	N24F	2007	
322	EU07FVM	Optare Solo M780	Optare	N24F	2007	
323	EU07FVN	Optare Solo M780	Optare	N24F	2007	
324	EU07FVO	Optare Solo M780	Optare	N24F	2007	
325	EU08WND	Optare Solo M880 SL	Optare	N28F	2008	
326	EU58AXX	Optare Solo M880 SL	Optare	N28F	2008	
411	SK02XGO	Dennis Dart SLF 8.8m	Plaxton Pointer MPD	N29F	2002	HAD, Shotts, 2004
412	SK02XGP	Dennis Dart SLF 8.8m	Plaxton Pointer MPD	N29F	2002	HAD, Shotts, 2004
413	EU03CFX	TransBus Dart SLF 8.8m	TransBus Mini Pointer	N29F	2003	
414	EU03CFY	TransBus Dart SLF 8.8m	TransBus Mini Pointer	N29F	2003	
416	SN56AYH	ADL Dart 8.8m	ADL Mini Pointer	N29F	2006	
417	EU07FRN	ADL Dart 8.8m	ADL Mini Pointer	N29F	2007	Excel, Stansted, 2007
450	P749HND	Dennis Dart SLF 10.6m	Plaxton Pointer 2	N37F	1997	Bryans, Denny, 2005
460	EU05CZA	ADL Dart 10.7m	ADL Pointer	N37F	2005	
501	EU58BRX	Scania OmniCity N230 UD	Optare Olympus	N51/29F	2008	
502		Scania OmniCity N230 UD	Optare Olympus	N51/29F	On order	
674	C174ECK	Leyland Olympian ONLXB/1R	Eastern Coach Works	B47/30F	1985	Stagecoach North West, 2003
676	C176ECK	Leyland Olympian ONLXB/1R	Eastern Coach Works	B47/30F	1985	Stagecoach, 2005
682	C382SAO	Leyland Olympian ONLXB/1RV	Alexander RL	B47/30F	1986	Stagecoach, 2004
683	C383SAO	Leyland Olympian ONLXB/1RV	Alexander RL	B47/30F	1986	Stagecoach, 2004
707	G707TCD	Leyland Olympian ON2R50G16Z4	Alexander RL	B51/34F	1990	Stagecoach, 2007
721	F621MSL	Leyland Olympian ONLXB/1R	Alexander RL	B51/36F	1988	Quantock MS, Wiveliscombe, '07
730	F630MSL	Leyland Olympian ONLXB/1R	Alexander RL	B51/36F	1988	Stagecoach, 2006
783	G183JHG	Leyland Olympian ON2R50G16Z4	Alexander RL	B51/31F	1990	Stagecoach, 2007
808	M408RVU	Volvo Olympian	Northern Counties Palatine 2	B47/30F	1995	Metrobus, Crawley, 2007
809	M409RVU	Volvo Olympian	Northern Counties Palatine 2	B47/30F	1995	Metrobus, Crawley, 2007
810	M410RVU	Volvo Olympian	Northern Counties Palatine 2	B47/30F	1995	Metrobus, Crawley, 2007
815	V415KMY	Dennis Trident	East Lancs Lolyne	N45/28F	1999	Low Fell Coaches, 2007
851	E851DPN	Leyland Olympian ONCL10/1RZ	Northern Counties	B47/30F	1988	East Yorkshire, 2007
852	E852DPN	Leyland Olympian ONCL10/1RZ	Northern Counties	B47/30F	1988	East Yorkshire, 2007
853	E856DPN	Leyland Olympian ONCL10/1RZ	Northern Counties	B47/30F	1988	East Yorkshire, 2007
855	H155BKH	Leyland Olympian ON2R50G13Z4	Northern Counties Palatine	B47/29F	1991	East Yorkshire, 2008
856	H156BKH	Leyland Olympian ON2R50G13Z4	Northern Counties Palatine	B47/29F	1991	East Yorkshire, 2008
858	H158BKH	Leyland Olympian ON2R50G13Z4	Northern Counties Palatine	B47/29F	1991	East Yorkshire, 2008
859	H159BKH	Leyland Olympian ON2R50G13Z4	Northern Counties Palatine	B47/29F	1991	East Yorkshire, 2008
862	J562HAT	Leyland Olympian ON2R50G13Z4	Northern Counties Palatine	B47/29F	1992	East Yorkshire, 2008

Stephensons' Mini Pointer Dart 411, SK02XGO, is seen in Southend where the company now has a strong presence.
David Longbottom

Re-registered after transfer from Northern Scottish to Stagecoach North West, Olympian 683, C383SAO, is seen in Southend on January 1st 2008, when a vehicle running day was underway. Stephensons of Essex can trace its origins back to 1975 and currently operates over fifty buses and coaches throughout south and mid-Essex.
Richard Godfrey

863	J563HAT	Leyland Olympian ON2R50G13Z4	Northern Counties Palatine	B47/29F	1992	East Yorkshire, 2008
864	J564HAT	Leyland Olympian ON2R50G13Z4	Northern Counties Palatine	B47/29F	1992	East Yorkshire, 2008
866	J566HAT	Leyland Olympian ON2R50G13Z4	Northern Counties Palatine	B47/29F	1992	East Yorkshire, 2008
869	J569HAT	Leyland Olympian ON2R50G13Z4	Northern Counties Palatine	B47/29F	1992	East Yorkshire, 2008
876	C62CHM	Leyland Olympian ONLXB/1RH	Eastern Coach Works	B42/30F	1986	Phoenix, Witham, 2003
881	C81CHM	Leyland Olympian ONLXB/1RH	Eastern Coach Works	B42/30F	1986	Phoenix, Witham, 2003
882	C82CHM	Leyland Olympian ONLXB/1RH	Eastern Coach Works	B42/30F	1986	Phoenix, Witham, 2003
892	C92CHM	Leyland Olympian ONLXB/1RH	Eastern Coach Works	B42/30F	1986	Phoenix, Witham, 2003
901	OFS683Y	Leyland Olympian ONTL11/2R	Eastern Coach Works	B50/35F	1983	Powell Bus, Hellaby, 2006
902	UAR219Y	Leyland Olympian ONTL11/2R	Eastern Coach Works	B50/35F	1982	Prentice, Haddington, 2003
947	H547VAT	Leyland Olympian ON2R50G13Z4	Northern Counties Palatine	B47/29F	1990	East Yorkshire, 2008
949	H549VAT	Leyland Olympian ON2R50G13Z4	Northern Counties Palatine	B47/29F	1990	East Yorkshire, 2008
951	H551VAT	Leyland Olympian ON2R50G13Z4	Northern Counties Palatine	B47/29F	1990	East Yorkshire, 2008
952	H552VAT	Leyland Olympian ON2R50G13Z4	Northern Counties Palatine	B47/29F	1990	East Yorkshire, 2008
953	L953MSC	Volvo Olympian	Alexander RH	B51/34F	1994	Lothian Buses, 2007
955	L955MSC	Volvo Olympian	Alexander RH	B51/34F	1994	Lothian Buses, 2007
962	L962MSC	Volvo Olympian	Alexander RH	B51/34F	1994	Lothian Buses, 2007
992	L762SNO	Leyland Olympian ON3R56C18Z5	Alexander RH	BC63/44F	1993	Fargo, Rayne, 2007
998	L808SNO	Leyland Olympian ON3R56C18Z5	Alexander RH	BC63/44F	1993	Fargo, Rayne, 2007
1017	G717FVX	Dennis Condor DDA1702	Duple Metsec	B63/43D	1989	Fargo, Rayne, 2007
1029	F329UJN	Dennis Condor DDA1702	Duple Metsec	B63/43D	1989	Fargo, Rayne, 2007

Special event vehicles:

424	424DCD	Leyland Titan PD3/4	Northern Counties	CO39/30F	1964	Southdown
1233	MRJ233W	Leyland Fleetline FE33ALR	Northern Counties	B49/31F	1981	Southend

Previous registrations:

424DCD	424DCD, AOR158B	G717FVX	EJ4042 (HK)
A17SOE	N45ARC	L762SNO	FS4861 (HK)
C382SAO	C473SSO, GSO3V	L808SNO	FS6392 (HK)
C383SAO	C474SSO, GSO4V	M408RVU	M408RVU, WLT838
F329UJN	EH2206 (HK)	UAR219Y	OFS693Y, SIL1075, TSO29X

Depots: South Street, Rochford; Bates Road, Maldon and Crittall Road, Witham.

SUPERTRAVEL

Supertravel Omnibus Ltd, STC House, Speke Hall Road, Speke, Liverpool, L24 9HD

	M110TVH	Mercedes-Benz 814D	Plaxton Beaver	BC33F	1995	Courtesy, Werneth, 1997
	P478CBV	Mercedes-Benz 410D	Frank Guy	M15	1997	Mencap, Ormskirk, 2006
	S7STM	Mercedes-Benz Atego O1120L	Ferqui Solera	C35F	1998	
004	V263BNV	Dennis Dart SLF 8.8m	Plaxton Pointer MPD	N29F	1999	
006	V264BNV	Dennis Dart SLF 8.8m	Plaxton Pointer MPD	N29F	1999	
016	V268BNV	Dennis Dart SLF 8.8m	Plaxton Pointer MPD	N29F	1999	
031	V331CVV	Dennis Dart SLF 8.8m	Plaxton Pointer MPD	N29F	2000	
033	V618EVU	Dennis Dart SLF 10.7m	Plaxton Pointer 2	N35F	2000	Pink Elephant Parking, 2006
001	W365ABD	Dennis Dart SLF 8.8m	Plaxton Pointer MPD	N29F	2000	
021	X712CCA	Dennis Dart SLF 8.8m	Plaxton Pointer MPD	N29F	2001	
014	Y184KNB	Dennis Dart SLF 8.9m	Alexander ALX200	N29F	2001	
009	Y185KNB	Dennis Dart SLF 8.9m	Alexander ALX200	N29F	2001	
015	Y942GEU	Renault Master	Rohill	B14F	2001	
	X313FVV	Mercedes-Benz Sprinter 313cdi	Mercedes-Benz	M16	2001	van, 2003
	X984SKH	Mercedes-Benz Sprinter 313cdi	Mercedes-Benz	M16	2001	van, 2006
	Y773OEE	Mercedes-Benz Atego 1223L	Ferqui/Optare Beluga	C39F	2001	Shaftesbury & District, 2004
020	KP51SYE	Dennis Dart SLF 8.8m	Plaxton Pointer MPD	N29F	2002	
019	KP51SYR	Dennis Dart SLF 8.8m	Plaxton Pointer MPD	N29F	2002	
032	PJ02RGZ	Dennis Dart SLF 10.7m	Plaxton Pointer 2	N37F	2002	Pete's Travel, W Bromwich, 2004
008	YN53ZWH	Optare Solo M920	Optare	N33F	2003	
007	YN53ZWJ	Optare Solo M920	Optare	N33F	2003	
	MX53SXO	Mercedes-Benz Vario 0814	Olympus	C24F	2003	McLean, Chapelhall, 2008
	MX53ZWC	Mercedes-Benz Vario 0814	Onyx	C24F	2003	McColl, Balloch, 2008
	YX04AWV	Mercedes-Benz Atego O1120L	Ferqui/Optare Solara	C39F	2004	Steel, Skipton, 2007
	VX54CLU	Mercedes-Benz Vario 0814	Mellor	BC33F	2005	
003	MV54KYJ	Optare Solo M920	Optare	N33F	2005	
037	MX05OSW	Optare Solo M780	Optare	N23F	2005	Leven Valley, South Bank, 2006
025	MX55BYP	Optare Solo M1020	Optare	N37F	2005	
026	MX55BYO	Optare Solo M1020	Optare	N37F	2005	
027	MX55BXN	Optare Solo M1020	Optare	N37F	2005	
024	MX55BYC	Optare Solo M920	Optare	N33F	2005	
022	MX55BYD	Optare Solo M920	Optare	N33F	2005	
023	MX55BYF	Optare Solo M920	Optare	N33F	2005	
034	MX56AAV	Optare Solo M920	Optare	N33F	2006	
035	MX56AAY	Optare Solo M920	Optare	N33F	2006	
036	MX56AAZ	Optare Solo M920	Optare	N33F	2006	

Previous registration: X712CCA X1STM

Liveried for the Runcorn Campus Link, Solo 025, MX55BYP is one of the longer 10.2 metre examples of the Solo. So far 3285 units of the standard Solo have been built along with 29 of the new SR models. *Ken Lansdowne*

SWANBROOK

Swanbrook Transport Ltd, Thomas House, St Margaret's Road, Cheltenham, GL50 4DZ

A900SUL	MCW Metrobus DR101/16	MCW	B43/32F	1984	London General, 2000
A958SYF	MCW Metrobus DR101/17	MCW	B43/28D	1984	London United, 1998
A703THV	MCW Metrobus DR101/18	MCW	B43/28D	1984	London United, 1998
B221WUL	MCW Metrobus DR101/17	MCW	B43/32F	1985	Arriva London, 2001
H97PVW	Leyland Olympian ON2R50C13Z4	Alexander RH	B47/33F	1991	Dublin Bus, 2004
H119PVW	Leyland Olympian ON2R50C13Z4	Alexander RH	B47/33F	1991	Dublin Bus, 2004
H137PVW	Leyland Olympian ON2R50C13Z4	Alexander RH	B47/33F	1991	Dublin Bus, 2004
H794PVW	Leyland Olympian ON2R50C13Z4	Alexander RH	B47/33F	1991	Dublin Bus, 2008
J908CEV	Leyland Olympian ON2R50C13Z4	Alexander RH	B47/33F	1991	Dublin Bus, 2008
UJI1761	Volvo B10M-60	Plaxton Paramount 3500 III	C53F	1991	Oxford Bus Company, 2000
UJI1762	Volvo B10M-60	Plaxton Paramount 3500 III	C53F	1991	Oxford Bus Company, 2000
L194OVO	Mercedes-Benz 811D	Plaxton Beaver	B33F	1994	City of Nottingham, 2001
L195OVO	Mercedes-Benz 811D	Plaxton Beaver	B33F	1994	City of Nottingham, 2001
M578UBA	Mercedes-Benz 814D	UVG CitiStar	BC29F	1995	MoD (04RN41), 2005
M588UBA	Mercedes-Benz 814D	UVG CitiStar	BC29F	1995	MoD (04RN39), 2005
M407BFG	Volvo B10M-60	Plaxton Première 350	C53F	1995	Stagecoach, 2007
M409BFG	Volvo B10M-60	Plaxton Première 350	C53F	1995	Stagecoach, 2007
N647JNO	Volvo Olympian	Alexander RH	B47/27D	1996	Dublin Bus, 2008
P255RUM	Mercedes-Benz 810D	Plaxton Beaver	B27F	1997	McColl, Balloch, 2007
R113NTA	Mercedes-Benz Vario O814	Alexander ALX100	B29F	1998	Stagecoach, 2007
T12SBK	Dennis Dart SLF	Marshall Capital	N43F	1999	
V844OOF	Optare Solo M850	Optare	N28F	2000	Zak's, Birmingham, 2004
V845OOF	Optare Solo M850	Optare	N28F	2000	Zak's, Birmingham, 2004
X469XUT	Volvo B6LE	East Lancs Spryte	N36F	2001	Kent Coach Tours, Ashford, 2002
BU51OWL	Optare Solo M920	Optare	N31F	2001	Browns, Builth Wells, 2005
SK51SBK	Dennis Dart SLF	Marshall Capital	N40F	2002	
YX05DWA	Mercedes-Benz Vario O814	Plaxton Beaver 2	BC33F	2005	Hutchinson, Easingwold, 2005
VX55FWE	Mercedes-Benz Vario O814	Plaxton Beaver 2	BC29F	2005	Hutchinson, Easingwold, 2006
MX06ABF	Optare Solo M850 SL	Optare	N25F	2006	Cross Gates Coaches, 2007

Previous registrations:

H97PVW	91D1097		M407BFG	M407BFG, SYC852
H119PVW	91D1096		M409BFG	M409BFG, CSU992
H137PVW	91D1095		N647JNO	96D278
H794PVW	91D1080		UJI1761	H957DRJ
J908CEV	92D139		UJI1762	H960DRJ

Depot: Pheasant Lane, Golden Valley, Staverton   **Web:** www.swanbrook.co.uk

An interesting applicaion of moulding changes the appearance of the Plaxton Paramount 3500 body. UJI1762 is a Volvo B10M that was new to Shearings Holidays. *Dave Heath*

TM TRAVEL

T M Travel Ltd, The Bus Depot, Station Road, Halfway, Sheffield, S20 3GZ

BSG550W	Leyland Tiger TRCTL11/3R	East Lancs EL2000 (1993)	BC57F	1981	Stagecoach, 2008
KSP329X	Leyland Tiger TRCTL11/3R	East Lancs EL2000 (1995)	BC57F	1982	Stagecoach, 2008
A111TRP	Leyland Tiger TRCTL11/3R	Plaxton Paramount 3200E	C55F	1983	Thompson, Parkgate, 2004
D229HMT	Leyland Tiger TRCTL11/3RZ	Van Hool Alizée	C57F	1987	Reynolds, Watford, 2007
F603CET	Leyland Tiger TRBTL11/2RP	Plaxton Derwent 2	B54F	1988	Arriva Midlands, 2007
G523VBB	Leyland Olympian ON2R50C13Z4	Northern Counties	B47/27D	1990	Arriva London, 2006
G524VBB	Leyland Olympian ON2R50C13Z4	Northern Counties	B47/27D	1990	Arriva London, 2005
G531VBB	Leyland Olympian ON2R50C13Z4	Northern Counties	B47/27D	1990	Arriva London, 2005
G532VBB	Leyland Olympian ON2R50C13Z4	Northern Counties	B47/27D	1990	Arriva London, 2005
G535VBB	Leyland Olympian ON2R50C13Z4	Northern Counties	B47/27D	1990	Arriva London, 2005
G370YUR	Leyland Olympian ONCL10/1RZ	Alexander AL	B47/30F	1990	Thompson, Parkgate, 2004
G472EKV	Leyland Tiger TRCL10/3ARZM	Plaxton Paramount 3200 III	C53F	1990	Morse, Sheriff Hutton, 2006
G435MWU	Leyland Tiger TRCL10/3ARZA	Plaxton Paramount 3200 III	C55F	1990	East Yorkshire, 2008
H803RWJ	Scania N113 DRB	Northern Counties Palatine	B47/33F	1990	Arriva Midlands, 2006
H784PVW	Leyland Olympian ON2R50C13Z4	Alexander AL	B47/31F	1991	Dublin Bus, 2008
J844TSC	Leyland Olympian ON2R56C13Z4	Alexander RH	B51/34F	1991	Lothian Buses, 2008
J850TSC	Leyland Olympian ON2R56C13Z4	Alexander RH	B51/34F	1991	Lothian Buses, 2008
J854TSC	Leyland Olympian ON2R56C13Z4	Alexander RH	B51/34F	1991	Lothian Buses, 2008
J619CEV	Leyland Olympian ON2R50C13Z4	Alexander RH	B47/31F	1992	Dublin Bus, 2006
J628CEV	Leyland Olympian ON2R50C13Z4	Alexander RH	B47/31F	1992	Dublin Bus, 2006
K482GNN	Leyland Olympian ON2R56C14Z5	East Lancs	B49/35F	1992	City of Nottingham, 2005
K482VVR	Volvo B10M-60	Van Hool Alizée H	S70F	1993	Draper, Tibshelf, 2008
L720SNO	Leyland Olympian ON2R50C13Z4	Alexander RH	B47/31F	1993	Dublin Bus, 2006
L726SNO	Leyland Olympian ON2R50C13Z4	Alexander RH	B47/31F	1993	Dublin Bus, 2006

Sheffield bus station is the location for this view of YJ56KAO, a VDL Bus SB120 with Plaxton Centro bodywork.
TM Travel has now moved into the Halfway depot once used by Sheffield Transport and latterly by First.
Mark Bailey

During 2008, TM Travel received its first Optare Versa bus. Pictured shortly after entering service in Chesterfield, is 408, YJ08PFN, is an example of the shorter 11-metre model. *Tony Wilson*

L351MRR	Scania N113DRB	East Lancs	B49/35F	1993	City of Nottingham, 2005
L352MRR	Scania N113DRB	East Lancs	B49/35F	1993	City of Nottingham, 2005
L353MRR	Scania N113DRB	East Lancs	B49/35F	1993	City of Nottingham, 2005
P487SWC	Volvo Olympian	Alexander RH	B47/31F	1996	Dublin Bus, 2008
P337ROO	DAF DB250	Northern Counties Palatine 2	B43/29F	1997	Courtney, Bracknell, 2005
P505OUG	Optare Excel L1070	Optare	N35F	1997	Rotala, Birmingham, 2008
R211DKG	Optare Excel L1150	Optare	N42F	1997	Rotala, Birmingham, 2008
R213DKG	Optare Excel L1150	Optare	N42F	1997	Rotala, Birmingham, 2008
R374DJN	Optare Excel L1070	Optare	N35F	1997	East Thames Buses, 2008
R379DJN	Optare Excel L1070	Optare	N35F	1997	East Thames Buses, 2008
S779RNE	Dennis Dart SLF 11.3m	Plaxton Pointer SPD	N41F	1998	Beaumont, Gloucester, 2005
T455HNH	Mercedes-Benz Vario O814	Alexander ALX100	B27F	1999	Menzies, Heathrow, 2007
T112AUA	DAF SB3000	Ikarus Blue Danube 396	C49FT	1999	Alfa, Chorley, 2008
V389KVY	Optare Excel L1150	Optare	N44F	1999	Rotala, Birmingham, 2008
W261EWU	Optare Solo M850	Optare	N23F	2000	Arriva Scotland, 2007
W284EYG	Optare Solo M850	Optare	N21F	2000	HAD, Shotts, 2007
W285EYG	Optare Solo M850	Optare	N23F	2000	Norfolk Green, King's Lynn, 2007
W286EYG	Optare Solo M850	Optare	N23F	2000	Arriva Scotland, 2007
W288EYG	Optare Solo M850	Optare	N28F	2000	HAD, Shotts, 2007
W289EYG	Optare Solo M850	Optare	N23F	2000	Courtney, Bracknell, 2007
W292EYG	Optare Solo M850	Optare	N23F	2000	Smith, Patna, 2007
W295EYG	Optare Solo M850	Optare	N23F	2000	HAD, Shotts, 2007
W923JNF	Dennis Dart SLF 10.8m	Alexander ALX200	N38F	2000	Connexions, Redhill, 2005
X685REC	DAF SB220	East Lancs Myllennium	N46F	2000	Lakeland, Hurst Green, 2006
X732FPO	Optare Solo M920	Optare	N27F	2000	Hampshire CC, 2007
X941NUB	Optare Solo M850	Optare	N23F	2000	Raddoneur, Birkenhead, 2007
X385VVY	Optare Solo M850	Optare	N30F	2001	APL, Crudwell, 2008
Y867PWT	Optare Solo M850	Optare	N29F	2001	Diamond Bus, Birmingham, 2007
Y198KNB	Optare Solo M920	Optare	N29F	2001	Beaumont, Gloucester, 2002
LA02WMZ	Optare Solo M850	Optare	N29F	2002	Bristol Airport, 2008
YL02FKU	Optare Solo M850	Optare	N27F	2002	*Operated for SouthYorkshire*
YL02FKV	Optare Solo M850	Optare	N27F	2002	*Operated for SouthYorkshire*
YL02FKW	Optare Solo M920	Optare	N31F	2002	NCP, Paisley, 2007
TM52BUS	Optare Solo M920	Optare	N33F	2002	
MW52PZD	Optare Solo M920	Optare	N33F	2003	Selwyns, Runcorn, 2004

YN03ZXE	Optare Solo M920	Optare	N33F	2003	Birchill & Steed, Whitstable, 2008
MX03YDF	Optare Solo M920	Optare	N33F	2003	Hut, Finstock, 2006
MX03YCN	Optare Solo M850	Optare	N29F	2003	K-Line, Honley, 2004
MX03YCP	Optare Solo M850	Optare	N29F	2003	Timeline, Melling, 2004
1294RU	VDL Bus DB250	East Lancs Lowlander	N51/29F	2003	
MX53FDD	Optare Solo M920	Optare	N33F	2003	Shuttle Buses, Kilwinning, 2004
YN53SSZ	Optare Solo M920	Optare	N33F	2003	
YN53EJX	Mercedes-Benz Vario O814	TransBus Beaver 2	B31F	2004	
MK04VMK	Mercedes-Benz Vario O814	TransBus Beaver 2	B31F	2004	Speedwell, Glossop, 2006
MX04VLM	Optare Solo M920	Optare	N33F	2004	McNee, Ratho Station, 2005
YN54WWU	Mercedes-Benz Vario O814	Plaxton Beaver 2	B31F	2004	
YN54WWV	Mercedes-Benz Vario O814	Plaxton Beaver 2	B31F	2004	
YN54SYG	VDL Bus DB250	East Lancs Lowlander	N45/24F	2004	
YT55TMT	ADL Dart	East Lancs Myllennium	N41F	2005	
YN55KMV	Enterprise Plasma	Plaxton Primo	N28F	2005	
YN55YSC	Enterprise Plasma	Plaxton Primo	N28F	2005	
YN06CYO	VDL Bus SB120	Plaxton Centro	N40F	2006	
YN06CYP	Mercedes-Benz Vario O814	Plaxton Cheetah	C29F	2006	
YN56AHY	Optare Solo M920	Optare	N33F	2006	
YN56OWP	Optare Solo M920	Optare	N33F	2006	
YJ56KAO	VDL Bus SB120	Plaxton Centro	N40F	2006	
YJ56JYA	VDL Bus SB120	Plaxton Centro	N40F	2007	
YN07JWA	VDL Bus SB200	Plaxton Centro	N45F	2007	
YJ07JWU	VDL Bus SB120	Plaxton Centro	N40F	2007	
YJ07JWW	VDL Bus SB120	Plaxton Centro	N40F	2007	
YJ07JNN	VDL Bus SB120	Plaxton Centro	N40F	2007	
YJ07JVC	VDL Bus SB120	Plaxton Centro	N40F	2007	
YJ07JVE	VDL Bus SB120	Plaxton Centro	N40F	2007	
YN07DVO	Volvo B12B	Plaxton Panther	C49FT	2007	
YN07SYS	Optare Solo M780 SL	Optare	N27F	2007	
YJ57EHB	Optare Solo M950	Optare	N33F	2007	
YJ57EHC	Optare Solo M950	Optare	N33F	2007	
YJ57KFE	Optare Solo M880 SL	Optare	N28F	2007	
YN57BXG	Mercedes-Benz Vario O816	Plaxton Cheetah 2	C33F	2007	
YJ57EHU	Optare Versa V1100	Optare	N38F	2007	Optare demonstrator, 2008
YJ08PFN	Optare Versa V1100	Optare	N40F	2008	
YN08DNU	Volvo B12M	Plaxton Paragon	C45FT	2008	
YN08NKW	Volvo B12M	Plaxton Paragon	C61F	2008	
YN08JWC	MAN 12.240	Plaxton Centro	N38F	2008	
YN08JWD	MAN 12.240	Plaxton Centro	N38F	2008	
YN08JWE	MAN 12.240	Plaxton Centro	N38F	2008	
YN58NDV	Optare Solo M880	Optare	N29F	2008	

Special event vehicles:

JRA635	Leyland Tiger PS1	Crossley	B35R	1947	Chesterfield
Q723GHG	Leyland RETL11	Eastern Coach Works	B51F	1985	Leyland Bus
FRN816W	Leyland Tiger TRCTL11/3R	Van Hool Alizée	C51F	1981	Leyland Bus
VOY182X	Leyland Tiger TRCTL11/2R	Plaxton Viewmaster IV	C49FT	1981	British Airways
CBV775Y	Leyland Royal Tiger B50	Leyland Doyen	C49FT	1983	Leyland Bus

Previous registrations:

1294RU	YJ53VBM	K482VVR	K482VVR, ONT46
BSG550W	BSG550W, MSV926, IUI5454	KSP329X	VSS6X, WLT759, IIL6440, PIW4457
CBV775Y	544UYD, XHG245Y	L720SNU	93D10165
D229HMT	D229HMT, FIL8614	L726SNU	93D10167
F603CET	F603CET, A19RBL, 49XBF	P487SWC	96D304
FRN816W	FRN816W, MSV927, BFW233W, NCT833, BFW428W		
G435MWU	G435MWU, 95EYM		
G472EKV	G878BKV, MIW5785, 89CE3071		
H784PVW	91D1089	C779RNE	S779HNE, 98D70567
J019CEV	92D127	T455HNH	T455HNH, J101750
J628CEV	92D126		

Web: www.tmtravel.co.uk

THAMES TRAVEL

Thames Travel (Wokingham) Ltd; Thames Travel (Wallingford) Ltd; Wyndham House, Lester Way, Wallingford, OX10 9TD

501	M501VJO	Dennis Dart 9.8m	Marshall C37	B36D	1995	City Line, Oxford, 2002
67	V67SVY	Volvo B10BLE	Alexander ALX300	NC44F	1999	First, 2004
103	BX02CMF	Mercedes-Benz Sprinter 411cdi	EvoBus	M13	2002	
153	KX53SGY	Optare Solo M850	Optare	N26F	2003	
104	OU04FMV	TransBus Dart 8.8m	TransBus Mini Pointer	N29F	2004	
154	OU54PGZ	Scania OmniDekka N94UD	East Lancs	N47/33F	2005	
105	AE05EUX	MAN 14.220	MCV Evolution	N40F	2005	
205	AE05EUZ	MAN 14.220	MCV Evolution	N40F	2005	
106	PF06ENO	Scania N94UB	East Lancs Esteem	N31F	2006	
206	PF06ENM	Scania N94UB	East Lancs Esteem	N31F	2006	
306	PF06ENN	Scania N94UB	East Lancs Esteem	N31F	2006	
406	PF06ENL	Scania N94UB	East Lancs Esteem	N31F	2006	
506	KX06LXN	ADL Dart	ADL Mini Pointer	N29F	2006	
606	KX06LXO	ADL Dart	ADL Mini Pointer	N29F	2006	
156	KX56HCZ	ADL Dart	ADL Mini Pointer	N29F	2006	
356	AE56OUF	MAN 14.220	MCV Evolution	N40F	2007	
157	OU57FHA	ADL Dart 4 8.9m	ADL Enviro 200	N29F	2007	
257	OU57FGV	ADL Dart 4 8.9m	ADL Enviro 200	N29F	2007	
357	OU57FGX	ADL Dart 4 8.9m	ADL Enviro 200	N29F	2007	
457	OU57FGZ	ADL Dart 4 8.9m	ADL Enviro 200	N29F	2007	

Thames Travel has expanded its services in the Berkshire and Oxfordshire area. Typical of the modern fleet is MCV-bodied MAN 108, AE08DLD, seen here in Wallingford while working route X39 to Oxford. *Mark Lyons*

Recent arrivals to fulfill the double-deck requirement at Thames Travel are three integral OmniCity N230 UDs asembled in Poland by Scania. Seen passing through Crowmarsh Gifford is 508, OU08HGN. *Mark Lyons*

557	AE57LYH	MAN 14.220	MCV Evolution	N40F	2007
657	AE57LYJ	MAN 14.220	MCV Evolution	N40F	2007
757	AE57LYK	MAN 14.220	MCV Evolution	N40F	2007
857	AE57LYP	MAN 14.220	MCV Evolution	N40F	2007
957	OU57FKB	ADL Dart 4 8.9m	ADL Enviro 200	N29F	2007
1057	AE57LYR	MAN 14.220	MCV Evolution	N40F	2007
1157	AE57LYS	MAN 14.220	MCV Evolution	N40F	2007
1257	AE57LYO	MAN 14.220	MCV Evolution	N40F	2007
108	AE08DLD	MAN 14.220	MCV Evolution	N40F	2008
208	AE08DKY	MAN 14.220	MCV Evolution	N40F	2008
308	AE08DKX	MAN 14.220	MCV Evolution	N40F	2008
408	OU08HGM	Scania OmniCity N230 UD	Scania	NC45/30F	2008
508	OU08HGN	Scania OmniCity N230 UD	Scania	NC45/30F	2008
608	OU08HGO	Scania OmniCity N230 UD	Scania	NC45/30F	2008
158	AJ58PZS	ADL Dart 4 8.9m	MCV Evolution	N29F	2008

Previous registrations:

OU04FMV	SN04EFW		V67SVY	V3JJL

Web: www.thames-travel.co.uk
Depots: Lester Way, Wallingford and Elms Farm, Grove, Wantage

THORNES / INDEPENDENT

Thornes Independent Ltd, The Coach Station, Hull Rd, Hemingbrough, Selby, YO8 6QG
Independent Coachways Ltd; Edwards Coachways Ltd, Low Fold Garage,
New Road Side, Horsforth, Leeds, LS18 4DR

85	WWW33	Volvo B10M-61	Plaxton Paramount 3500	S67F	1984	
93	VFN53	Volvo B9M	Plaxton Paramount 3200 II	C43F	1987	Courtlands, Horley, 1989
131	W20VHO	Mercedes Benz Vario O814	Autobus Nouvelle	C33F	2000	
132	W200VHO	DAF SB220	East Lancashire Myllennium	N52F	2000	
133	T200VHO	Volvo B9M	Plaxton Première 320	C43F	1999	
136	SHO800	Volvo B10M-66SE	Plaxton Première 320	C53F	1996	
138	S200VHO	Volvo B7R	Plaxton Prima	C57F	1999	Compass Royston, Stockton, 2000
139	P20VHO	Volvo B10M-62	Plaxton Première 320	C53F	1997	Halpenny, Blackrock, 2002
142	V200CHO	Volvo B10M-62	Plaxton Première 320	C53F	2000	Airport Parking, Copthorne, 2003
143	SMV24	Volvo Citybus B10M-61	East Lancs	B44/30D	1985	Forest, Aintree, 2003
144	VHO2OOH	Volvo B12B	Van Hool T9 Alizée	C47FT	2002	Blackburn Buses, 2003
145	MBO1F	Dennis Javelin 12m	Plaxton Première 320	S70F	1997	Baker, Enstone, 2004
146	ABT1K	Dennis Javelin 12m	Plaxton Première 320	S70F	1997	Baker, Enstone, 2004
148	R200VHO	DAF SB220	Optare Delta	B52F	1997	Duff, Sutton-on-the-Forest, 2005
151	YN06RVT	Volvo B7R	Plaxton Profile	C53F	2006	
152	SAC500	Volvo B12M	Plaxton Paragon	C53F	2003	Park's of Hamilton, 2007
153	YN55KLX	ADL Javelin 10M	Plaxton Profile	C41F	2005	Bicknell, Godalming, 2007
154	URN172Y	Leyland Atlantean AN68C/2R	East Lancashire	B50/36F	1982	Preston Bus, 2007
156	SF06WEC	Volvo B12M	Plaxton Panther	C53F	2006	Park's of Hamilton, 2008
157	SED232	Mercedes Benz Vario O814	Plaxton Cheetah	C20FT	2000	Logan, Dunloy, 2008

Representing the Thornes' fleet for this edition is 143, SMV24, a Volvo Citybus with East Lancs bodywork.
David Longbottom

Several operators have found a market for using vintage vehicles particularly for weddings and festivals. Thornes operates six, including OKP980, a Beadle-bodied Leyland that carries its original Maidstone & District livery. *David Longbottom*

Special event vehicles

32	OKP980	Beadle-Leyland	Beadle	C35F	1952	Maidstone & District
35	VHO200	Seddon Mk19	Harrington Wayfarer	C41F	1959	Liss & District, 1961
74	LBT380N	Bristol LHS6L	Plaxton Supreme III	C35F	1975	
141	KVY789N	Bedford YRQ	Duple Dominant	C45F	1975	Gorwood, East Cottingwith, 2002

Previous registrations:

ABT1K	R772WOB	SF06WEC	LSK496
LBT380N	LBT380N, KKH101N, WRL270, JCV385N	SHO800	N200VHO, OKP980, N232LAG
MBO1F	R771WOB	SMV24	B868XYR
OKP980	OKP980, VSJ556	V200CHO	V59KWO
P20VHO	97LH686, P232JBC	VFN53	D866YPH
R200VHO	R26GNW	VHO2OOH	PJ02UXR
S200VHO	S121OEF	VHO200	From new
SAC500	KSK954, SG03ZEM	WWW33	B385JVY
SED232	W906RET		

Depot: Hull Road, Hemingbrough (Thornes) and Low Fold Garage, Horsforth (Independent).

TRAVEL DE COURCEY

Mike De Courcey Travel Ltd, Rowley Drive, Coventry, CV3 4FG

508	POG508Y	MCW Metrobus DR102/27	MCW	B43/30F	1982	Travel West Midlands, 2003
513	POG513Y	MCW Metrobus DR102/27	MCW	B43/30F	1982	Travel West Midlands, 2003
536	POG536Y	MCW Metrobus DR102/27	MCW	B43/30F	1982	Travel West Midlands, 2003
567	POG567Y	MCW Metrobus DR102/27	MCW	B43/30F	1982	Travel West Midlands, 2003
630	ROX630Y	MCW Metrobus DR102/27	MCW	B43/30F	1983	Travel West Midlands, 2003
656	ROX656Y	MCW Metrobus DR102/27	MCW	B43/30F	1983	Travel West Midlands, 2003
663	ROX663Y	MCW Metrobus DR102/27	MCW	B43/30F	1983	Travel West Midlands, 2005
666	ROX666Y	MCW Metrobus DR102/27	MCW	B43/30F	1983	Travel West Midlands, 2003
671	A671UOE	MCW Metrobus DR102/27	MCW	B43/30F	1983	Travel West Midlands, 2004
684	A684UOE	MCW Metrobus DR102/27	MCW	B43/30F	1983	Travel West Midlands, 2003
706	A706UOE	MCW Metrobus DR102/27	MCW	B43/30F	1983	Travel West Midlands, 2004
712	A712UOE	MCW Metrobus DR102/27	MCW	B43/30F	1984	Travel West Midlands, 2003
101	A101WVP	MCW Metrobus GR133/1	MCW	B43/30F	1984	Travel West Midlands, 2003
107	A107WVP	MCW Metrobus GR133/1	MCW	B43/30F	1984	Travel West Midlands, 2003
740	A740WVP	MCW Metrobus DR102/27	MCW	B43/30F	1984	Travel West Midlands, 2005
748	A748WVP	MCW Metrobus DR102/27	MCW	B43/30F	1984	Travel West Midlands, 2003
110	G110NGN	Volvo Citybus B10M-50	Northern Counties	B47/38F	1989	Diamond Bus, Birmingham, 2006
113	G113NGN	Volvo Citybus B10M-50	Northern Counties	B47/38F	1989	Diamond Bus, Birmingham, 2006
114	G114NGN	Volvo Citybus B10M-50	Northern Counties	B47/38F	1989	Diamond Bus, Birmingham, 2006
122	G122NGN	Volvo Citybus B10M-50	Northern Counties	B47/38F	1989	Diamond Bus, Birmingham, 2006
51	MJI3751	Leyland Olympian ONCL10/1RZ	Alexander RL	B47/30F	1990	Holmeswood Coaches, 2007
68	MJI3768	Leyland Olympian ONCL10/1RZ	Alexander RL	B47/30F	1990	Holmeswood Coaches, 2007
65	MJI2365	Volvo B10B	Plaxton Verde	B51F	1995	Diamond Bus, Birmingham, 2008
66	MJI2366	Volvo B10B	Plaxton Verde	B51F	1995	Diamond Bus, Birmingham, 2008
453	N453VOD	Mercedes-Benz 709D	Alexander Sprint	B23F	1996	Lancashire United, Blackburn, '02
454	N464VOD	Mercedes-Benz 709D	Alexander Sprint	B23F	1996	Lancashire United, Blackburn, '02
861	P861VFG	Scania N113 DRB	East Lancs Cityzen	BC47/31F	1997	Brighton & Hove, 2008
862	P862VFG	Scania N113 DRB	East Lancs Cityzen	BC47/31F	1997	Brighton & Hove, 2009
863	P863VFG	Scania N113 DRB	East Lancs Cityzen	BC47/31F	1997	Brighton & Hove, 2009
864	P864VFG	Scania N113 DRB	East Lancs Cityzen	BC47/31F	1997	Brighton & Hove, 2009
791	R791YFL	LDV Convoy	LDV	M8	1997	Oxford Contract Vehicles, 2004

Travel de Courcey operates several Metrobuses that were new to West Midlands Travel on its routes in the area. Seen in Coventry is A706UOE.
Alan Blagburn

Travel de Courcey has been a keen user of the MAN 14.220 chassis fitted with MCV Evolution bodywork. Around thirty are now in service including 568, AE57CNN, seen on the service from Coventry to Rugby. *Alan Blagburn*

70	MJI2370	Volvo B10M-62	Plaxton Première 350	C53F	1998	North Kent Express, 2003
63	MJI7863	Volvo B10M-62	Plaxton Première 350	C53F	1998	North Kent Express, 2003
72	MJI7472	Volvo B10M-62	Van Hool Alizée HE	C53F	1998	Parks of Hamilton, 2003
61	MJI7861	Volvo B10M-62	Van Hool Alizée HE	C53F	1998	Parks of Hamilton, 2003
23	Y23OXF	Volvo B10M-62	Plaxton Excalibur	C45FT	2001	Oxford Bus Company, 2008
24	Y24OXF	Volvo B10M-62	Plaxton Excalibur	C45FT	2001	Oxford Bus Company, 2008
64	MJI2364	Volvo B12M	Jonckheere Mistral 50	C49FT	2002	Compass Royston, Stockton, '06
71	BW52SVA	Volvo B10M-62	Jonckheere Mistral 45	C53F	2002	Glynn, Ennis, 2002
67	MJI2367	Mercedes-Benz Touro OC500	Mercedes-Benz	C53F	2003	Cummer, Galway, 2007
29	MJI3529	Mercedes-Benz Touro OC500	Mercedes-Benz	C53F	2003	Cummer, Galway, 2007
62	MJI7862	Mercedes-Benz Touro OC500	Mercedes-Benz	C53F	2003	Cummer, Galway, 2007
541	AE54JRO	MAN 14.220	MCV Evolution	N40F	2004	
542	AE54JRU	MAN 14.220	MCV Evolution	N40F	2004	
543	AE54JRV	MAN 14.220	MCV Evolution	N40F	2004	
544	AE54JRX	MAN 14.220	MCV Evolution	N40F	2004	
545	AE54MVW	MAN 14.220	MCV Evolution	N40F	2004	
546	AE54MVX	MAN 14.220	MCV Evolution	N40F	2004	
547	AE54MVY	MAN 14.220	MCV Evolution	N40F	2004	
548	AE54NUH	MAN 14.220	MCV Evolution	N40F	2004	
549	AE54MWA	MAN 14.220	MCV Evolution	N40F	2004	
550	AE05EVD	MAN 14.220	MCV Evolution	N40F	2005	
551	AE55EHR	MAN 14.220	MCV Evolution	N40F	2005	
552	AE55EHP	MAN 14.220	MCV Evolution	N40F	2005	
553	AE55VGM	MAN 14.220	MCV Evolution	N40F	2006	
554	AE06OPJ	MAN 14.220	MCV Evolution	N40F	2006	
555	AE06OPK	MAN 14.220	MCV Evolution	N40F	2006	
556	AE06OPH	MAN 14.220	MCV Evolution	N40F	2006	
557	AE56LWH	MAN 14.220	MCV Evolution	N40F	2006	
558	AE05OVC	MAN 14.220	MCV Evolution	N40F	2005	MCV demonstrator, 2006
73	PN06KTG	MAN 18.360	Marcopolo Viaggio 350	C51FT	2006	
74	PN06TKD	MAN 18.360	Marcopolo Viaggio 350	C51FT	2006	
75	DC07MDC	MAN 18.360	Marcopolo Viaggio III	C53FT	2007	
76	CY08COV	MAN 18.360	Marcopolo Viaggio III	C51FT	2007	
559	AE07DZJ	MAN 14.220	MCV Evolution	N40F	2007	
560	AE07DZP	MAN 14.220	MCV Evolution	N40F	2007	

English Bus Handbook: Notable Independents

Pictured in Coventry, 569, AE57FBC shows the offside styling of the MCV Evolution. *Alan Blagburn*

561	AE07DZR	MAN 14.220	MCV Evolution	N40F	2007	
562	AE07DZX	MAN 14.220	MCV Evolution	N40F	2007	
563	AE07NYR	MAN 14.220	MCV Evolution	N40F	2007	
564	AE57CMX	MAN 14.220	MCV Evolution	N40F	2007	
565	AE57CMY	MAN 14.220	MCV Evolution	N40F	2007	
566	AE57CMZ	MAN 14.220	MCV Evolution	N40F	2007	
568	AE57CNN	MAN 14.220	MCV Evolution	N40F	2007	
569	AE57FBC	MAN 14.220	MCV Evolution	N40F	2008	

Special event vehicles.

239	VWK239	Daimler CVG6DD	Metro-Cammell	B33/27R	1958	Coventry Corporation
246	JHA246L	Leyland Leopard PSU3B/2R	Marshall	BC49F	1973	Midland Red

Previous registrations:

BW52SWA	97CE920, MJI2365	MJI2370	S756CTO
G114NGN	G114NGN, MJI2366	MJI3529	03G1441, BW52OUH
MJI2364	NV51XCS	MJI3751	G369YUR
MJI2365	N604FJO	MJI7472	KSK952, R902JGA
MJI2366	N607FJO	MJI7861	KSK953, R903JGA
MJI2367	03G2481, BW52OZJ	MJI7862	03G2487, BW52OVC
MJI2368	G364YUR, A4HWD	MJI7863	S755CTO

Web: www.traveldecourcey.com

TRUSTYBUS

Galleon Travel - Trustybus

Galleon Travel Ltd, Enterprise House, 17 Perry Road, Harlow, CM18 7PN

	Reg	Chassis	Body	Seating	Year	Origin
	A860KFP	Volvo B10M-61	Plaxton Paramount 3200	C53F	1983	Cape, Marlow, 2008
	J51EDM	Dennis Dart 9m	Plaxton Pointer	B35F	1991	Chesterbus, 2007
	K92BNY	Dennis Dart 9m	Plaxton Pointer	B35F	1993	Ensignbus, Purfleet, 2007
	K58LLG	Dennis Dart 9m	Plaxton Pointer	B35F	1993	Chesterbus, 2007
	K59LLG	Dennis Dart 9m	Plaxton Pointer	B35F	1993	Chesterbus, 2007
	L61PDM	Dennis Dart 9m	Plaxton Pointer	B35F	1993	Chesterbus, 2007
	L63SFM	Dennis Dart 9m	Plaxton Pointer	B35F	1994	Chesterbus, 2007
	L64SFM	Dennis Dart 9m	Plaxton Pointer	B35F	1994	Chesterbus, 2007
	M150HPL	Dennis Dart 9.8m	Plaxton Pointer	B40F	1994	Metrobus, Crawley, 2004
	P844KOT	Dennis Dart SLF 10.7m	UVG Urbanstar	N39F	1996	Low Fell Coaches, 2007
	P10TLS	MAN 11.190	Optare Vecta	B39F	1996	Pink Elephant, Heathrow, 2002
	P501OUG	Optare Excel L1070	Optare	N35F	1996	Diamond Bus, Birmingham, 2007
	P504OUG	Optare Excel L1070	Optare	N35F	1996	Diamond Bus, Birmingham, 2007
	P506OUG	Optare Excel L1070	Optare	N35F	1996	Diamond Bus, Birmingham, 2007
	P508OUG	Optare Excel L1070	Optare	N35F	1996	Diamond Bus, Birmingham, 2007
	R711MEW	Dennis Dart SLF 8.8m	Marshall Capital	N28F	1998	Metroline, Harrow, 2007
518	S518KFL	Dennis Dart SLF 8.8m	Marshall Capital	N27F	1998	Metroline, Harrow, 2007
521	S521KFL	Dennis Dart SLF 8.8m	Marshall Capital	N27F	1998	Metroline, Harrow, 2007
522	S522KFL	Dennis Dart SLF 8.8m	Marshall Capital	N32F	1998	Metroline, Harrow, 2008
528	S528KFL	Dennis Dart SLF 8.8m	Marshall Capital	N32F	1998	Metroline, Harrow, 2008
	X216HCD	Mercedes-Benz Vario 0814	Plaxton Cheetah	C25F	2001	OFJ Connections, Heathrow, 2007
DPL1	LN03AYL	TransBus Dart 9.5m	TransBus Pointer	N37F	2003	
DPL2	LN03AYM	TransBus Dart 9.5m	TransBus Pointer	N37F	2003	
	YN53CEX	Scania K114 EB4	Irizar Century	C49FT	2003	Scania demonstrator, 2004
	YN53GHH	Scania K124 EB6	VDL Berkhof Axial 100	C55/19DT	2004	
DPL3	EU04BZY	TransBus Dart 9.5m	TransBus Pointer	N37F	2004	
	WX05NFY	Ayats Atlantis A24-13	Ayats	C59/16FT	2005	
DML6	AE55VGD	ADL Dart 4 10.7m	MCV Evolution	N41F	2006	
DML	AE55MVL	ADL Dart 4 10.7m	MCV Evolution	N43F	2006	Central Connect, Birmingham, '08
	YN06CKY	Scania K114 EB4	Irizar PB	C49FT	2006	
DMA6	AE06HBU	ADL Dart 4 9.5m	MCV Evolution	N33F	2006	
DMA8	AE06HBX	ADL Dart 4 9.5m	MCV Evolution	N33F	2006	
DES9	LK07GTF	ADL Dart 4 8.9m	ADL Enviro 200	N29F	2007	
	YN07LFT	Scania K340 EB4	Irizar PB	C49FT	2007	
DES	KX58GTF	ADL Dart 4 8.9m	ADL Enviro 200	N29F	2008	

Previous registrations:

A860KFP	A941VMH, TFJ694	K92BNY	K92BNY, 392MBF

Mutton Lane in Potters Bar is the location for this view of Marshall-bodied R711MEW, a Dart new to Metroline.
Richard Godfrey

TYRER BUS

Tyrer Tours Ltd , 16 Kirby Road, Lomeshaye Ind Est, Nelson, BB7 6RS

F387GVO	Leyland Lion LDTL11/1R	East Lancs	B47/31F	1988	City of Nottingham, 2003
L944HTM	Mercedes-Benz 811D	Plaxton Beaver	B31F	1994	Gibson Direct, Renfrew, 2003
L313AUT	Mercedes-Benz 709D	Alexander Sprint	B25F	1994	Arriva Fox County, 2002
UDZ7580	Mercedes-Benz 709D	Plaxton Beaver	B25F	1994	Ulsterbus, 2004
N748YVR	Mercedes-Benz 711D	Plaxton Beaver	B27F	1995	First PMT, 2000
R849JGD	Mercedes-Benz Vario O814	Mellor	B33F	1998	Coakley Bus, Motherwell, 2002
W556JVV	Dennis Dart SLF 8.8m	Plaxton Pointer MPD	N29F	2000	A1A, Birkenhead, 2005
W563JVV	Dennis Dart SLF 8.8m	Plaxton Pointer MPD	N29F	2000	A1A, Birkenhead, 2005
KX03HZC	TransBus Dart 8.8m	TransBus Mini Pointer	N29F	2003	
KX03HZD	TransBus Dart 8.8m	TransBus Mini Pointer	N29F	2003	
MX03YCK	Optare Solo M850	Optare	N29F	2003	*Operated for Lancashire CC*
MX03YCL	Optare Solo M850	Optare	N29F	2003	*Operated for Lancashire CC*
YJ54UBL	Optare Solo M780	Optare	N23F	2004	*Operated for Lancashire CC*
MX05OTM	Optare Solo M920	Optare	N33F	2005	Speedwell, Hyde, 2008
YJ05WCM	Optare Solo M780	Optare	N23F	2005	*Operated for Lancashire CC*
YJ05WCN	Optare Solo M780	Optare	N23F	2005	*Operated for Lancashire CC*
YJ05WCO	Optare Solo M780	Optare	N23F	2005	*Operated for Lancashire CC*
PN05RNU	Optare Alero	Optare	N13C	2005	*Operated for Lancashire CC*

Tyrer Bus operates several Optare Solo buses, several of which are owned by Lancashire County Council. Recently, Tyrer Bus has commenced operating services into Bolton, with MX57CCA seen on the Hall-I'-Th'-Wood and Tonge Moor routes. *Alan Blagburn*

One of a pair of Mini Pointer Darts from Dennis with Tyrer Bus is W556JBV, seen here passing the bus station in Burnley. A further pair by Dennis' successor, TransBus, is also used. *Dave Heath*

YJ56AOW	Optare Solo M780	Optare	N24F	2006
YJ56AOX	Optare Solo M780	Optare	N24F	2006
YJ56AOZ	Optare Solo M780	Optare	N24F	2006
MX57CBO	Optare Solo M880	Optare	N28F	2007
MX57CBU	Optare Solo M880	Optare	N28F	2007
MX57CBV	Optare Solo M880	Optare	N28F	2007
MX57CBY	Optare Solo M880	Optare	N28F	2007
MX57CCA	Optare Solo M880	Optare	N28F	2007
MX08DGE	Optare Solo M880	Optare	N28F	2008
MX08DGO	Optare Solo M880	Optare	N28F	2008
MX08DGU	Optare Solo M880	Optare	N28F	2008
MX08DGV	Optare Solo M880	Optare	N28F	2008
MX08DHA	Optare Solo M950	Optare	N32F	2008
MX08DHK	Optare Solo M950	Optare	N32F	2008
MX08DHM	Optare Solo M950	Optare	N32F	2008
MX08DHN	Optare Solo M950	Optare	N32F	2008
MX08MYM	Optare Solo M950	Optare	N29F	2008
MX08MYO	Optare Solo M950	Optare	N29F	2008
MX08MYT	Optare Solo M950	Optare	N29F	2008

Depots: Lomeshaye Industrial Estate, Nelson and Cuthbert Street, Bolton

English Bus Handbook: Notable Independents

UNŌ

Unō Buses Ltd; Universitybus Ltd, Gypsy Moth Avenue, Hatfield, AL10 9BS

100	KF52NBN	Dennis Dart SLF	Plaxton Pointer 2	N33F	2002	
101	KF52NBM	Dennis Dart SLF	Plaxton Pointer 2	N33F	2002	
102	FM52GEY	Dennis Dart SLF 8.8m	Plaxton Pointer MPD	N29F	2003	Centrebus, Dunstable, 2008
103	FM52GFA	Dennis Dart SLF 8.8m	Plaxton Pointer MPD	N29F	2003	Centrebus, Dunstable, 2008
104	KC03PGK	TransBus Dart 8.8m	TransBus Mini Pointer	N29F	2003	Centrebus, Dunstable, 2008
105	KC03PGU	TransBus Dart 8.8m	TransBus Mini Pointer	N29F	2003	Centrebus, Dunstable, 2008
111	KE03TWC	TransBus Dart	TransBus Pointer	N37F	2003	
112	KE03TWA	TransBus Dart	TransBus Pointer	N37F	2003	
113	KC03OSE	TransBus Dart	TransBus Pointer	N37F	2003	
114	KC03OSF	TransBus Dart	TransBus Pointer	N37F	2003	
115	KE53LMY	TransBus Dart	TransBus Pointer	N37F	2003	
116	V895DNB	Dennis Dart SLF	Plaxton Pointer	N39F	1999	Munro's of Jedburgh, 2003
117	V896DNB	Dennis Dart SLF	Plaxton Pointer	N39F	1999	Munro's of Jedburgh, 2003
118	KE04UMA	TransBus Dart	TransBus Pointer	N37F	2004	
119	KE04UMB	TransBus Dart	TransBus Pointer	N37F	2004	
120	KE04UMC	TransBus Dart	TransBus Pointer	N37F	2004	
121	KE04UMD	TransBus Dart	TransBus Pointer	N37F	2004	
144	K434OKH	Dennis Dart 9m	Plaxton Pointer	B34F	1992	Centrebus, Dunstable, 2008
146	M146VVS	Dennis Dart 9.8m	Wright HandyBus	B42F	1995	
148	M148VVS	Dennis Dart 9.8m	Wright HandyBus	B42F	1995	
151	FJ54MMO	TransBus Dart	Caetano Compass	N36F	2004	
152	FJ54MMU	TransBus Dart	Caetano Compass	N36F	2004	
153	UN55UNO	TransBus Dart	Caetano Compass	N38F	2005	
154	UO55UNO	TransBus Dart	Caetano Compass	N38F	2005	
201	YJ53VBN	DAF DB250	Optare Spectra	N50/27F	2003	
202	YJ53VBO	DAF DB250	East Lancs Myllennium	N45/29F	2003	
203	UH55UNO	Scania OmniDekka N94UD	East Lancs	N47/33F	2005	
204	UJ55UNO	Scania OmniDekka N94UD	East Lancs	N47/33F	2005	
205	UL55UNO	Scania OmniDekka N94UD	East Lancs	N47/33F	2005	
206	PN08SVZ	Scania N230 UD	East Lancs Olympus	N51/29F	2008	
207	PN08SWF	Scania N230 UD	East Lancs Olympus	N51/29F	2008	
250	N429FKK	Volvo Olympian	Alexander RH	B47/27D	1996	Dublin Bus, 2007
251	N430FKK	Volvo Olympian	Alexander RH	B47/27D	1996	Dublin Bus, 2007
252	N431FKK	Volvo Olympian	Alexander RH	B47/27D	1996	Dublin Bus, 2007
301	BX56VSJ	Mercedes-Benz Citaro O530	Mercedes-Benz	N27D	2006	
302	BX56VSK	Mercedes-Benz Citaro O530	Mercedes-Benz	N27D	2006	
303	BX56VSL	Mercedes-Benz Citaro O530	Mercedes-Benz	N27D	2006	
304	BX56VSM	Mercedes-Benz Citaro O530	Mercedes-Benz	N27D	2006	
305	BX56VSN	Mercedes-Benz Citaro O530	Mercedes-Benz	N27D	2006	
306	BV57VHE	Mercedes-Benz Citaro O530	Mercedes-Benz	N42F	2007	
307	BV57VHF	Mercedes-Benz Citaro O530	Mercedes-Benz	N42F	2007	
308	BV57VHG	Mercedes-Benz Citaro O530	Mercedes-Benz	N42F	2007	
309	BV57VHH	Mercedes-Benz Citaro O530	Mercedes-Benz	N42F	2007	
310	BV57VHJ	Mercedes-Benz Citaro O530	Mercedes-Benz	N42F	2007	
311	BN58BJJ	Mercedes-Benz Citaro O530	Mercedes-Benz	N42F	2008	
312	BN58BJK	Mercedes-Benz Citaro O530	Mercedes-Benz	N42F	2008	
313	BN58BJO	Mercedes-Benz Citaro O530	Mercedes-Benz	N42F	2008	
351	YN56NRY	Scania OmniCity N94 UB	Scania	N42F	2007	
352	YN56NRZ	Scania OmniCity N94 UB	Scania	N42F	2007	
353	YN57FZA	Scania OmniCity K230 UB	Scania	N45F	2007	
354	YN57FZB	Scania OmniCity K230 UB	Scania	N45F	2007	
401	YJ05XNB	Optare Solo M880	Optare	N25F	2005	Centrebus, Dunstable, 2008
402	YJ05CCY	Optare Solo M880	Optare	N25F	2005	Centrebus, Dunstable, 2008
403	MX56BZY	Optare Solo M880	Optare	N27F	2007	Centrebus, Dunstable, 2008
404	YJ58PHN	Optare Solo M880 SL	Optare	N28F	2008	
405	YJ58PHO	Optare Solo M880 SL	Optare	N28F	2008	
429	Y129TBF	Mercedes-Benz Vario 0814	Plaxton Beaver 2	B31F	2001	Centrebus, Dunstable, 2008
451	L951NBH	Mercedes-Benz 811D	Plaxton Beaver	B31F	1994	Centrebus, Dunstable, 2008
458	M458UUR	Mercedes-Benz 811D	Plaxton Beaver	B31F	1995	Centrebus, Dunstable, 2008
563	V563JBH	Dennis Dart SLF 10.7m	Plaxton Pointer 2	N40F	1999	Centrebus, Dunstable, 2008
564	V564JBH	Dennis Dart SLF 10.7m	Plaxton Pointer 2	N40F	1999	Centrebus, Dunstable, 2008
565	V565JBH	Dennis Dart SLF 10.7m	Plaxton Pointer 2	N40F	1999	Centrebus, Dunstable, 2008
567	W567XRO	Dennis Dart SLF 10.7m	Plaxton Pointer 2	N40F	2000	Centrebus, Dunstable, 2008
568	W568XRO	Dennis Dart SLF 10.7m	Plaxton Pointer 2	N40F	2000	Centrebus, Dunstable, 2008

UNŌ operates three Scania OmniDekka buses alongside a pair of the later Scania chassis with East Lancs body, the Olympus, including 203, UH55UNO. The OmniDekka was a collaborative product from Scania that has been replaced by the integral Scania OmniCity double-deck. However, Scania is still supplying chassis to the British market for Alexander Dennis and East Lancs (now Optare) to complete. *Tony Wilson*

569	W569XRO	Dennis Dart SLF 10.7m	Plaxton Pointer 2	N40F	2000	Centrebus, Dunstable, 2008
571	W571XRO	Dennis Dart SLF 10.7m	Plaxton Pointer 2	N40F	2000	Centrebus, Dunstable, 2008
572	W572XRO	Dennis Dart SLF 10.7m	Plaxton Pointer 2	N40F	2000	Centrebus, Dunstable, 2008
573	W573XRO	Dennis Dart SLF 10.7m	Plaxton Pointer 2	N40F	2000	Centrebus, Dunstable, 2008
574	W574XRO	Dennis Dart SLF 10.7m	Plaxton Pointer 2	N40F	2000	Centrebus, Dunstable, 2008
575	W575XRO	Dennis Dart SLF 10.7m	Plaxton Pointer 2	N40F	2000	Centrebus, Dunstable, 2008
576	W576XRO	Dennis Dart SLF 10.7m	Plaxton Pointer 2	N40F	2000	Centrebus, Dunstable, 2008
577	W577XRO	Dennis Dart SLF 10.7m	Plaxton Pointer 2	N40F	2000	Centrebus, Dunstable, 2008
578	W578XRO	Dennis Dart SLF 10.7m	Plaxton Pointer 2	N40F	2000	Centrebus, Dunstable, 2008
647	P647SBH	Dennis Dart SLF	Wright Crusader	N38F	1997	
648	R648VMJ	Dennis Dart SLF	Wright Crusader	N41F	1998	
649	R649VBM	Dennis Dart SLF	Wright Crusader	N41F	1997	
650	R650VBM	Dennis Dart SLF	Wright Crusader	N41F	1997	
651	R651VBM	Dennis Dart SLF	Wright Crusader	N41F	1997	
652	R652VBM	Dennis Dart SLF	Wright Crusader	N41F	1997	
653	R653VBM	Dennis Dart SLF	Wright Crusader	N41F	1997	
654	R654VBM	Dennis Dart SLF	Wright Crusader	N41F	1997	
655	YDZ2082	Dennis Dart SLF	Wright Crusader	N39F	1998	Wright demonstrator, 1998
664	P664PNM	Dennis Dart SLF	Wright Crusader	N41F	1996	
665	P665PNM	Dennis Dart SLF	Wright Crusader	N41F	1996	
667	P667PNM	Dennis Dart SLF	Wright Crusader	N41F	1996	
668	P668PNM	Dennis Dart SLF	Wright Crusader	N41F	1996	
773	X773EVS	DAF SB120	Wright Cadet	N41F	2000	
774	X774EVS	DAF SB120	Wright Cadet	N41F	2000	
776	X776EVS	DAF SB120	Wright Cadet	N41F	2000	
777	KE51VJD	DAF SB120	Wrightbus Cadet	N39F	2001	
778	KE51VJC	DAF SB120	Wrightbus Cadet	N39F	2001	

Previous registrations:

N429FKK	96D269		V895DNB	99D72545
N430FKK	96D277		V896DNB	99D72547
N431FKK	96D276			

Web: www.universitybus.co.uk

VELVET

Black Velvet Travel Ltd, Suite A Binning House, 4a High Street, Eastleigh, SO50 5LA

302	F302MYJ	Volvo Citybus B10M-50	Northern Counties	BC43/33F	1989	Stagecoach, 2008
303	F303MYJ	Volvo Citybus B10M-50	Northern Counties	BC43/33F	1989	Stagecoach, 2008
309	F309MYJ	Volvo Citybus B10M-50	Northern Counties	BC43/33F	1989	Stagecoach, 2008
507	V7GMT	DAF SB220	East Lancs Myllennium	N35F	1999	Go-Ahead London, 2008
511	V11GMT	DAF SB220	East Lancs Myllennium	N35F	1999	Go-Ahead London, 2008
512	V12GMT	DAF SB220	East Lancs Myllennium	N35F	1999	Go-Ahead London, 2008
514	V14GMT	DAF SB220	East Lancs Myllennium	N35F	1999	Go-Ahead London, 2008
841	J841TSC	Leyland Olympian ON2R56C13Z4	Alexander RH	B51/34F	1991	Lothian Buses, 2008
843	J843TSC	Leyland Olympian ON2R56C13Z4	Alexander RH	B51/34F	1991	Lothian Buses, 2008
851	J851TSC	Leyland Olympian ON2R56C13Z4	Alexander RH	B51/34F	1991	Lothian Buses, 2008

Depot: Barton Park Industrial Estate, Eastleigh

Black Velvet Travel operates just three vehicle types. Four of the then new low-floor version of the DAF SB220 built with East Lancs Myllennium bodywork and new to London Central in 1999 are now employed. Number 514, V14GMT, was pictured at Hedge End in April. *Richard Godfrey*

WARDLE

Wardle Transport, 105 Ford Green Road, Smallthorne, Stoke-on-Trent, ST6 1NT

1	K345OFM	Leyland Olympian ON3R42C18Z4	Alexander RH	BC49/37F	1993	Citybus, Hong Kong, 2004
2	K346OFM	Leyland Olympian ON3R42C18Z4	Alexander RH	BC49/37F	1993	Citybus, Hong Kong, 2004
3	K347OFM	Leyland Olympian ON3R42C18Z4	Alexander RH	BC49/37F	1993	Citybus, Hong Kong, 2004
4	B13STA	Volvo B10M-62	Jonckheere Deauville	C53F	1995	Hawkins, Dewsbury, 2004
5	UKZ5466	Mercedes-Benz Vario O814	Plaxton Cheetah	C33F	2001	Rennies, Dunfermline, 2003
7	DG52TYP	Dennis Dart SLF 8.8m	Plaxton Pointer MPD	N29F	2002	D&G, Adderley Green, 2006
8	DG52TYS	Dennis Dart SLF 8.8m	Plaxton Pointer MPD	N29F	2002	D&G, Adderley Green, 2006
9	DG52TYT	Dennis Dart SLF 8.8m	Plaxton Pointer MPD	N29F	2002	D&G, Adderley Green, 2006
10	DG52TYU	Dennis Dart SLF 8.8m	Plaxton Pointer MPD	N29F	2002	D&G, Adderley Green, 2006
11	DA51XTC	Dennis Dart SLF 8.8m	Plaxton Pointer MPD	N29F	2002	D&G, Adderley Green, 2006
12	W216JND	Mercedes-Benz 312D	Concept	M12	2000	Janeway, Wythenshaw, 2002
14	IHZ8821	Mercedes-Benz Sprinter 413cdi	Excel	C16F	2001	Thorpe, Gosforth, 2007
15	P175UAD	Mercedes-Benz 508D	UVG	M5	1996	
16	REZ8516	Mercedes-Benz Sprinter 411cdi	Mercedes-Benz	M5	2002	
17	NEZ9506	Mercedes-Benz Sprinter 411cdi	Mercedes-Benz	M12	2002	
18	OUI3925	Mercedes-Benz Sprinter 413cdi	Concept	M16	2001	Halls, Stafford, 2002
19	DA51XTE	Dennis Dart SLF 8.8m	Plaxton Pointer MPD	N29F	2002	D&G, Adderley Green, 2007
20	S2WMS	Dennis Dart SLF 10.7m	Plaxton Pointer 2	N39F	1999	Worth, Enstone, 2007
21	ELZ2362	Mercedes-Benz Sprinter 411cdi	Minibus Options	M16	2000	Ellesmere Port Community, 2004
22	BV51ENL	Mercedes-Benz Sprinter 413cdi	Excel	M16	2002	Brookfleet, Blackheath, 2003
23	YX53CYZ	Mercedes-Benz Sprinter 413cdi	Onyx	M5	2003	
25	OEZ2159	Mercedes-Benz Sprinter 411cdi	Stanway	M11	2001	
26	X2PCC	Mercedes-Benz Vario O814	Plaxton Beaver 2	BC33F	2002	PC Coaches, Lincoln, 2007
27	XIL8793	Mercedes-Benz 412D	Olympus	M14	1999	First West Yorkshire, 2004
28	RUI2486	Mercedes-Benz Sprinter 411cdi	Advanced	M4	2002	Edsor, Billinghurst, 2004
29	FY03WZV	Mercedes-Benz Sprinter 413cdi	Ferqui	C16F	2003	McNee, Ratho Station, 2008
30	PCZ6034	Mercedes-Benz Sprinter 413cdi	Onyx	M16	2001	Plant, Cheadle, 2004
31	FHZ2128	Mercedes-Benz 412D	Olympus	M14	2000	First West Yorkshire, 2004
32	IJZ2331	Mercedes-Benz 612D	Mellor	B16F	1998	Nightingale, Maidenhead, 2004
33	OLZ7276	Mercedes-Benz 308D	Frank Guy	M4	1995	Derbyshire CC, 2005
34	S260JUG	Mercedes-Benz 614D	UVG	B14F	1998	Leeds City Council, 2005
35	FCZ3413	Mercedes-Benz Sprinter 416cdi	Mercedes-Benz	M16	2000	Tinsley, Belfast, 2005
36	SF04RGY	Mercedes-Benz Sprinter 413cdi	Onyx	M16	2004	Osman, Leyton, 2005

Wardle's fleet mostly comprises minibuses or smaller midibuses as they are one of the first 'City Rider On Call' providers in the UK, a Government initiative to encourage greater use of public transport. Seen leaving Hanley bus station is 7, DG52TYP. *Alan Blagburn*

Wardle's local service is designed to make public transport more accessible and flexible, offering daily non-stop bus services on fixed routes with the ability to divert into designated areas (off route) to drop off and collect passengers. Collections are arranged via a dedicated call centre based in the city. Mini Pointer 19, DA51XTE, came from fellow local operator D&G. *Alan Blagburn*

37	V939ECU	Mercedes-Benz 512D	Crest	M4	1999	St Clare's Hospice, 2005
38	YX55BGY	Mercedes-Benz Sprinter 413cdi	Onyx	M16	2005	
39	OKZ9847	Mercedes-Benz 410D	Mellor	M16	1999	Burnt Tree Hire, 2005
40	HX55EZF	Mercedes-Benz Sprinter 416cdi	Driveline	M16	2005	
41	W478EUB	Mercedes-Benz 614D	Oughtred & Harrison	B13F	2000	Leeds City Council, 2007
42	W482EUB	Mercedes-Benz 614D	Oughtred & Harrison	B10F	2000	Leeds City Council, 2007
43	W483EUB	Mercedes-Benz 614D	Oughtred & Harrison	B13F	2000	Leeds City Council, 2007
44	S268JUG	Mercedes-Benz 614D	UVG	B14F	1998	Leeds City Council, 2006
45	YX56DJD	Mercedes-Benz Sprinter 416cdi	Yorkshire Conversions	M16	2006	
46	YX56DJE	Mercedes-Benz Sprinter 416cdi	Yorkshire Conversions	M16	2006	
47	YU06KFA	Mercedes-Benz Sprinter 413cdi	Onyx	M16	2006	
48	X613HNV	Mercedes-Benz Sprinter 311cdi	Mercedes-Benz	M5	2000	
49	KY51SXD	Mercedes-Benz Sprinter 311cdi	Mercedes-Benz	M5	2001	
50	T20CCH	Volvo B10M-62	Berkhof Axial 50	C50F	1999	Len Wright, Watford, 2007
51	DF02OPR	Mercedes-Benz Vito 108cdi	Mercedes-Benz	M7	2002	St Mary's Hospital, 2007
	WBZ8737	Volvo Olympian	East Lancs	B49/35F	1996	City of Nottingham, 2008
	VLZ9237	Volvo Olympian	East Lancs	B49/35F	1998	City of Nottingham, 2008
	MNZ1138	Leyland Olympian ONCL10/1RZ	Alexander RH	B45/30F	1988	Central Connect, Birmingham, '08
	L754VNL	Dennis Dart 9.8m	Plaxton Pointer	B40F	1993	Bankfoot, Perth, 2008
	M113BMR	Dennis Dart 9.8m	Plaxton Pointer	B40F	1994	Bankfoot, Perth, 2008
	BU54ALL	Optare Solo M780	Optare	N23F	2004	*operated for Stoke-on-Trent CC*
	YK54AWH	Mercedes-Benz Sprinter 411cdi	Oughtred & Harrison	M16	2005	Bradford Community, 2008
	YK54AWJ	Mercedes-Benz Sprinter 411cdi	Oughtred & Harrison	M16	2005	Bradford Community, 2008
	YJ05JXP	Optare Solo M850	Optare	N27F	2005	*operated for Stoke-on-Trent CC*
	R107TKO	Mercedes-Benz Vario O810	Plaxton Beaver 2	B27F	1998	Bysiau Cwm Taf, Whitland, 2008
	WT08BUS	Optare Versa V1100	Optare	N38F	2008	
	WT58BUS	Optare Versa V1100	Optare	N38F	2008	
	WT58SOT	Optare Versa V1100	Optare	N38F	2008	

Previous registrations:

B13STA	M314KRY, 95WD1	OUI3925	Y181KNE
ELZ2362	X798ULG	PCZ6034	DK51LTU
FHZ2128	V903WUB	REZ8516	DG52WJU
IHZ8821	Y2UTO, Y668RVY	RUI2486	LV02ODW
IJZ2331	R876DCA	UKZ5466	Y477TSU
MNZ1138	E219WBG	VLZ9237	R468RRA
NEZ9506	DG52WJY	WBZ8737	P490CVO
OEZ2159	Y569HPK	XIL8793	T59BUB
OKZ9847	T160RWK	X613HNV	X613HNV, UXT629
OLZ7276	N731BWG		

WEARDALE

Weardale Motor Services Ltd, Shittlehope Burn Garage, Stanhope, Bishop Auckland, DL13 2YQ

	Registration	Chassis	Body	Config	Year	Notes
	OGR625T	Leyland Leopard PSU3E/4R	Plaxton Supreme IV	C53F	1979	
u	A633BCN	MCW Metrobus DR102/43	MCW	B46/31F	1984	Go North East, 2006
	C778SFS	Leyland Olympian ONTL11/2R	Eastern Coach Works	B51/36F	1985	Jay, Greengairs, 2004
	C779SFS	Leyland Olympian ONTL11/2R	Eastern Coach Works	B51/36F	1985	Jay, Greengairs, 2004
u	C787SFS	Leyland Olympian ONTL11/2R	Eastern Coach Works	B51/36F	1985	Jay, Greengairs, 2004
	C793SFS	Leyland Olympian ONTL11/2R	Eastern Coach Works	B51/36F	1986	Jay, Greengairs, 2004
	C777OCN	MCW Metrobus DR102/55	MCW	B46/31F	1986	Go North East, 2006
	GIB8666	Volvo B10M-61	Plaxton Paramount 3200 II	S70F	1986	
	NIL9777	Leyland Tiger TRCTL11/2RH	Plaxton Paramount (1985)	C53F	1986	Anchor Training, 2000
	D6WMS	Leyland Olympian ONLXB/2RZ	Alexander RL	B51/36F	1988	Stagecoach, 2006
	RIB6199	Leyland Olympian ONLXB/2RZ	Alexander RL	B51/38F	1988	Stagecoach, 2006
	G16WMS	Leyland Olympian ONCL10/1RZ	Alexander RH	B45/29F	1989	Go North East, 2006
	G18WMS	Leyland Olympian ONCL10/1RZ	Alexander RH	B45/29F	1989	Go North East, 2006
	D16WMS	Leyland Olympian ONCL10/2RZ	Alexander RH	B55/34F	1989	Lothian Buses, 2008
	G640EVV	Leyland Olympian ONLXB/2RZ	Alexander RL	B51/34F	1989	Stagecoach, 2006
	C6WMS	Leyland Olympian ON2R50C13Z4	Alexander RH	B47/31F	1991	Dublin Bus, 2006
	E6WMS	Leyland Olympian ON2R50C13Z4	Alexander RH	B47/31F	1991	Dublin Bus, 2006
	J640CEV	Leyland Olympian ON2R50C13Z4	Alexander RH	B47/31F	1991	Dublin Bus, 2006
	J600WMS	Leyland Olympian ON2R56C13Z4	Alexander RH	B51/34F	1991	Lothian Buses, 2007
	CHZ1751	DAF SB2700	Van Hool Alizée	C51FT	1992	Gardiners, Spennymoor, 2008
	L666WMS	Leyland Olympian ON3R56C18Z5	Alexander RH	B57/45F	1993	Citybus, Hong Kong, 2006
u	M6WMS	Mercedes-Benz 709D	Plaxton Beaver	B25F	1994	
	GIB6197	DAF SB3000	Ikarus Blue Danube	C53F	1998	Alfa, Chorley, 2007
	W6WMS	DAF SB220	Ikarus Citibus	N44F	2000	
	W666WMS	DAF SB220	Ikarus Citibus	N44F	2000	

High capacity buses initially used in Hong Kong are chosen by some operators who specialise in school work. Seen between journeys, L666WMS was new to Stagecoach subsidiary, Citybus in Hong Kong and seats 102 following refurbishment. *David Longbottom*

A pair of low-floor DAF SB220s with Ikarus Citibus bodies joined Weardale in 2000. W6WMS shows the special fleet name applied to this pair as it heads for Stanhope. *David Longbottom*

V6WMS	Mercedes-Benz Vario O814	Plaxton Beaver	B31F	2000	Brijan, Bishops's Waltham, 2006
P6WMS	Mercedes-Benz 614D	Crest	C22F	2000	LB Hackney, 2006
Y6WMS	Mercedes-Benz Sprinter 411cdi	Koch	N13	2002	*Operated for Durham CT*
P666WMS	Mercedes-Benz Sprinter 411cdi	Koch	N13	2002	*Operated for Durham CT*
GO05WMS	Volvo B12M	Plaxton Paragon	C53F	2005	
GIB976	Scania K114 EB4	VDL Berkhof Axial 50	C55FT	2005	
LG06HSF	LDV Maxus	LDV	M9	2006	van, 2008
YN06FVR	Neoplan Euroliner N316 SHD	Neoplan	C61FT	2006	
YN56BGV	Neoplan Euroliner N316 SHDL	Neoplan	C61FT	2007	
YN07CCC	Neoplan Tourliner N2216 SHD	Neoplan	C59FT	2007	
YJ07EGC	Optare Solo M950	Optare	N33F	2007	
MX07NTJ	Optare Solo M880	Optare	N28F	2007	Britannia, Bournemouth, 2008
YN07KHC	Irisbus EuroRider 397E.12.31A	Plaxton Panther	C57F	2007	
YN57AAY	Irisbus EuroRider 397E.12.43A	Beulas Cygnus	C55FT	2007	
YN57AAF	Irisbus EuroRider 397E.12.43A	Beulas Cygnus	C59FT	2008	
SF57JSX	Mercedes-Benz Sprinter 616cdi	Koch	N25F	2008	
YN08DME	Mercedes-Benz Vario O816	Plaxton Cheetah	C33F	2008	

Special event vehicle:

AEF680A	Leyland Titan PD2/10	Roe	B37/28R	1958	Stanhope MS

Previous registrations:

C6WMS	91D10112, J639CEV	GO03WMS	LK03GXX
CHZ1751	J799KHD	J600WMS	J836TSC
D6WMS	F624MSL	J640CEV	91D10113
E6WMS	91D1060, H592PVW	L666WMS	FS5611(HK), L662SNO
G16WMS	G676TCN	P6WMS	W993VGU
G18WMS	G677TCN	RIB6199	F622MSL
GIB6197	R119GNW	V6WMS	V505FSF, V66BJT

Depots: Shittlehope Burn, Stanhope; Frosterley; Copeland Lane, Crook.

WEAVAWAY

S Weaver, 169 Main Street, New Greenham Park, Newbury, RG19 6HN

	NMW329X	Bedford YNT	Duple Dominant IV	C53F	1982	preservation, 2007
	OIL4201	Leyland Olympian ONTL11/2R	Eastern Coach Works	B50/31D	1983	Bennett's, Gloucester, 2008
	OFS702Y	Leyland Olympian ONTL11/2R	Eastern Coach Works	B50/31D	1983	Bennett's, Gloucester, 2008
ND	KW02DSE	ADL E300	ADL Enviro 300	N44F	2002	Minsterley Motors, 2008
	BKZ70	Bova Futura FHD 12.340	Bova	S70F	2003	
	NBZ70	Bova Futura FHD 12.340	Bova	S70F	2003	
	NDZ70	Bova Futura FHD 12.340	Bova	S70F	2003	
	X70OXF	Bova Futura FHD 12.340	Bova	S70F	2003	
CT	CW51TOP	Bova Futura FHD 12.340	Bova	S70F	2003	
OX	CW52TOP	Bova Futura FHD 12.340	Bova	S70F	2003	
OX	WA05DFG	VDL Bova Futura FHD 13.340	VDL Bova	C53FT	2005	
OX	WA05DFJ	VDL Bova Futura FHD 13.340	VDL Bova	C53FT	2005	
WW	OU05AVB	Volvo B9TL	East Lancs Nordic	BC63/39F	2005	
WW	OU05AVD	Volvo B9TL	East Lancs Nordic	BC63/39F	2005	
WW	OU05AVY	Volvo B9TL	East Lancs Nordic	BC63/39F	2005	
WW	OU05AWJ	Volvo B9TL	East Lancs Nordic	BC63/39F	2005	
WW	OU05KKB	Volvo B9TL	East Lancs Nordic	BC63/39F	2005	
WW	OU05KLA	Volvo B9TL	East Lancs Nordic	BC63/39F	2005	
OX	YN06CYX	Mercedes-Benz Vario O814	Plaxton Cheetah	C33F	2006	
CT	WA06GSU	VDL Bova Magiq XHD122.340	VDL Bova	C53FT	2006	
CT	WA06GSV	VDL Bova Magiq XHD122.340	VDL Bova	C53FT	2006	
CT	CW56TOP	VDL Bova Magiq XHD122.340	VDL Bova	C32FT	2006	
WW	WA56ENV	VDL Bova Magiq XHD122.340	VDL Bova	C32FT	2006	
CT	CW53TOP	Scania K114EB4	VDL Berkhof Axial 50	C57FT	2006	
CT	CW54TOP	Scania K340 EB4	VDL Berkhof Axial 50	C51FT	2007	
CT	CW55TOP	Scania K340 EB4	VDL Berkhof Axial 50	C51FT	2007	

One of the Weavaway high capacity double-decks from East Lancs with the Nordic style of body. These buses feature coach seating for over a hundred passengers. Seen in the all-white scheme os OU05AVB.
Ken Landsdown

Weavaway's operation in Newbury uses the Newbury & District name. AE57FBB is one of three MAN buses used on this network, two of which have MCV Evolution bodywork, the third carrying the Plaxton product.
Richard Godfrey

ND	AE57FBA	MAN 14.220	MCV Evolution	NC40F	2007
ND	AE57FBB	MAN 14.220	MCV Evolution	NC40F	2007
ND	KX08OLO	MAN 12.240	Plaxton Centro	N38F	2008
WW	KX08OMB	Volvo B12B	Plaxton Panther	C49FT	2008
WW	KX08OMC	Volvo B12B	Plaxton Panther	C49FT	2008
WW	KX08OMD	Volvo B12B	Plaxton Panther	C49FT	2008
OX	KX08OME	Volvo B12B	Plaxton Panther	C53F	2008
OX	KX08OMF	Volvo B12B	Plaxton Panther	C53F	2008
OX	KX08OMG	Volvo B12B	Plaxton Panther	C53F	2008
ND	YX58FPP	Enterprise Plasma EB01	Plaxton Primo	N28F	2008
ND	RX58KHF	ADL E300	ADL Enviro 300	NC53F	2008
ND	RX58KHG	ADL E300	ADL Enviro 300	NC53F	2008
ND	RX58KHH	ADL E300	ADL Enviro 300	NC53F	2008

Previous registrations:

BKZ70	YJ03GXP	NDZ70	YJ03GXN
CW51TOP	WA03UDY, NBZ70	OIL4201	OFS701Y
CW52TOP	WA03UDZ, NDZ70	WA06GSU	WA06GSU, CW54TOP
CW53TOP	WA56EZW	WA06GSV	WA06GSV, CW55TOP
CW54TOP	YN07LKJ	WA56ENV	WA56ENV, CW56TOP
CW55TOP	TN07LKK	X70OXF	YJ03GXR
CW56TOP	WA56ENW	YN06CYX	YN06CYX, CW53TOP
LUI9693	H251JJT		
NBZ70	YJ03GXM		

Depots: Station Road, Culham; New Greenham Park, Newbury and Great Knollys Street, Reading
Liveries: Countrywide Top Travel (CT); Haywards (HS); Newbury & District (ND); Oxford International (OX) and Weavaway (WW),

WEST END TRAVEL

West End Travel - Rutland Travel

Romdrive Ltd, 17 Norfolk Drive, Melton Mowbray, LE13 0AJ

RT	PIL7971	Volvo B58-61	Plaxton Supreme IV	C53F	1981	Tucker, Melton Mowbray, 1999
RT	C366LFF	Scania K112 CRS	Plaxton Bustler	BC49F	1986	Tucker, Melton Mowbray, 1999
RT	PIL7972	Leyland Tiger TRCL10/3ARZA	Plaxton Paramount 3200 III	C53F	1989	Duff, Sutton-on-the-Forest, 2001
RT	WET1K	Volvo B10M-60	Plaxton Paramount 3500 III	C53F	1991	George Young, Ross-on-Wye, '04
WE	NEL1F	Mercedes-Benz 709D	Alexander Sprint	BC27F	1994	Penniston, Melton Mowbray, 1999
WE	M465LPG	Mercedes-Benz 709D	Alexander Sprint	B29F	1995	Arriva Fox County, 2002
WE	SNR1H	Mercedes-Benz 709D	Alexander Sprint	B25F	1996	Penniston, Melton Mowbray, 1999
WE	D1WET	Setra S250 Special	Setra	C48FT	1997	Miller, Dagenham, 2008
WE	AV51AVA	Mercedes-Benz Vario 0810	Plaxton Beaver 2	B27F	2001	Repton, New Haw, 2004
WE	T1WET	Mercedes-Benz Sprinter 413cdi	Crest	M16	2002	Norwich City, Acle, 2005
WE	Y1WET	Dennis Dart SLF 8.8m	Plaxton Pointer MPD	N29F	2002	
RT	L1WET	Volvo B7R	Jonckheere Modulo	C55F	2002	Bebb, Llantwit Fardre, 2004
RT	M1WET	Volvo B7R	Plaxton Profile	C53F	2003	Streamline, West Malling, 2006
WE	EU03EUD	TransBus Dart 8.8m	TransBus Mini Pointer	N29F	2003	Localink, Bishop's Stortford, 2008
WE	BU53AXB	Mercedes-Benz Sprinter 411cdi	Koch	N16F	2003	
WE	JP54JON	Volkswagen Transporter	Volkswagen	M8	2004	private owner, 2007
WE	JP55WET	Setra S315 GT-HD	Setra	C49FT	2005	
WE	FN06FLL	Enterprise Plasma EB01	Plaxton Primo	N28F	2006	
WE	MX56HYO	ADL Dart 8.8m	ADL Mini Pointer	N29F	2006	NCP, Heathrow, 2008
WE	MX56HYP	ADL Dart 8.8m	ADL Mini Pointer	N29F	2006	NCP, Heathrow, 2008
WE	H1WET	Setra S416 GT	Setra	C49FT	2006	Evobus demonstrator, 2008
WE	BV57VGP	Mercedes-Benz Sprinter 515cdi	Mercedes-Benz	N16	2008	
WE	BV57VHD	Setra S415 GT-HD	Setra	C49FT	2008	
WE	BN58BKA	Setra S415 GT-HD	Setra	C49FT	2008	

Special event vehicle:

WE	VUB396H	Leyland Leopard PSU3A/4R	Plaxton Paramount Elite	C53F	1970	Wallace Arnold

Previous registrations:

AV51AVA	AJ51REP	PIL7972	F702ENE
C366LFF	C938VLB, GSU498	SNR1H	N845ASF
D1WET	P210FRC	T1WET	AD52HMY
H1WET	BU06HSX	VUB396H	VUB396H, OO1908, VRY562H, TIB8702
JP54JON	GL54BPV	VUT1X	-
L1WET	CE02YRV	WET1K	J706CWT, A10GYC
M1WET	GK53HTV	XLC1S	-
NEL1F	M673RAJ, XLC1S	Y1WET	FE52KNS
PIL7971	UGG916W, CCC257, ULA485W		

Representing West End Travel is FN06FLL, an Enterprise Plasma with Plaxton Primo bodywork. The bus is seen in London Road in Leicester.
A P Rodgers

WESTERN GREYHOUND

Western Greyhound Ltd, Western House, St Austell Street, Summercourt, TR8 5DR

101	WK05FJE	Mercedes-Benz Sprinter 413cdi	Plaxton Pronto	N16F	2005	Liskeard & District, 2007
111	WK02XLN	Mercedes-Benz Sprinter 411cdi	G&M	M12	2002	Liskeard & District, 2007
112	WK02XLO	Mercedes-Benz Sprinter 411cdi	G&M	M12	2002	Liskeard & District, 2007
121	WK02XLR	Mercedes-Benz Vito 110cdi	Mercedes-Benz	M7	2002	Liskeard & District, 2007
201	WK08ESU	Mercedes-Benz Citaro O530	Mercedes-Benz	N39F	2008	
202	WK08CZV	Mercedes-Benz Citaro O530	Mercedes-Benz	N39F	2008	
203	WK08ESV	Mercedes-Benz Citaro O530	Mercedes-Benz	N39F	2008	
204	WK08ESY	Mercedes-Benz Citaro O530	Mercedes-Benz	N39F	2008	
205	WK08ETA	Mercedes-Benz Citaro O530	Mercedes-Benz	N39F	2008	
450	S450ATV	Volvo Olympian	East Lancs Pyoneer	B49/35F	1998	City of Nottingham, 2008
452	S452ATV	Volvo Olympian	East Lancs Pyoneer	B49/35F	1998	City of Nottingham, 2008
453	S453ATV	Volvo Olympian	East Lancs Pyoneer	B49/35F	1998	City of Nottingham, 2008
454	S454ATV	Volvo Olympian	East Lancs Pyoneer	B49/35F	1998	City of Nottingham, 2008
455	S455ATV	Volvo Olympian	East Lancs Pyoneer	B49/35F	1998	City of Nottingham, 2008
467	R467RRA	Volvo Olympian	East Lancs Pyoneer	B49/35F	1997	City of Nottingham, 2008
472	R472RRA	Volvo Olympian	East Lancs Pyoneer	B49/35F	1997	City of Nottingham, 2008
473	R473RRA	Volvo Olympian	East Lancs Pyoneer	B49/35F	1998	City of Nottingham, 2008
475	R475RRA	Volvo Olympian	East Lancs Pyoneer	B49/35F	1997	City of Nottingham, 2008
476	R476RRA	Volvo Olympian	East Lancs Pyoneer	B49/35F	1997	City of Nottingham, 2008
478	R478RRA	Volvo Olympian	East Lancs Pyoneer	B49/35F	1997	City of Nottingham, 2006
498	UWR498	Volvo Olympian	East Lancs Pyoneer	B49/35F	1998	City of Nottingham, 2006
499	499YMK	Volvo Olympian	East Lancs Pyoneer	B49/35F	1998	City of Nottingham, 2006
500	S100PAF	Mercedes-Benz Vario O814	Plaxton Beaver 2	BC31F	1998	Hambly's of Kernow, 2004
501	S501SRL	Mercedes-Benz Vario O814	Plaxton Beaver 2	B27F	1999	
502	S502SRL	Mercedes-Benz Vario O814	Plaxton Beaver 2	B27F	1999	
503	S503SRL	Mercedes-Benz Vario O814	Plaxton Beaver 2	B27F	1999	

One of the less usual buses operated by Western Greyhound is forward-entrance Routemaster NMY648E. New to London Transport for its BEA service to Heathrow, it is seen at Newquay. *Richard Godfrey*

The mainstay of the Western Greyhound fleet is the Mercedes-Benz Vario fitted with Beaver 2 bodywork of which some sixty are operated. One of the first to arrive in the fleet was 502, S502SRL, seen at Sennen Cove close to Land's End. Many of the type feature high-back seating although this example is fitted with normal bus seating. *Steve Rice*

507	R807HWS	Mercedes-Benz Vario 0814	Plaxton Beaver 2	BC33F	1998	Andybus, Tetbury, 2003
508	R808HWS	Mercedes-Benz Vario 0814	Plaxton Beaver 2	BC33F	1998	Andybus, Tetbury, 2003
509	R809HWS	Mercedes-Benz Vario 0814	Plaxton Beaver 2	BC33F	1998	Andybus, Tetbury, 2003
510	R810HWS	Mercedes-Benz Vario 0814	Plaxton Beaver 2	BC33F	1998	Andybus, Tetbury, 2003
518	S18RED	Mercedes-Benz Vario 0814	Plaxton Beaver 2	B27F	1999	Hazell, Northlew, 2003
520	S20WGL	Mercedes-Benz Vario 0814	Plaxton Beaver 2	BC33F	1998	Cerbydau Carreglefn, 2003
521	T961ACC	Mercedes-Benz Vario 0814	Plaxton Beaver 2	BC31F	1999	Sunset,Sancreed, 2007
522	V22WGL	Mercedes-Benz Vario 0814	Plaxton Beaver 2	B27F	2000	Mike Halford, Bridport, 2007
528	Y28WGL	Mercedes-Benz Vario 0814	Plaxton Beaver 2	B27F	2001	Blennarhassett, Ormskirk, 2005
529	W29WGL	Mercedes-Benz Vario 0814	Plaxton Beaver 2	BC33F	2001	Brylaine, Boston, 2002
530	S30ARJ	Mercedes-Benz Vario 0814	Plaxton Beaver 2	BC33F	1998	Andybus, Tetbury, 2001
531	V31WGL	Mercedes-Benz Vario 0814	Plaxton Beaver 2	BC28F	1999	Countryliner, Guildford, 2003
533	X33WGL	Mercedes-Benz Vario 0814	Plaxton Beaver 2	BC33F	2001	Wilson, Rhu, 2003
534	S34BMR	Mercedes-Benz Vario 0814	Plaxton Beaver 2	BC33F	1998	Andybus, Tetbury, 2001
537	WSV537	Mercedes-Benz Vario 0814	Plaxton Beaver 2	B31F	2000	Autowater, Bathgate, 2006
540	V40WGL	Mercedes-Benz Vario 0814	Plaxton Beaver 2	B31F	1999	Countryliner, Guildford, 2002
541	Y371TGE	Mercedes-Benz Vario 0814	Plaxton Beaver 2	B31F	2001	Autowater, Bathgate, 2006
544	W44WGL	Mercedes-Benz Vario 0814	Plaxton Beaver 2	BC33F	2000	Brylaine, Boston, 2002
545	W385WGE	Mercedes-Benz Vario 0814	Plaxton Beaver 2	B31F	2000	Autowater, Bathgate, 2006
548	W348WCS	Mercedes-Benz Vario 0814	Plaxton Beaver 2	B31F	2000	Rowe, Kilmarnock, 2006
549	W349WCS	Mercedes-Benz Vario 0814	Plaxton Beaver 2	B31F	2000	Rowe, Kilmarnock, 2006
550	W50MBH	Mercedes-Benz Vario 0814	Plaxton Beaver 2	BC33F	2000	Brylaine, Boston, 2002
551	WK51AVP	Mercedes-Benz Vario 0814	Plaxton Beaver 2	BC33F	2001	
552	WK51HNF	Mercedes-Benz Vario 0814	Plaxton Beaver 2	BC29F	2002	
553	WK53BNA	Mercedes-Benz Vario 0814	TransBus Beaver 2	BC33F	2003	
554	WK53BNB	Mercedes-Benz Vario 0814	TransBus Beaver 2	BC33F	2003	
555	WK53BND	Mercedes-Benz Vario 0814	TransBus Beaver 2	BC33F	2003	
556	WK53BUS	Mercedes-Benz Vario 0814	TransBus Beaver 2	BC33F	2003	McGowan, Neilston, 2005
557	WK04BUS	Mercedes-Benz Vario 0814	TransBus Beaver 2	BC33F	2004	
558	WK02SAT	Mercedes-Benz Vario 0814	Plaxton Beaver 2	BC29F	2002	Fairline, Glasgow, 2005
559	WK02SUN	Mercedes-Benz Vario 0814	Plaxton Beaver 2	BC31F	2002	Fairline, Glasgow, 2006
560	WK02BUS	Mercedes-Benz Vario 0814	Plaxton Beaver 2	BC33F	2002	Mitchell, Plean, 2004
561	WK51BUS	Mercedes-Benz Vario 0814	Plaxton Beaver 2	BC33F	2001	Hutchinson, Easingwold, 2003
562	WK53EUT	Mercedes-Benz Vario 0814	TransBus Beaver 2	BC33F	2003	
563	WK53EUU	Mercedes-Benz Vario 0814	TransBus Beaver 2	BC33F	2003	

English Bus Handbook: Notable Independents

564	WK04CUA	Mercedes-Benz Vario 0814	TransBus Beaver 2	BC33F	2004	
565	WK04CUC	Mercedes-Benz Vario 0814	TransBus Beaver 2	BC33F	2004	
566	WK54BHL	Mercedes-Benz Vario 0814	Plaxton Beaver 2	BC33F	2004	
567	WK54BHN	Mercedes-Benz Vario 0814	Plaxton Beaver 2	BC33F	2004	
568	WK54BHO	Mercedes-Benz Vario 0814	Plaxton Beaver 2	BC33F	2004	
569	WK54BHP	Mercedes-Benz Vario 0814	Plaxton Beaver 2	BC33F	2004	
570	WK05CFD	Mercedes-Benz Vario 0814	Plaxton Beaver 2	BC33F	2005	
571	WK05CFE	Mercedes-Benz Vario 0814	Plaxton Beaver 2	BC33F	2005	
572	WK05CFF	Mercedes-Benz Vario 0814	Plaxton Beaver 2	BC33F	2005	
573	WK05AOX	Mercedes-Benz Vario 0814	Plaxton Beaver 2	BC33F	2005	
574	WK02TUE	Mercedes-Benz Vario 0814	Plaxton Beaver 2	BC31F	2002	Fairline, Glasgow, 2006
575	SUO575	Mercedes-Benz Vario 0814	Plaxton Beaver 2	BC33F	2005	
576	WK03BUS	Mercedes-Benz Vario 0814	Plaxton Beaver 2	B33F	2003	Autowater, Bathgate, 2006
577	WK04HSD	Mercedes-Benz Vario 0814	TransBus Beaver 2	BC33F	2004	Hambley, Pelynt, 2006
578	578PTA	Mercedes-Benz Vario 0814	Plaxton Beaver 2	BC33F	2004	Frome Minibuses, 2007
579	WK04KUO	Mercedes-Benz Vario 0814	Plaxton Beaver 2	B31F	2004	Eurotaxis, Bristol, 2007
580	WK04OTA	Mercedes-Benz Vario 0814	TransBus Beaver 2	BC33F	2004	Rowe & Tudhope, Kilmarnock, '07
581	WK04CCC	Mercedes-Benz Vario 0814	TransBus Beaver 2	BC33F	2004	Stonehouse Coaches, 2007
582	WK04DAD	Mercedes-Benz Vario 0814	Plaxton Beaver 2	BC29F	2003	Fairline, Glasgow, 2007
591	WK56SUN	Mercedes-Benz Vario 0814	Plaxton Beaver 2	BC29F	2003	Sunset Travel, Sancreed, 2007
592	WK56SET	Mercedes-Benz Vario 0814	Plaxton Beaver 2	BC29F	2003	Sunset Travel, Sancreed, 2007
599	YN53VBO	Mercedes-Benz Vario 0814	Plaxton Beaver 2	BC33F	2003	Graham's of Tredegar, 2007
701	WK58EAE	Optare Solo M710 SE	Optare	NC23F	2009	
702	WK58EAF	Optare Solo M710 SE	Optare	NC23F	2009	
703	WK58EAG	Optare Solo M710 SE	Optare	NC23F	2009	
704	WK58EAJ	Optare Solo M710 SE	Optare	NC23F	2009	
910	WK55BUS	Optare Solo M920	Optare	N33F	2006	Fairline, Glasgow, 2008
911	WK06SUN	Optare Solo M920	Optare	N33F	2006	Fairline, Glasgow, 2008
912	WK06SET	Optare Solo M920	Optare	N33F	2006	Fairline, Glasgow, 2008
913	WK57TAF	Optare Solo M920	Optare	N33F	2007	Fairline, Glasgow, 2008
956	WK56PEN	Optare Solo M920	Optare	N33F	2007	Fairline, Glasgow, 2008

Special event vehicles:

	MLL528	AEC Regal IV 9821LT	Metro Cammell	B39F	1952	London Transport
259	JWV259W	Bristol VRT/SL3/6LXB	Eastern Coach Works	B43/31F	1981	Southdown
282u	LUA282V	Leyland Leopard PSU5D/4R	Plaxton Supreme IV	C53F	1980	Wallace Arnold
347u	MHJ347F	Leyland Titan PD3/4	East Lancs	B41/32R	1967	Southend
RM1783	783DYE	AEC Routemaster R2RH	Park Royal	O36/28R	1964	London Transport
RM1062	62CLT	AEC Routemaster R2RH	Park Royal	B36/28R	1962	London Transport
RMA11	NMY648E	AEC Routemaster R2RH	Park Royal	B32/24R	1967	London Transport (BEA)
RML2737	SMK737F	AEC Routemaster R2RH/1	Park Royal	B40/32R	1967	London Transport
2748	EOO590	Bristol FLF6B	Eastern Coach Works	B38/32R	1962	Eastern National

Ancillary vehicle:

DSU107	Dennis Javelin 10m	Wadham Stringer Vanguard III	B40F	1995	

Previous registrations:

62CLT	62CLT, 794UXA	WK02TUE	SK02NZA
499YMK	R480RRA	WK03BUS	SF03SCX
578PTA	YK53HSN	WK04BUS	YX04GNZ
DSU107	M492GCD	WK04CCC	SJ04LLA
M492GCD	74KK68, CHZ5937	WK04DAD	SN03NLL
MLL528	MLL589, XKE146A, UVE678	WK04HSD	WAF156
S20WGL	S135UEY	WK04KUO	MX04VMJ
SUO575	YX54GYY	WK04OTA	SF04HXM
T961ACC	T961ACC, T5TAF	WK06SET	YJ06FZF
TFO319	HSK646, L656ADS	WK06SUN	YJ06FZE
UWR498	R479RRA	WK51BUS	PX51ELJ
V22WGL	V384HGG, CJZ3881	WK53BUS	SN53KYH
V31WGL	V451FOT	WK55BUS	YJ55BJZ
V40WGL	V991DNB	WK56PEN	MX56NCT
W29WGL	W968JNF	WK56SET	YN56DZE
W44WGL	W965JNF	WK56SUN	YN56DVM
W50MBH	W969JNF	WK57TAF	YJ57EJC
WK02BUS	SL02COU	WSV557	W389WGE
WK02SAT	SK02NZS	X33WGL	X437JHS
WK02SUN	SK02NZC	Y28WGL	Y909BFS

Web: westerngreyhound.com - **Depot:** St Austell Street, Summercourt.
Outstations: Bodmin (Priory Road); Camelford (Tregarth Farm); Hayle (Marsh Lane); Liskeard (Clemo Road); Looe (Polperro Road); Padstow (Jury Park); Penzance (Castle Horneck) and Wadebridge (Kestle Quarry, Sladesbridge).

WHITE BUS SERVICES

C E Jeatt & Sons Ltd, North Street Garage, North Street, Winkfield, Windsor, SL4 4TF

Reg	Chassis	Body	Seats	Year	History
L606ASU	Dennis Javelin 12m	Wadham Stringer Vanguard II	S70F	1994	MoD (47KL53), 2005
L321XTC	Dennis Javelin 12m	Wadham Stringer Vanguard II	S70F	1995	MoD (75KK27), 2006
L561ASU	Dennis Javelin 12m	Wadham Stringer Vanguard II	S70F	1995	MoD (77KK25), 2005
L523MJB	Dennis Javelin 12m	Wadham Stringer Vanguard II	S70F	1995	MoD (47KL48), 2005
P131RWR	DAF SB220	Optare Delta	B49F	1987	Luton Airport, 2004
R89GNW	DAF SB220	Optare Delta	B49F	1998	Flyerbus, Dublin, 2003
R674OEB	Dennis Javelin 12m	Berkhof Excellence 1000LD	C57F	1998	Greys of Ely, 2005
R714KGK	Dennis Javelin 12m	Berkhof Radial	C53F	1998	Lodge, High Easter, 2006
S158JUA	DAF SB220	Optare Delta	B49F	1998	National Express, W Drayton, '03
V917TAV	Dennis Javelin 12m	Berkhof Radial	C57F	2000	Grey, Witchford, 2007
CB53MTB	Volvo B7R	TransBus Profile	C53F	2003	Baker, Enstone, 2008
DB53MTB	Volvo B7R	TransBus Profile	C53F	2003	Baker, Enstone, 2008
SN55DVB	ADL E300	ADL Enviro 300	S60F	2005	Courtney, Bracknell, 2008
YJ07EGD	Optare Tempo X1260	Optare	N46F	2007	Optare demonstrator, 2007
YN57EHV	Optare Tempo X1260	Optare	N47F	2007	

Previous registrations:

L321XTC	L860EPM	R89GNW	R89GNW, 98D61842
R674OEB	ESU308, G18ELY, ESU350	V917TAV	ESU350, G12ELY

YEOMANS

Yeomans - Lugg Valley Primrose

Yeomans Canyon Travel Ltd, Old School Lane, Hereford, HR1 1EX
Lugg Valley Primose Travel Ltd; Lugg Valley Travel Ltd,
The New Bus Station, Westbury Street, Leominster, HR6 8PX

1	YO	YP52VEO	Scania K114 IB4	Van Hool T9 Alizée	C49FT	2002	
2	YO	H17GOW	Dennis Javelin 12m	Plaxton Paramount 3200 III	S70F	1991	Whittle, Kidderminster, 2000
3	YO	DX53YYG	BMC 850	BMC	B35F	2004	
4	YO	FJ07TKD	Scania K340 EB4	Caetano Levanté	C49FT	2007	
5	YO	YK04GWL	Optare Solo M850	Optare	N25F	2004	Optare demonstrator, 2005
6	YO	YN53EKZ	Optare Solo M920	Optare	N31F	2003	
7	YO	FN03DXH	Volvo B12M	Transbus Paragon	C49FT	2003	
9	YO	J74CVJ	Dennis Dart 8.5m	Alexander Dash	B30F	1991	
10	YO	YP52KSK	Volvo B7R	Plaxton Profile	S70F	2002	
11	YO	YCT187	Neoplan Cityliner N516 SHD	Neoplan	C44FT	2000	Parry's, Cheslyn Hay, 2003
12	YO	N801SJU	Dennis Javelin GX 12m	Marcopolo Explorer	C51FT	1996	Top Kat, Bedwas, 2006
14	YO	X534NWT	Optare Solo M920	Optare	N35F	2000	
15	YO	YJ54UBG	Optare Solo M990	Optare	N35F	2005	
16	YO	E277VJW	Volvo B10M-61	Plaxton Paramount 3500 III	C53F	1988	Empress, Bethnal Green, 1993
17	YO	VIB5072	Dennis Javelin 12m	Marcopolo Explorer	C53F	1996	Ludlows, Halesowen, 2005
19	YO	LUI1687	Dennis Lance 11m	Duple 320	C53F	1988	Huckle, Sutton Coldfield, 2002
20	YO	W443CWX	Optare Solo M920	Optare	N35F	2000	
21	YO	YCT502	Volvo B12B	TransBus Paragon	C49FT	2003	

Taking a break on its travels away from the Lugg Valley is 129, FN03DXG a Volvo B12B with Plaxton Paragon bodywork. *Mark Doggett*

A growing number of the services centred on Leominster and Hereford is using Optare Solo buses. Pictured leaving Leominster is 20, W443CWX, a 9.2 metre example. *Alan Blagburn*

22	YO	YP52CTV	Scania K114IB4	Van Hool T9 Alizée	C49FT	2002	
23	YO	YJ56WVW	Optare Solo M990	Optare	N35F	2006	
24	YO	RJI8611	Dennis Javelin 12m	Plaxton Paramount 3200 III	C53F	1988	Carmel Coaches, Northlew, 2003
25	YO	M208BGK	Dennis Javelin 12m	Plaxton Première 320	C53F	1994	Sullivan Bus, Potters Bar, 2007
	YO	M941JBO	Dennis Javelin 11m	Plaxton Première Interurban	BC49F	1994	R&D, Formby, 2008
28	YO	B995YTC	Leyland Tiger TRCTL11/3LZ	Wadham Stringer Vanguard II	S70F	1985	MoD (37KC30), 2000
29	YO	H457MEY	Dennis Dart 9.8m	Carlyle Dartline	B40F	1991	Padarn, Caernarfon, 1995
30	YO	Y687HPG	Dennis Javelin GX	Berkhof Axial 50	C51F	2001	
31	YO	E477VDA	Volvo B10M-61	Plaxton Paramount 3500 III	C53F	1988	Flights, Birmingham, 1992
33	YO	G28HDW	Dennis Javelin 11m	Duple 300	B55F	1985	East End, Clydach, 2002
36	YO	J78VOE	Ford Transit	Ford	M8	1992	private owner, 1999
40	YO	R144PCJ	Ford Transit	Ford	M8	1997	
41	YO	FJ08KNK	Scania K340 EB6	Caetano Levanté	C49FT	2007	
42	YO	FJ08KNX	Scania K340 EB6	Caetano Levanté	C49FT	2007	
43	YO	FJ08KNZ	Scania K340 EB6	Caetano Levanté	C49FT	2007	
44	YO	FJ08KOB	Scania K340 EB6	Caetano Levanté	C49FT	2007	
46	YO	G41TGW	Dennis Dart 8.5m	Carlyle Dartline	B28F	1990	Connex, 2002
55	YO	M341SCJ	Volvo B10M-62	Plaxton Expressliner 2	C46FT	1995	
56	YO	YOI2747	Volvo B10M-62	Plaxton Expressliner 2	C44FT	1995	
57	YO	YOI298	Volvo B10M-62	Plaxton Expressliner 2	C46FT	1995	
64	YO	YCV834	Volvo B10M-62	Plaxton Expressliner 2	C49FT	2000	
102	LV	H902AHS	Volvo B10M-60	Plaxton Paramount 3500 III	C53F	1991	Wallace Arnold, 1994
103	LV	F992HGE	Volvo B10M-61	Plaxton Paramount 3200 III	C53F	1989	Park's of Hamilton, 1991
104	LV	G57RGG	Volvo B10M-60	Plaxton Paramount 3500 III	C49FT	1990	Park's of Hamilton, 1996
107	LV	X533NWT	Optare Excel L1180	Optare	N45F	2000	
109	LV	YK05ENU	Optare Solo M990	Optare	N35F	2000	
111	LV	YN53EKX	Optare Solo M920	Optare	N31F	2003	
112	LV	YN53EKY	Optare Solo M920	Optare	N31F	2003	
116	LV	YG02DGU	Optare Solo M920	Optare	N31F	2002	
117	LV	AF51DXH	Dennis Dart SLF	Marshall Capital	N35F	2001	
118	LV	N211HGO	Dennis Javelin 12m	Plaxton Première 320	C53F	1995	Metrobus, Crawley, 2007
119	LV	G25HDW	Dennis Javelin 11m	Duple 300	B55F	1990	South Lancs, Atherton, 2003
120	LV	DJS203	Volvo B10M-62	Jonckheere Mistral 50	C53F	2000	
121	LV	YN05HFV	Scania K94 IB4	Irizar S-kool	S70F	2005	Weardale, Stanhope, 2005
122	LV	YJ58CCX	Optare Solo M950	Optare	N31F	2008	

An additon to the fleet in 2001 was Dennis Dart 117, AF51DXH, which has a Marshall Capital body. It is seen operating the main link between the two operating centres, route 426. *David Longbottom*

122	YO	BU04CAX	BMC 1100	BMC	B40F	2004	
123	LV	YJ57EHE	Optare Versa V1100	Optare	N37F	2007	Optare demonstrator, 2008
124	LV	G525VYE	Dennis Dart 8.5m	Duple Dartline	B28F	1990	Connex, 2002
125	LV	YJ58CCY	Optare Solo M950	Optare	N31F	2008	
126	LV	YJ58CCZ	Optare Solo M950	Optare	N31F	2008	
127	LV	YG02DGX	Optare Solo M920	Optare	N31F	2002	
128	YO	YG02DGV	Optare Solo M920	Optare	N31F	2002	
129	LV	FN03DXG	Volvo B12B	Plaxton Paragon	C53FT	2003	
131	LV	Y59HBT	Optare Excel L1180	Optare	N45F	2001	
132	LV	G216LGK	Dennis Dart 9m	Duple Dartline	B36F	1990	Metroline London Northern, 2001
133	LV	YG02DGZ	Optare Solo M920	Optare	N31F	2002	
134	LV	GIL2633	Dennis Javelin GX	Berkhof Axial 50	C51F	2001	
135	LV	YG02DGY	Optare Solo M920	Optare	N31F	2002	
138	LV	P71MOV	Dennis Lance 11m	Northern Counties Paladin	B46F	1997	Serveverse, Mile Oak, 1999
	YO	YJ56WVW	Optare Solo M920	Optare	N35F	2006	
	LV	YN07LGG	Scania K114 IB4	Irizar PB	C49FT	2007	
	YO	FJ08KNK	Scania K340 EB6	Caetano Levanté	C49FT	2007	

Previous registrations:

DJS203	W866AAY	M341SCJ	M341SCJ, YCT502
E277VJW	E277VJW, YOI298	RJI8611	F129TRU
E477VDA	E477VDA, YCV834	VIB5072	N791SJU
F992HGE	F992HGE, DJS203	YCT187	W905MDT
G57RGG	G57RGG, GIL2633	YCT502	YN03NJJ
GIL2633	Y688HPG	YCV834	W634MKY
		YOI298	M343SCJ
LXI9357	E589UHS	YOI2747	M342SCJ

Depots: Old School Lane, Hereford (YO); Baron's Cross Garage, Leominster (LV) and Worcester Road, Leominster (LV).

Index to vehicles

Reg	Operator	Reg	Operator	Reg	Operator	Reg	Operator
1BLU	Bluebird	851FNN	Quantock	9383MX	Simonds	A860KFP	Trustybus
62CLT	Western Greyhound	864DYE	Ensignbus	9423RU	Bakerbus	A867SUL	Nu-Venture
131ASV	Beestons	890ADV	Quantock	9595RU	Bakerbus	A873SUL	Nu-Venture
157TYB	Quantock	930FDV	Hopley's	9983PW	Simonds	A883SUL	Emsworth & D
159FCG	Safeguard	956CCE	Hopley's	A3YRR	Marshalls	A900SUL	Swanbrook
166UMB	Simonds	969EHW	Abus	A9AVN	Avon Buses	A905SUL	Nu-Venture
187NKN	Chalkwell	1294RU	TM Travel	A10AVN	Avon Buses	A911SYE	Go Whippet
196BLU	Bluebird	1497RU	Bakerbus	A14SOE	Stephensons	A933SYE	Ham's Travel
196FCG	Safeguard	1513RU	Bakerbus	A17AVN	Avon Buses	A958SYF	Swanbrook
200APB	Safeguard	1901HE	Aintree Coachlines	A17SOE	Stephensons	A971YSX	Carousel
214CLT	Halifax JC	2091PW	Simonds	A18HOF	Compass Bus	A984SYE	Nu-Venture
217OPO	Beestons	2732RH	Hornsby Travel	A19AVN	Avon Buses	A984SYF	Speedwell
219GRA	Beestons	3093RU	Bakerbus	A240VL	Quantock	AA05BLU	Bluebird
221GRA	Beestons	3102RU	Bakerbus	A33MRN	Preston	ABT1K	Thornes
222GRA	Beestons	3275RU	Bakerbus	A101WVP	Travel de Courcey	ACH441	Quantock
224ENG	Simonds	3353RU	Bakerbus	A102SUU	Nu-Venture	ACT540L	Delaine
229LRB	Beestons	3471RU	Bakerbus	A107WVP	Travel de Courcey	AD03OCT	Delaine
247FCG	Safeguard	354TRT	Sanders Cs	A111TRP	TM Travel	AD04OCT	Delaine
256JPA	Simonds	3558RU	Nu-Venture	A163HLV	Aintree Coachlines	AD05OCT	Delaine
259VYC	Sanders Cs	3563RU	Bakerbus	A250SVW	Ensignbus	AD07DBL	Delaine
270BLU	Bluebird	3566RU	Bakerbus	A305MKJ	Ham's Travel	AD08DBL	Delaine
277FCG	Safeguard	3601RU	Bakerbus	A431VNY	Carousel	AD09DBL	Delaine
304CLT	Ensignbus	3747RH	Halifax JC	A441UUV	Ensignbus	AD56DBL	Delaine
360BLU	Bluebird	3990ME	Sanders Cs	A462LFV	Fishwick	AD57BDY	Anglian
378BNG	Simonds	4085RU	Bakerbus	A499MHG	Jim Stones	AD57EXA	Anglian
424DCD	Stephensons	4512UR	Simonds	A633BCN	Weardale	AD58DBL	Delaine
499YMK	Western Greyhound	4614RU	Bakerbus	A671UOE	Travel de Courcey	AE05EUX	Thames Travel
508AHU	Hopley's	4940VF	Simonds	A681KDV	Country Liner	AE05EUY	Hulleys
515FCG	Safeguard	5280NW	Ensignbus	A684UOE	Travel de Courcey	AE05EUZ	Thames Travel
524FN	Beestons	5621RU	Bakerbus	A703THV	Swanbrook	AE05EVD	Travel de Courcey
532FN	Beestons	5777RU	Bakerbus	A703YFS	Blackpool	AE05OVC	Travel de Courcey
536FN	Sanders Cs	5946RU	Bakerbus	A704YFS	Blackpool	AE06HBU	Trustybus
537FN	Sanders Cs	6280RU	Bakerbus	A706UOE	Travel de Courcey	AE06HBX	Trustybus
538ELX	Simonds	6542FN	Sanders Cs	A707DAU	Blackpool	AE06OPH	Travel de Courcey
538FCG	Safeguard	6546FN	Sanders Cs	A708DAU	Blackpool	AE06OPJ	Travel de Courcey
538FN	Sanders Cs	7025RU	Bakerbus	A709DAU	Blackpool	AE06OPK	Travel de Courcey
577HLU	Chalkwell	7092RU	Bakerbus	A710DAU	Blackpool	AE06VPY	Country Liner
578PTA	Western Greyhound	7236PW	Simonds	A711YFS	Cedar Cs	AE06VPZ	Country Liner
640UAF	Hopley's	7345FM	Regal Busways	A712UOE	Travel de Courcey	AE06XSA	Simonds
713WAF	Cedar Cs	7455RH	Hornsby Travel	A719YFS	Blackpool	AE06XSB	Simonds
776WME	Country Liner	8150RU	Bakerbus	A720YFS	Blackpool	AE06ZBR	Reading
783DYE	Western Greyhound	8399RU	Bakerbus	A721YFS	Blackpool	AE06ZBT	Reading
784EYB	Anglian	8421RU	Nu-Venture	A740WVP	Travel de Courcey	AE07DZD	Country Liner
805EVT	Quantock	8439RU	Bakerbus	A748WVP	Travel de Courcey	AE07DZE	Country Liner
831HKA	Anglian	8955RH	Hornsby Travel	A857SUL	Nu-Venture	AE07DZF	Country Liner

Reg	Operator	Reg	Operator	Reg	Operator	Reg	Operator
AE07DZG	Simonds	AE57CNN	Travel de Courcey	AU04JKN	Anglian	B203WUL	Halifax JC
AE07DZH	Simonds	AE57FAJ	Halton	AU06BOV	Anglian	B217WUL	Halifax JC
AE07DZJ	Travel de Courcey	AE57FAK	Halton	AU06BPE	Anglian	B221WUL	Swanbrook
AE07DZP	Travel de Courcey	AE57FAM	Halton	AU06BPF	Anglian	B285WUL	Ensignbus
AE07DZR	Travel de Courcey	AE57FAO	Halton	AU06BPK	Anglian	B522HAM	Ham's Travel
AE07DZS	Avon Buses	AE57FBA	Weavaway	AU06BPO	Anglian	B542HAM	Ham's Travel
AE07DZT	Avon Buses	AE57FBB	Weavaway	AU07KMK	Anglian	B565SOW	TM Travel
AE07DZX	Travel de Courcey	AE57FBC	Travel de Courcey	AU07KMM	Anglian	B625DWF	Nu-Venture
AE07NYR	Travel de Courcey	AE57LYH	Thames Travel	AU08DKL	Anglian	B740GSC	Blackpool
AE07NYS	Avon Buses	AE57LYJ	Thames Travel	AU08DKN	Anglian	B863XYR	Shamrock
AE07NYT	Avon Buses	AE57LYK	Thames Travel	AU08GLY	Anglian	B865XYR	Shamrock
AE07NYU	Avon Buses	AE57LYO	Thames Travel	AU53GWC	Anglian	B866XYR	Shamrock
AE08DKO	Avon Buses	AE57LYP	Thames Travel	AU53KSO	Norfolk Green	B892UAS	Abus
AE08DKU	Avon Buses	AE57LYR	Thames Travel	AU54ENY	Anglian	B893UAS	Abus
AE08DKV	Avon Buses	AE57LYS	Thames Travel	AU54EOA	Anglian	B896UAS	Abus
AE08DKX	Thames Travel	AF51DXH	Yeomans	AU54JRV	Simonds	B899UAS	Abus
AE08DKY	Thames Travel	AF51JZU	Beestons	AU57EZL	Anglian	B995YTC	Yeomans
AE08DLD	Thames Travel	AF52VMD	Go Whippet	AU57EZM	Anglian	BAS563	Quantock
AE08DLF	Stansted Transit	AFB592V	Abus	AU58AKK	Anglian	BB05BLU	Bluebird
AE54JRO	Travel de Courcey	AFB597V	South Lancs	AU58AKN	Anglian	BBW213Y	Heddingham
AE54JRU	Travel de Courcey	AIG8286	Hopley's	AU58AUV	Anglian	BBW218Y	Heddingham
AE54JRV	Travel de Courcey	AJ05BUS	Andybus	AV51AVA	West End Travel	BCJ710B	Ensignbus
AE54JRX	Travel de Courcey	AJ07BUS	Andybus	AV52GHZ	Beestons	BCP671	Halifax JC
AE54MVW	Travel de Courcey	AJ08BUS	Andybus	AY51EFG	Beestons	BD02HDG	Go Whippet
AE54MVX	Travel de Courcey	AJ54AMJ	Bluebird	B1BUS	Jim Stones	BD02HDJ	Go Whippet
AE54MVY	Travel de Courcey	AJ55BUS	Andybus	B1JYM	Jim Stones	BD03YFW	Premiere
AE54MWA	Travel de Courcey	AJ58PZH	Halton	B4AVN	Avon Buses	BD52HKN	Premiere
AE54NUH	Travel de Courcey	AJ58PZK	Halton	B5AVN	Avon Buses	BED729C	N Warrington
AE55EHM	Regal Busways	AJ58PZL	Halton	B10JYM	Jim Stones	BEZ7262	Cedar Cs
AE55EHP	Travel de Courcey	AJ58PZM	Halton	B10PTL	Premiere	BGR684W	Minsterley Ms
AE55EHR	Travel de Courcey	AJ58PZS	Thames Travel	B11JYM	Jim Stones	BHZ8675	D & G
AE55MVL	Trustybus	AJ58WAA	Bodmans	B12PTL	Premiere	BIG5013	Bakerbus
AE55VGD	Trustybus	AJA132	Quantock	B13STA	Wardle	BIG9704	Hopley's
AE55VGM	Travel de Courcey	ALD978B	Aintree Coachlines	B15PTL	Premiere	BIG9705	Hopley's
AE56LWH	Travel de Courcey	ANZ8798	Renown	B16TYG	Jim Stones	BIL4419	N I B S
AE56LWJ	Avon Buses	ANZ8799	Renown	B17PTL	Premiere	BIL4539	N I B S
AE56LWK	Avon Buses	AO02LVC	Anglian	B44HAM	Ham's Travel	BIL4710	N I B S
AE56LWL	Avon Buses	AO52LJF	Anglian	B49PJA	Shamrock	BIL6538	N I B S
AE56MBX	Country Liner	AO52NXD	Simonds	B111WUV	Go Whippet	BIL9406	N I B S
AE56MBY	Country Liner	AO57HCC	Anglian	B124BOO	Heddingham	BK04MZU	Rossendale
AE56MDF	Country Liner	AO57HCD	Anglian	B135WNB	Shamrock	BK04MZV	Rossendale
AE56MDJ	Country Liner	AP02XOO	Norfolk Green	B162WUL	Halifax JC	BK04MZW	Rossendale
AE56MDK	Country Liner	AP03BUS	Abus	B175VDV	Plymouth	BK04MZZ	Rossendale
AE56MDO	Country Liner	AP03BUZ	Abus	B176VDV	Plymouth	BKZ70	Weavaway
AE56OUF	Thames Travel	AP04BUS	Abus	B192WUL	Speedwell	BN08OOW	Marshalls
AE57CMX	Travel de Courcey	AP53BUS	Abus	B196WUL	Halifax JC	BN58BJJ	Uno
AE57CMY	Travel de Courcey	ATV11B	Beestons	B201EFM	Aintree Coachlines	BN58BJK	Uno
AE57CMZ	Travel de Courcey	AU03FSA	Simonds	B203DTU	Regal Busways	BN58BJO	Uno

BN58BKA	West End Travel	BX56VTM	Safeguard	C779SFS	Weardale	D84UTF	Shamrock
BOK68V	Ensignbus	BX56VTP	Bennetts	C787SFS	Weardale	D138FYM	Ipswich
BT08SZR	Bennetts	BXI2563	Nu-Venture	C793SFS	Weardale	D149FYM	Blackpool
BU03LXV	Country Liner	BXI2573	South Lancs	C900JGA	Sargeants	D150FYM	Ipswich
BU03LXW	Country Liner	BYW430V	Pennine	CAZ6602	Midland Rider	D162FYM	Renown
BU04CAX	Yeomans	BYW432V	Pennine	CAZ6603	Chalkwell	D167FYM	Renown
BU04UTP	Country Liner	BYX217V	Halifax JC	CAZ6604	Chalkwell	D168FYM	Renown
BU05EEO	Brylaine	BYX248V	Halifax JC	CAZ6617	Midland Rider	D169FYM	Renown
BU05EEP	Brylaine	C1WYC	Carousel	CAZ6619	Midland Rider	D173FYM	Blackpool
BU05EER	Brylaine	C2WYC	Carousel	CAZ6641	Midland Rider	D180FYM	Renown
BU06CVA	Bennetts	C3WYC	Carousel	CB51BUS	Carousel	D183FYM	Ensignbus
BU51OWL	Swanbrook	C6WMS	Weardale	CB52BUS	Carousel	D194FYM	Stansted Transit
BU52LEE	Jim Stones	C8LEA	Emsworth & D	CB53BUS	Carousel	D201FYM	Ipswich
BU53AXB	West End Travel	C25CHM	Renown	CB53MTB	White Bus	D211FYM	Renown
BU53AYB	Bennetts	C51CHM	N I B S	CB54BUS	Carousel	D213FYM	Blackpool
BU54AJP	Abus	C59HOM	Premiere	CBV775Y	TM Travel	D214FYM	Renown
BU54ALL	Wardle	C62CHM	Stephensons	CC05BLU	Bluebird	D219FYM	Stansted Transit
BU57HAM	Ham's Travel	C73CHM	N I B S	CCX777	Quantock	D229HMT	TM Travel
BUS1N	Jim Stones	C81CHM	Stephensons	CDB206	Quantock	D235FYM	Ensignbus
BUS1S	Jim Stones	C82CHM	Stephensons	CHG541	Quantock	D238FYM	Ensignbus
BUS1T	Jim Stones	C92CHM	Stephensons	CHZ1751	Weardale	D245FYM	Stansted Transit
BUS51T	Jim Stones	C97CHM	Ipswich	CHZ8960	D & G	D251FYM	Ensignbus
BV51ENL	Wardle	C100CHM	Ipswich	CL07PTL	Premiere	D257FYM	Blackpool
BV55UAZ	Reliance	C100UBC	N Warrington	CN04NBY	Country Liner	D258FYM	Blackpool
BV57VGP	West End Travel	C101CHM	Ipswich	CNZ2250	D & G	D367JJD	Blackpool
BV57VHD	West End Travel	C101UBC	N Warrington	CRU184C	Ensignbus	D602RGJ	Emsworth & D
BV57VHE	Uno	C102UBC	N Warrington	CSK282	Anglian	DA02PUX	Halton
BV57VHF	Uno	C103UBC	N Warrington	CSU960	Chalkwell	DA02PUY	Halton
BV57VHG	Uno	C174ECK	Stephensons	CU04AMV	Rossendale	DA51XTC	Wardle
BV57VHH	Uno	C176ECK	Stephensons	CU04AMX	Rossendale	DA51XTE	Wardle
BV57VHJ	Uno	C178ECK	Shamrock	CU04AOP	Rossendale	DA52ZVK	N Warrington
BW52SVA	Travel de Courcey	C264XEF	Regal Busways	CUD223Y	Heddingham	DA52ZVL	N Warrington
BX02CMF	Thames Travel	C329HWJ	Renown	CUD224Y	Heddingham	DA52ZVM	N Warrington
BX03BKU	Rossendale	C334HWJ	Hornsby Travel	CUL162V	Ham's Travel	DAZ1558	Olympus
BX04DVV	Hornsby Travel	C345BUV	Carousel	CUV220C	Ensignbus	DAZ1559	Olympus
BX54EBD	Bennetts	C356BUV	Carousel	CUV226C	Ensignbus	DAZ4300	Kime's
BX54EBF	Bennetts	C366LFF	West End Travel	CW51TOP	Weavaway	DAZ4301	Kime's
BX54EBG	Bennetts	C382SAO	Stephensons	CW52TOP	Weavaway	DAZ4302	Kime's
BX54EBP	Country Liner	C383SAO	Stephensons	CW53TOP	Weavaway	DB53MTB	White Bus
BX54ECV	Bennetts	C386BUV	Carousel	CW54TOP	Weavaway	DBV132Y	Shamrock
BX55FYH	Chambers	C424BUV	Halifax JC	CW55TOP	Weavaway	DC07MDC	Travel de Courcey
BX55FYJ	Chambers	C432BUV	Carousel	CW56TOP	Weavaway	DCK219	Quantock
BX55OFL	Cedar Cs	C481YWY	Shamrock	CY08COV	Travel de Courcey	DD56BLU	Bluebird
BX56VSJ	Uno	C521DND	Ham's Travel	D1WET	West End Travel	DE02URX	N Warrington
BX56VSK	Uno	C521LJR	Country Liner	D6WMS	Weardale	DE04YNB	N Warrington
BX56VSL	Uno	C722NNN	Ipswich	D16WMS	Weardale	DE04YNC	N Warrington
BX56VSM	Uno	C777OCN	Weardale	D82UTF	Shamrock	DE04YND	N Warrington
BX56VSN	Uno	C778SFS	Weardale	D83UTF	Shamrock	DE04YNF	N Warrington

Reg	Operator	Reg	Operator	Reg	Operator	Reg	Operator
DE04YNG	N Warrington	DK54JPU	Halton	DY52DZF	Minsterley Ms	EK51JAU	Heddingham
DE04YNH	N Warrington	DK55HMF	N Warrington	E6WMS	Weardale	EK51JBE	Heddingham
DE52URZ	Halton	DK55HMG	N Warrington	E50TYG	Quantock	ELP223	Ensignbus
DE52USB	Halton	DK55HMH	N Warrington	E100AFW	Delaine	ELZ2362	Wardle
DE52USC	Halton	DK55HMJ	N Warrington	E1760EW	Go Whippet	ENZ2127	D & G
DF02EHY	Halton	DK55HMO	N Warrington	E215WBG	Geldards	ENZ4635	D & G
DF02EKC	Halton	DK55HMU	N Warrington	E277VJW	Yeomans	EOO590	Western Greyhound
DF02OPR	Wardle	DK550MM	N Warrington	E323MSG	Preston	ER05BUS	Somerbus
DF03NTE	Compass Bus	DK550MO	N Warrington	E324MSG	Preston	ESK697	Bakerbus
DF52ABU	N Warrington	DK550MP	N Warrington	E325MSG	Preston	ESK841	Bakerbus
DF52AXG	N Warrington	DK550MR	N Warrington	E326MSG	Preston	EU03BZK	Heddingham
DG02WXT	Halton	DK550PL	N Warrington	E327MSG	Preston	EU03CFX	Stephensons
DG02WXU	Halton	DK550PM	N Warrington	E328MSG	Preston	EU03CFY	Stephensons
DG02WXV	Halton	DK56MLJ	N Warrington	E329MSG	Preston	EU03EUD	West End Travel
DG02WYB	D & G	DK56MLL	N Warrington	E330MSG	Preston	EU03XOJ	Ham's Travel
DG52TYP	Wardle	DK56MLN	N Warrington	E331MSG	Preston	EU04BZY	Trustybus
DG52TYS	Wardle	DK56MLO	N Warrington	E332MSG	Preston	EU04CPV	Ensignbus
DG52TYT	Wardle	DK56MLU	N Warrington	E441ADV	Go Whippet	EU04CUW	Ensignbus
DG52TYU	Wardle	DK56MLV	N Warrington	E477VDA	Yeomans	EU04CZR	Ensignbus
DG53FJU	N Warrington	DK56MLX	N Warrington	E749SKR	Halifax JC	EU04CZS	Ensignbus
DG53FJV	N Warrington	DLJ111L	Shamrock	E767LBT	Ensignbus	EU05AUR	Heddingham
DG53FJX	N Warrington	DM05GSM	Simonds	E848DPN	Shamrock	EU05AUT	Heddingham
DG53FJY	N Warrington	DM51BUS	DRM	E849DPN	Shamrock	EU05BZM	Ensignbus
DG53FLH	N Warrington	DM55DRM	DRM	E850DPN	Shamrock	EU05CLJ	Heddingham
DG53FLJ	N Warrington	DPV65D	Quantock	E851DPN	Stephensons	EU05CZA	Stephensons
DJS203	Yeomans	DSK558	Safeguard	E852DPN	Stephensons	EU05ECV	N I B S
DK03NTD	Halton	DSK559	Safeguard	E853DPN	Shamrock	EU05VBG	Ensignbus
DK03NTE	Halton	DSK560	Safeguard	E854DPN	Shamrock	EU05VBJ	Ensignbus
DK03TNL	Halton	DSK648	Simonds	E855DPN	Shamrock	EU05VBK	Ensignbus
DK03TNN	Halton	DSU107	Western Greyhound	E856DPN	Shamrock	EU05VBL	Ensignbus
DK04MKE	Halton	DTG370V	Halifax JC	E857DPN	Shamrock	EU05VBM	Ensignbus
DK04MKF	Halton	DUF179	Quantock	E858DPN	Shamrock	EU05VBN	Ensignbus
DK04MKG	Halton	DW02HAM	Ham's Travel	E889CDS	Cedar Cs	EU05VBO	Ensignbus
DK04MKJ	Halton	DW05HAM	Ham's Travel	E911DRD	Go Whippet	EU05VBP	Ensignbus
DK07EZG	N Warrington	DW07HAM	Ham's Travel	E912DRD	Go Whippet	EU05VBT	Ensignbus
DK07EZH	N Warrington	DW52HAM	Ham's Travel	E921KYR	Shamrock	EU06KCX	Heddingham
DK07EZJ	N Warrington	DW54HAM	Ham's Travel	E925CDS	Marshalls	EU06KHK	Olympus
DK07EZL	N Warrington	DW55HAM	Ham's Travel	E927KYR	Shamrock	EU06KOW	Stansted Transit
DK07EZM	N Warrington	DX04MVR	D & G	E935CDS	Aintree Coachlines	EU06KOX	Stansted Transit
DK07EZN	N Warrington	DX04XMS	D & G	EAZ2576	Olympus	EU06KPA	Stansted Transit
DK07EZO	N Warrington	DX050MB	D & G	EAZ2588	Olympus	EU07FRN	Stephensons
DK07EZP	N Warrington	DX07WEF	Bakerbus	EE56BLU	Bluebird	EU07FVL	Stephensons
DK07EZR	N Warrington	DX07WEH	Bakerbus	EG04ZHT	Simonds	EU07FVM	Stephensons
DK07FWH	N Warrington	DX07WEJ	Bakerbus	EGP33J	Ensignbus	EU07FVN	Stephensons
DK07FWJ	N Warrington	DX53YYG	Yeomans	EIG1357	Beestons	EU07FVO	Stephensons
DK07FWL	N Warrington	DX57JXS	Minsterley Ms	EIG1358	Beestons	EU07GVY	Heddingham
DK54JPJ	Halton	DX57REU	Minsterley Ms	EJ02KYY	Heddingham	EU07XGK	Olympus
DK54JPO	Halton	DX57TVW	Minsterley Ms	EJZ2291	D & G	EU07XGL	Olympus

EU07XGM	Olympus	F167SMT	Nu-Venture	F973HGE	Plymouth	FIG6292	Cedric
EU08FHB	Olympus	F168SMT	Nu-Venture	F988HGE	Plymouth	FIL4166	Beestons
EU08FHC	Olympus	F196FFM	Country Liner	F992HGE	Yeomans	FIL8615	Beestons
EU08FHD	Olympus	F212YHG	Preston	FC02DRX	Chalkwell	FJ03AAZ	Cedric
EU08WND	Stephensons	F234YTJ	Halifax JC	FC02DRZ	Chalkwell	FJ06URR	Red Rose
EU53MVZ	Heddingham	F246HNE	Chambers	FCZ3413	Wardle	FJ07TKD	Yeomans
EU55BWC	Heddingham	F254HAM	Ham's Travel	FD02SDX	Nottingham	FJ08KNK	Yeomans
EU56FLM	Heddingham	F259YTJ	Stansted Transit	FD02SDY	Nottingham	FJ08KNK	Yeomans
EU56FLN	Heddingham	F292NHJ	Ensignbus	FD02SEY	Nottingham	FJ08KNX	Yeomans
EU56FLP	Heddingham	F302MYJ	Velvet	FD02SFE	Nottingham	FJ08KNZ	Yeomans
EU56FLR	Heddingham	F303MYJ	Velvet	FD02SFF	Nottingham	FJ08KOB	Yeomans
EU58AXX	Stephensons	F309MYJ	Velvet	FD02SFJ	Nottingham	FJ54MMO	Uno
EU58BLJ	N I B S	F329UJN	Stephensons	FD02SFK	Nottingham	FJ54MMU	Uno
EU58BRX	Stephensons	F352WSC	Fishwick	FD02SFN	Nottingham	FJ54ZCL	Minsterley Ms
EVD406	Quantock	F353WSC	Fishwick	FD02SFO	Nottingham	FJ54ZCO	Minsterley Ms
EX02RYR	Heddingham	F355WSC	Fishwick	FD05SDZ	Nottingham	FJZ4196	D & G
EY57FZE	Heddingham	F368AFR	Blackpool	FD05YDV	Nottingham	FM52GEY	Uno
EYE336V	Carousel	F369AFR	Blackpool	FD05YDW	Nottingham	FM52GFA	Uno
F31XOF	Ensignbus	F370AFR	Blackpool	FD51EYR	Nottingham	FN02VBK	Olympus
F44YHB	Cedar Cs	F371AFR	Blackpool	FD51EYS	Nottingham	FN03DXG	Yeomans
F45YHB	Ensignbus	F372AFR	Blackpool	FD51EYT	Nottingham	FN03DXH	Yeomans
F49ENF	Red Rose	F373AFR	Blackpool	FD51EYU	Nottingham	FN04FSG	Premiere
F52XOF	Ensignbus	F387GVO	Tyrer Bus	FD51EYV	Nottingham	FN06FLL	West End Travel
F85MJH	Shamrock	F389GVO	Aintree Coachlines	FD51EYW	Nottingham	FN54FLC	Renown
F86MJH	Shamrock	F406DUG	Aintree Coachlines	FD51EYX	Nottingham	FNZ1052	Quantock
F87MJH	Shamrock	F440AKB	Cedar Cs	FD51EYY	Nottingham	FP02XMA	Nottingham
F95STB	N Warrington	F510NJE	Carousel	FD54EOO	Olympus	FP02XMB	Nottingham
F96STB	N Warrington	F511NJE	Carousel	FD54EOP	Olympus	FP02XMC	Nottingham
F97STB	N Warrington	F600GVO	Plymouth	FE02AKV	Nottingham	FP51AOH	Nottingham
F98STB	N Warrington	F601GVO	Plymouth	FE02LWD	Felix	FP51AOJ	Nottingham
F99STB	N Warrington	F602GVO	Plymouth	FE51RAU	Go Whippet	FP51EXN	Nottingham
F103XEM	N Warrington	F603CET	TM Travel	FE51RBU	Go Whippet	FP51EXO	Nottingham
F104XEM	N Warrington	F603GVO	Plymouth	FE51RCU	Go Whippet	FRN816W	TM Travel
F117PHM	Go Whippet	F603RPG	Speedwell	FE51RDU	Go Whippet	FSK598	Country Liner
F120PHM	Richardson	F604GVO	Plymouth	FF56BLU	Bluebird	FSK868	Cedric
F121XEM	N Warrington	F605GVO	Plymouth	FFY401	Halifax JC	FSV428	Premiere
F122XEM	N Warrington	F606GVO	Plymouth	FFY403	Halifax JC	FW5696	Marshalls
F123PHM	Geldards	F607GVO	Plymouth	FG52WFT	Nottingham	FWL778Y	Heddingham
F127PHM	Marshalls	F621MSL	Stephensons	FG52WFU	Nottingham	FWL779Y	Heddingham
F129PHM	Geldards	F630MSL	Stephensons	FG52WFV	Nottingham	FWL780Y	Heddingham
F130PHM	Geldards	F653OFG	Country Liner	FG52WFW	Nottingham	FWL781Y	Heddingham
F131PHM	Geldards	F660RTL	Premiere	FG52WFX	Nottingham	FX04TJY	Kime's
F139PHM	Richardson	F809TLV	Geldards	FG52WFY	Nottingham	FX06JVF	Hornsby Travel
F140PHM	Richardson	F814TLV	Geldards	FG52WUC	Felix	FX06JVG	Hornsby Travel
F144PHM	Richardson	F815YLV	Ensignbus	FH51LTX	Nottingham	FX08REU	Hornsby Travel
F148SPV	Heddingham	F824TLV	Geldards	FHJ565	Beestons	FX54LLE	Hornsby Travel
F150LTW	Heddingham	F825TLV	Geldards	FHZ2128	Wardle	FXT183	Ensignbus
F166SMT	Nu-Venture	F865LCU	Ham's Travel	FIG2843	Olympus	FY02VHF	Brylaine

English Bus Handbook: Notable Independents

Reg	Operator	Reg	Operator	Reg	Operator	Reg	Operator
FY02VHG	Brylaine	G372RTO	Cedar Cs	G645BPH	Shamrock	GU52HJY	Coastal
FY03WZN	Safeguard	G376RTO	Cedar Cs	G703TCD	Go Whippet	GU52HKA	Reading
FY03WZV	Wardle	G401DPD	Anglian	G707TCD	Stephensons	GU52HKB	Reading
G18WMS	Weardale	G435MWU	TM Travel	G711LKW	Geldards	GU52HKC	Coastal
G25HDW	Yeomans	G472EKV	TM Travel	G714LKW	Geldards	GU52HKD	Coastal
G28HDW	Yeomans	G501SFT	Speedwell	G717FVX	Stephensons	GU52HXM	Courtney
G34OCK	Preston	G503VYE	Halifax JC	G727RGA	Emsworth & D	GX02CGY	Compass Bus
G35OCK	Preston	G505XLO	Emsworth & D	G738RTY	Midland Rider	GX03AZJ	Compass Bus
G36HKY	Geldards	G516VYE	Emsworth & D	G756UYT	Cedric	GX03AZL	Compass Bus
G36OCK	Preston	G518VBB	Renown	G760VRT	Geldards	GX04ASU	Coastal
G37OCK	Preston	G520VBB	Ensignbus	G767CDU	Emsworth & D	GX04AWR	Chalkwell
G39TGW	Emsworth & D	G521WJF	Speedwell	G802GSX	Fishwick	GX04AZA	Country Liner
G41TGW	Yeomans	G523VBB	TM Travel	G804GSX	Shamrock	GX04BXN	Compass Bus
G57RGG	Yeomans	G523WJF	Speedwell	G806GSX	Fishwick	GX05AOP	Coastal
G99NBD	Olympus	G524VBB	TM Travel	G809GSX	Shamrock	GX06AOE	Southdown PSV
G101NBV	Blackpool	G525VBB	Renown	G810GSX	Shamrock	GX07AVO	Coastal
G102NBV	Blackpool	G525VYE	Yeomans	G817GSX	Shamrock	GX07BYO	Coastal
G103NBV	Blackpool	G530VBB	Carousel	G823UMU	Go Whippet	GX54AWH	Compass Bus
G104NBV	Blackpool	G531VBB	TM Travel	G824UMU	Go Whippet	GX57AFV	Compass Bus
G105NBV	Blackpool	G532VBB	TM Travel	G828GSX	Shamrock	GX57BXG	Southdown PSV
G106NBV	Blackpool	G534VBB	Carousel	G830GSX	Shamrock	GX57BXH	Southdown PSV
G107NBV	Blackpool	G535VBB	TM Travel	G832GSX	Shamrock	GYE524W	Carousel
G108NBV	Blackpool	G537VBB	Renown	G864XDX	Aintree Coachlines	GZ2248	Green Bus
G110FJW	Geldards	G540VBB	Renown	G957KJX	Stansted Transit	H1JYM	Jim Stones
G110NGN	Travel de Courcey	G541TBD	D & G	G994VWV	Renown	H1WET	West End Travel
G112ENV	Premiere	G541VBB	Renown	G995VWV	Quantock	H2FBT	Blackpool
G113ENV	Premiere	G542VBB	Renown	GB03ACL	Aintree Coachlines	H3FBT	Blackpool
G113NGN	Travel de Courcey	G543VBB	Renown	GCA747	Green Bus	H3YRR	Marshalls
G114ENV	Premiere	G547VBB	Halifax JC	GIB6197	Weardale	H28YBV	Preston
G114NGN	Travel de Courcey	G556VBB	Renown	GIB8666	Weardale	H36YCW	Emsworth & D
G114PGT	D & G	G601KTX	N I B S	GIB976	Weardale	H37YCW	Emsworth & D
G122NGN	Travel de Courcey	G603KTX	N I B S	GIL2633	Yeomans	H48NDU	Heddingham
G122PGI	D & G	G604KTX	N I B S	GJ02LVT	Chalkwell	H64CCK	Fishwick
G147TYT	Geldards	G605KTX	N I B S	GJ52LUY	Chalkwell	H65CCK	Fishwick
G148TYT	Richardson	G607KTX	N I B S	GK04NZU	Kime's	H97PVW	Swanbrook
G16WMS	Weardale	G611CEF	Premiere	GL05BUS	Bennetts	H101BFR	Preston
G183JHG	Stephensons	G612OTV	Plymouth	GN53YUF	Nu-Venture	H102BFR	Preston
G184JHG	Go Whippet	G614CEF	Nu-Venture	GN54SVF	Chalkwell	H103BFR	Preston
G215KRN	Preston	G614OTV	Plymouth	GN54SVG	Chalkwell	H104BFR	Preston
G216LGK	Yeomans	G615CEF	Nu-Venture	GNH258F	Quantock	H109YHG	Blackpool
G218KRN	Preston	G615OTV	Plymouth	GNU750	Quantock	H110YHG	Blackpool
G285UMJ	Cedric	G620CEF	Premiere	GNZ3462	D & G	H112YHG	Blackpool
G304UYK	Rossendale	G621OTV	Plymouth	GNZ3561	D & G	H113YHG	Blackpool
G307UYK	Rossendale	G623OTV	Plymouth	GO05WMS	Weardale	H114ABV	Marshalls
G312UYK	Rossendale	G640CHF	Plymouth	GRN895W	Fishwick	H114YHG	Blackpool
G340KKW	Go Whippet	G640EVV	Weardale	GU51UVE	Ham's Travel	H115THE	Renown
G365YUR	Geldards	G641BPH	Cedar Cs	GU52HAO	Coastal	H115YHG	Blackpool
G370YUR	TM Travel	G643CHF	Plymouth	GU52HAX	Coastal	H116YHG	Blackpool

H118CHG	Blackpool	H549VAT	Stephensons	HX52WTE	Reading	J508GCD	Norfolk Green
H119CHG	Blackpool	H551VAT	Stephensons	HX52WTF	Reading	J530GCD	Norfolk Green
H119PVW	Swanbrook	H552GKX	Cedric	HX55EZF	Wardle	J535GCD	Norfolk Green
H122CHG	Blackpool	H552VAT	Stephensons	IAZ2314	Hopley's	J539GCD	Emsworth & D
H124THE	Ham's Travel	H554GKX	Carousel	IHZ8821	Wardle	J541GCD	Norfolk Green
H137PVW	Swanbrook	H556GKX	Carousel	IJZ2331	Wardle	J545GCD	D & G
H138MOB	Halifax JC	H563GKX	Carousel	INZ2296	Chalkwell	J548GCD	Emsworth & D
H146PVW	N Warrington	H564GKX	Carousel	IUI2138	Nu-Venture	J562HAT	Stephensons
H155BKH	Stephensons	H577MOC	Safeguard	J5BUS	Jim Stones	J563HAT	Stephensons
H156BKH	Stephensons	H620ACK	Red Rose	J7JFS	Fishwick	J564HAT	Stephensons
H158BKH	Stephensons	H705PTW	Ipswich	J14JFS	Fishwick	J564URW	Felix
H159BKH	Stephensons	H712LOL	Country Liner	J51EDM	Trustybus	J566HAT	Stephensons
H160HJN	Heddingham	H751PVW	Halifax JC	J54EDM	Aintree Coachlines	J569HAT	Stephensons
H177GTT	Plymouth	H764PVW	Halifax JC	J74CVJ	Yeomans	J599DUV	Compass Bus
H178GTT	Plymouth	H775PTW	Ipswich	J78VOE	Yeomans	J600WMS	Weardale
H17GOW	Yeomans	H784PVW	TM Travel	J100SOU	Compass Bus	J605XHL	Regal Busways
H187PVW	N Warrington	H794PVW	Swanbrook	J102DUV	Stansted Transit	J607KCU	Halifax JC
H191PVW	N Warrington	H803RWJ	TM Travel	J105DUV	Regal Busways	J615KCU	Pennine
H201DVM	Emsworth & D	H805AHA	Cedar Cs	J107KCW	Preston	J619CEV	TM Travel
H204DVM	Emsworth & D	H811WKH	Cedar Cs	J108KCW	Preston	J628CEV	TM Travel
H206LOM	Cedar Cs	H839PVW	Halifax JC	J109KCW	Preston	J629CEV	Halifax JC
H212LOM	Cedar Cs	H882LOX	APL Travel	J110KCW	Preston	J630CEV	Halifax JC
H215PVW	N Warrington	H902AHS	Yeomans	J112KCW	Preston	J636KCU	Midland Rider
H221LOM	Cedar Cs	H908PTW	Brylaine	J113KCW	Preston	J640CEV	Weardale
H225LOM	Ensignbus	H970XHR	Thamesdown	J114KCW	Preston	J642CEV	Halifax JC
H228LOM	N I B S	H971XHR	Thamesdown	J116WSC	Country Liner	J669LGA	Go Whippet
H229LOM	N I B S	H972XHR	Thamesdown	J117DUV	Central	J670LGA	Go Whippet
H237LOM	N I B S	H973XHR	Thamesdown	J123GRN	Blackpool	J687LGA	Go Whippet
H241LOM	N I B S	HFR507E	Blackpool	J124GRN	Blackpool	J688LGA	Go Whippet
H243LOM	N I B S	HG05URB	Ham's Travel	J125GRN	Blackpool	J689LGA	Go Whippet
H245LOM	N I B S	HIB644	Simonds	J126GRN	Blackpool	J697CEV	Cedar Cs
H247LOM	N I B S	HIJ6931	Anglian	J127LHC	Anglian	J702HMY	Cedar Cs
H258MFX	N I B S	HIL7391	Sanders Cs	J129GMP	Somerbus	J702KCU	Renown
H264MFX	N I B S	HIL7644	Ross Travel	J140DUV	Regal Busways	J717CEV	Brylaine
H303CAV	Go Whippet	HJA965E	Quantock	J155EDM	Aintree Coachlines	J722KBC	Go Whippet
H348SWA	Go Whippet	HKF151	Aintree Coachlines	J160LPV	Ipswich	J723KBC	Go Whippet
H367XGC	D & G	HKO169	Quantock	J201BVO	Sanders Cs	J724KBC	Heddingham
H457MEY	Yeomans	HLJ44	Ensignbus	J202VHN	Stansted Transit	J734CWT	Heddingham
H479PVW	Hornsby Travel	HOD55	Cedar Cs	J229HGY	Ham's Travel	J786KHD	Reading
H513RWX	N Warrington	HUI4165	Olympus	J248LLK	Somerbus	J808WFS	Go Whippet
H514RWX	N Warrington	HV52WSJ	Rossendale	J266SPR	N I B S	J810KHD	Sanders Cs
H515RWX	N Warrington	HV52WSK	Rossendale	J269SPR	N I B S	J811HMC	Geldards
H516RWX	N Warrington	HV52WSL	Rossendale	J295TWK	Heddingham	J811KHD	Compass Bus
H517RWX	N Warrington	HV52WSN	Rossendale	J377AWT	Midland Rider	J813HMC	Ensignbus
H519RWX	N Warrington	HV52WSO	Rossendale	J418PRW	Green Bus	J814NKK	Abus
H547GKX	Cedric	HWR449T	Pennine	J502GCD	Emsworth & D	J815HMC	Geldards
H547VAT	Stephensons	HX52WTC	Reading	J503GCD	Emsworth & D	J816HMC	Geldards
H548GKX	Cedric	HX52WTD	Reading	J504GCD	Emsworth & D	J817HMC	Geldards

English Bus Handbook: Notable Independents

Reg	Operator	Reg	Operator	Reg	Operator	Reg	Operator
J837TSC	Marshalls	JSK261	Plymouth	K346OFM	Wardle	K891CSF	Thamesdown
J838TSC	Heddingham	JSK262	Plymouth	K347OFM	Wardle	K892CSF	Thamesdown
J841TSC	Velvet	JSK263	Plymouth	K357DWJ	Heddingham	K893CSF	Thamesdown
J843TSC	Velvet	JSK264	Plymouth	K360DWJ	Cedric	K955PBG	Regal Busways
J844TSC	TM Travel	JSK265	Plymouth	K405FHJ	D & G	K984SCU	Pennine
J845TSC	Hornsby Travel	JSK492	Cedar Cs	K408FHJ	Olympus	K985CBO	Halifax JC
J850TSC	TM Travel	JTE546	Quantock	K414FHJ	Regal Busways	K986SCU	Pennine
J851TSC	Velvet	JUI4236	Simonds	K414MGN	Halifax JC	K989SCU	Pennine
J853TSC	Konectbus	JWV259W	W Greyhound	K430OKH	Country Liner	K993CBO	Halifax JC
J854TSC	TM Travel	JXC194	Ensignbus	K432OKH	Speedwell	K995CBO	Halifax JC
J855TSC	Konectbus	JXC432	Ensignbus	K434OKH	Uno	KAZ4127	Compass Bus
J908CEV	Swanbrook	K2SLT	South Lancs	K439DRW	APL Travel	KC03OSE	Uno
J923MKC	Halton	K3SLT	South Lancs	K479JHJ	Country Liner	KC03OSF	Uno
J924MKC	Halton	K16SLT	South Lancs	K482GNN	TM Travel	KC03PGK	Uno
J926MKC	Halton	K17SLT	South Lancs	K482VVR	TM Travel	KC03PGU	Uno
J927MKC	Halton	K18SLT	South Lancs	K508ESS	Heddingham	KDX108	Simonds
J954MFT	Pennine	K33GOW	Quantock	K518ESS	Heddingham	KDZ5805	Midland Rider
J976PRW	Preston	K40SLT	South Lancs	K574NHC	D & G	KE03TWA	Uno
J992XKU	Geldards	K50APL	APL Travel	K578NHC	Norfolk Green	KE03TWC	Uno
JAM145E	Thamesdown	K50SLT	South Lancs	K623UFR	Green Bus	KE04UMA	Uno
JB51BUS	Jim Stones	K58LLG	Trustybus	K640FAU	Sanders Cs	KE04UMB	Uno
JDZ2349	Midland Rider	K59LLG	Trustybus	K699ERM	Go Whippet	KE04UMC	Uno
JFJ875	Quantock	K60SLT	South Lancs	K702ERM	Go Whippet	KE04UMD	Uno
JFR10W	Shamrock	K70SLT	South Lancs	K703BBL	Ensignbus	KE51VJC	Uno
JFR11W	Shamrock	K92BNY	Trustybus	K703PCN	Renown	KE51VJD	Uno
JFR13W	Shamrock	K100ACL	Aintree Coachlines	K710KGU	Stansted Transit	KE53LMY	Uno
JG9938	Quantock	K100BLU	Reliance	K710PCN	Renown	KE53VDP	Stansted Transit
JHA246L	Travel de Courcey	K100SDC	Central	K711ASC	N Warrington	KE53VDY	Stansted Transit
JIL2426	Pennine	K102SFJ	Plymouth	K712ASC	N Warrington	KF02ZXX	Simonds
JIL2428	Pennine	K103SFJ	Plymouth	K713ASC	Go Whippet	KF52NBM	Uno
JIL2705	Pennine	K104SFJ	Plymouth	K716PCN	Regal Busways	KF52NBN	Uno
JIL3959	Bodmans	K105SFJ	Plymouth	K814HMV	Cedric	KFM767	Quantock
JIL7416	Pennine	K107JWJ	Heddingham	K815HMV	Cedric	KGK708	Ensignbus
JIL7417	Pennine	K107SFJ	Plymouth	K816HMV	Cedric	KGK758	Ensignbus
JJ58BLU	Bluebird	K109SFJ	Plymouth	K825NKH	Speedwell	KHU323P	South Lancs
JJD405D	Ensignbus	K110SFJ	Plymouth	K827NKH	Speedwell	KIA891	Simonds
JJD406D	Country Liner	K127UFV	Blackpool	K852MTJ	Halton	KK07BLU	Bluebird
JJZ3437	D & G	K128UFV	Preston	K853MTJ	Halton	KKZ7562	Olympus
JJZ5248	D & G	K129UFV	Blackpool	K859PCN	Midland Rider	KM51BFN	Country Liner
JJZ5250	D & G	K130UFV	Blackpool	K860PCN	Midland Rider	KM51BFZ	Stansted Transit
JJZ5278	D & G	K132TCP	Brylaine	K864LGN	Country Liner	KN04CSU	Ham's Travel
JJZ5289	D & G	K136ARE	Green Bus	K877UDB	Regal Busways	KP51SXU	Stansted Transit
JJZ5291	D & G	K159PGO	Abus	K879ODY	Norfolk Green	KP51SXV	Country Liner
JJZ5312	D & G	K198EVW	Heddingham	K882CSF	Thamesdown	KP51SXZ	Stansted Transit
JJZ5368	D & G	K250CBA	Midland Rider	K884CSF	Thamesdown	KP51SYA	Country Liner
JP54JON	West End Travel	K301WTA	Plymouth	K886CSF	Thamesdown	KP51SYC	Country Liner
JP55WET	West End Travel	K321YKG	Green Bus	K887CSF	Thamesdown	KP51SYE	Supertravel
JRA635	TM Travel	K345OFM	Wardle	K889CSF	Thamesdown	KP51SYR	Supertravel

KP51UFL	Renown	KX53SBZ	Premiere	L54CNY	Sargeants	L302YOD	Plymouth
KR1728	Ensignbus	KX53SDO	Premiere	L56UNS	Go Whippet	L308YDU	Green Bus
KR8385	Ensignbus	KX53SGY	Thames Travel	L61PDM	Trustybus	L313AUT	Tyrer Bus
KSP329X	TM Travel	KX54NLA	Courtney	L63SFM	Trustybus	L321XTC	White Bus
KTA356V	Minsterley Ms	KX54NLC	Courtney	L64SFM	Trustybus	L328CHB	Green Bus
KTF594	Quantock	KX54NLD	Courtney	L67UNG	Heddingham	L330CHB	D & G
KTJ204C	Ensignbus	KX56HCZ	Thames Travel	L69UNG	Heddingham	L351MRR	TM Travel
KTL780	Delaine	KX57BWE	Premiere	L70ARK	Aintree Coachlines	L352MRR	TM Travel
KU02YBG	Country Liner	KX57FMA	Stansted Transit	L91WBX	Sargeants	L353MRR	TM Travel
KU52RXX	Country Liner	KX57FMC	Stansted Transit	L100ACL	Aintree Coachlines	L452UEB	Emsworth & D
KUI2269	Shamrock	KX57FMD	Stansted Transit	L108HHV	Heddingham	L455YAC	Go Whippet
KUI9951	Quantock	KX57FME	Stansted Transit	L109LHL	Heddingham	L456YAC	Go Whippet
KV03ZFJ	Simonds	KX57FMF	Stansted Transit	L110HHV	Regal Busways	L463RDN	Bennetts
KV03ZFK	Simonds	KX57FMG	Stansted Transit	L112YOD	Plymouth	L476CFT	Compass Bus
KV03ZGL	Premiere	KX57OVR	JPT	L113YOD	Plymouth	L478CFT	Compass Bus
KV03ZGM	Premiere	KX57OVT	JPT	L114YOD	Plymouth	L481CFT	Compass Bus
KV03ZGN	Premiere	KX57OVU	JPT	L115YOD	Plymouth	L488CFT	Compass Bus
KV03ZGP	Stansted Transit	KX57OVY	JPT	L116YOD	Plymouth	L517EHD	Sanders Cs
KV03ZGR	Stansted Transit	KX58GTF	Trustybus	L117YOD	Plymouth	L519EHD	Bennetts
KV51KZC	Stansted Transit	KXW435	Ensignbus	L118YOD	Plymouth	L523MJB	White Bus
KV51KZD	Beestons	KY51SXD	Wardle	L119YOD	Plymouth	L561ASU	White Bus
KV51KZF	Beestons	KYN300X	Go Whippet	L120YOD	Plymouth	L606ASU	White Bus
KV51KZJ	Country Liner	KYO624X	Halifax JC	L121YOD	Plymouth	L657MYG	South Lancs
KVY789N	Thornes	KYV312X	Ham's Travel	L122YOD	Plymouth	L666LMT	APL Travel
KW02DSE	Weavaway	KYV347X	Nu-Venture	L123YOD	Plymouth	L666WMS	Weardale
KX03HYW	Premiere	KYV348X	Nu-Venture	L124YOD	Plymouth	L686CDD	Norfolk Green
KX03HZC	Tyrer Bus	KYV415X	Nu-Venture	L125YOD	Plymouth	L717OMV	Country Liner
KX03HZD	Tyrer Bus	KYV455X	Nu-Venture	L126YOD	Plymouth	L718OMV	Country Liner
KX04HRA	Courtney	KYV481X	Ham's Travel	L141VOM	Aintree Coachlines	L719OMV	Country Liner
KX06LXN	Thames Travel	KYV633X	Ensignbus	L149WAG	JPT	L720OMV	Renown
KX06LXO	Thames Travel	KYV643X	Halifax JC	L162ADX	Ipswich	L720SNO	TM Travel
KX06LYP	Country Liner	KYV670X	Ensignbus	L192MAU	Olympus	L726SNO	TM Travel
KX07HDG	Stansted Transit	KYV703X	Carousel	L194OVO	Swanbrook	L754VNL	Wardle
KX07HDH	Stansted Transit	KYV730X	Halifax JC	L195OVO	Swanbrook	L762SNO	Stephensons
KX07HDJ	Stansted Transit	KYV737X	Carousel	L202HYE	Heddingham	L780GMJ	Bodmans
KX08DNC	Premiere	KYV758X	Carousel	L202SKD	Aintree Coachlines	L808SNO	Stephensons
KX08HLR	Stansted Transit	KYY961	Ensignbus	L204SKD	Geldards	L833YDS	Heddingham
KX08HLU	Stansted Transit	L1JFS	Fishwick	L205SKD	Geldards	L86VDM	Aintree Coachlines
KX08HMD	Courtney	L1SLT	Aintree Coachlines	L207RNO	Heddingham	L881YVK	Renown
KX08HME	Courtney	L1WET	West End Travel	L207SKD	Geldards	L944HTM	Tyrer Bus
KX08OLO	Weavaway	L2SLT	Aintree Coachlines	L208RNO	Heddingham	L950MSC	Bennetts
KX08OMB	Weavaway	L4YTB	Regal Busways	L211YAG	Premiere	L951MSC	Bennetts
KX08OMC	Weavaway	L6HAM	Ham's Travel	L212TWM	Beestons	L951NBH	Uno
KX08OMD	Weavaway	L8YCL	Geldards	L215SKD	Geldards	L952MSC	Bennetts
KX08OME	Weavaway	L9YCL	Geldards	L255NFA	Green Bus	L953MSC	Stephensons
KX08OMF	Weavaway	L26FNE	Rossendale	L273LHH	Avon Buses	L954MSC	Thamesdown
KX08OMG	Weavaway	L27FNE	Rossendale	L287MJH	Compass Bus	L955MSC	Stephensons
KX51UDG	Country Liner	L51UNS	Go Whippet	L289ETG	Ross Travel	L957MSC	Thamesdown

English Bus Handbook: Notable Independents

L958MSC	Bennetts	M6WMS	Weardale	M157RBH	Olympus	M379SCK	Blackpool
L962MSC	Stephensons	M20HAM	Ham's Travel	M166LNC	South Lancs	M407BFG	Swanbrook
L966MSC	Thamesdown	M26XEH	Quantock	M167LNC	South Lancs	M408RVU	Stephensons
L967MSC	Thamesdown	M33ARJ	Andybus	M185UAN	Avon Buses	M409BFG	Swanbrook
L968MSC	Thamesdown	M41EPV	Ipswich	M191UAN	Avon Buses	M409RVU	Stephensons
L969MSC	Thamesdown	M42EPV	Ipswich	M197UAN	Avon Buses	M410RVU	Stephensons
L979MSC	Cedar Cs	M50APL	APL Travel	M203EGF	Bodmans	M437PUY	Hulleys
L981MSC	Cedar Cs	M51HOD	Plymouth	M206VSX	Thamesdown	M441CVG	Heddingham
LA02WMZ	TM Travel	M52HOD	Plymouth	M207VSX	Thamesdown	M452LLJ	Emsworth & D
LBT380N	Thornes	M53HOD	Plymouth	M208BGK	Yeomans	M453LLJ	Premiere
LDS279A	Ensignbus	M54PRA	South Lancs	M210VEV	Heddingham	M453VCW	Stephensons
LFM302	Quantock	M65CYJ	Renown	M211VSX	Thamesdown	M456EDH	D & G
LG06HSF	Weardale	M67HHB	D & G	M211WHJ	Heddingham	M456LLJ	Emsworth & D
LIL2493	Sanders Cs	M68CYJ	Renown	M212WHJ	Heddingham	M458UUR	Uno
LJH665	Quantock	M69CYJ	Regal Busways	M213VSX	Thamesdown	M465LPG	West End Travel
LJX198	Halifax JC	M74VJO	Nu-Venture	M216VSX	Thamesdown	M501VJO	Thames Travel
LK03PZW	Olympus	M79CYJ	Red Rose	M225VSX	Bennetts	M517NCG	Heddingham
LK07GTF	Trustybus	M81MYM	Chambers	M231VSX	Thamesdown	M530VWT	Olympus
LK55ABU	Nu-Venture	M84MYM	Ham's Travel	M233TBV	Stephensons	M571XKY	Heddingham
LK55ABV	Nu-Venture	M85MYM	Geldards	M239YCM	N Warrington	M578UBA	Swanbrook
LK55ABX	Nu-Venture	M87MYM	Geldards	M240YCM	N Warrington	M588UBA	Swanbrook
LL07BLU	Bluebird	M88SLT	South Lancs	M241XWS	Bodmans	M593HKH	N Warrington
LLZ3249	D & G	M91MYM	Ham's Travel	M241YCM	N Warrington	M604TTV	Heddingham
LN03AYL	Trustybus	M92JHB	Green Bus	M242YCM	N Warrington	M612RCP	Sanders Cs
LN03AYM	Trustybus	M97WBW	Renown	M243YCM	N Warrington	M640EPV	Ipswich
LN54VXL	Ham's Travel	M99SLT	South Lancs	M246YWM	N Warrington	M646RCP	Ensignbus
LNN353	Marshalls	M101BLE	Premiere	M247YWM	N Warrington	M647RCP	Ensignbus
LSU413	Beestons	M107UWY	Heddingham	M248YWM	N Warrington	M655KVU	Chambers
LUA255V	Emsworth & D	M110TVH	Supertravel	M261KWK	Heddingham	M711BMR	APL Travel
LUA282V	Western Greyhound	M112BMR	Bodmans	M262KWK	Heddingham	M718WUD	D & G
LUI1687	Yeomans	M112UWY	Bodmans	M273HOD	Plymouth	M733AOO	Olympus
LUI6233	Premiere	M113BMR	Wardle	M273UKN	Heddingham	M736AOO	Olympus
LUI7627	Premiere	M115BMR	Bodmans	M274HOD	Plymouth	M763RCP	Geldards
LUI7662	Premiere	M115XLV	Speedwell	M276UKN	Heddingham	M764RCP	Sanders Cs
LUI7665	Premiere	M118BMR	APL Travel	M282UKN	Heddingham	M774XHW	Bodmans
LUI7668	Premiere	M127HOD	Plymouth	M284HRH	N Warrington	M802PRA	Hulleys
LUI7672	Premiere	M128HOD	Plymouth	M294UKN	Heddingham	M803PRA	Hulleys
LUI8478	Premiere	M129HOD	Plymouth	M301KRY	Felix	M806RCP	Sanders Cs
LUI9633	D & G	M130HOD	Plymouth	M304KOD	Plymouth	M828RCP	Pennine
LUI9649	D & G	M131HOD	Plymouth	M305KOD	Plymouth	M832CVG	Heddingham
LV02LKD	Olympus	M132HOD	Plymouth	M311KRY	Sargeants	M832RCP	Sanders Cs
LX03KPE	Country Liner	M144UKN	Heddingham	M341SCJ	Yeomans	M834RCP	Sanders Cs
LXI9357	Yeomans	M146UKN	Heddingham	M370LAX	Green Bus	M836RCP	Sanders Cs
M1BUS	Jim Stones	M146VVS	Uno	M374SCK	Blackpool	M890GBB	Premiere
M10CT	Delaine	M148VVS	Uno	M375SCK	Blackpool	M941JBO	Yeomans
M1WET	West End Travel	M150HPL	Trustybus	M376SCK	Blackpool	M960VWY	Sargeants
M20CT	Delaine	M153RBH	Olympus	M377SCK	Blackpool	M966SDP	Avon Buses
M3KFC	Emsworth & D	M153XHW	Bodmans	M378SCK	Blackpool	M988NAA	Heddingham

MA02BLU	Bluebird	MRT9P	Ipswich	MX08DGO	Tyrer Bus	MX56HYA	Reading
MAZ6907	Beestons	MTD235	Pennine	MX08DGU	Tyrer Bus	MX56HYO	West End Travel
MB02BLU	Bluebird	MUH287X	Ipswich	MX08DGV	Tyrer Bus	MX56HYP	West End Travel
MBO1F	Thornes	MUI7843	Olympus	MX08DHA	Tyrer Bus	MX56NLD	Speedwell
MC02HAM	Ham's Travel	MV03AJY	Olympus	MX08DHF	Brylaine	MX56NLU	Speedwell
MC06HAM	Ham's Travel	MV54BLU	Bluebird	MX08DHK	Tyrer Bus	MX56NLZ	Country Liner
MC07HAM	Ham's Travel	MV54KYJ	Supertravel	MX08DHM	Tyrer Bus	MX56WWA	Bluebird
MC52HAM	Ham's Travel	MW52PZD	TM Travel	MX08DHN	Tyrer Bus	MX56WWB	Bluebird
MC56HAM	Ham's Travel	MW54BLU	Bluebird	MX08MYM	Tyrer Bus	MX56WWC	Bluebird
MCO658	Plymouth	MX03EHD	Premiere	MX08MYO	Tyrer Bus	MX57CBO	Tyrer Bus
MD02BLU	Bluebird	MX03YCK	Tyrer Bus	MX08MYT	Tyrer Bus	MX57CBU	Tyrer Bus
MDJ918E	Halifax JC	MX03YCL	Tyrer Bus	MX08PZH	JPT	MX57CBV	Tyrer Bus
ME52BLU	Bluebird	MX03YCN	TM Travel	MX08TCU	JPT	MX57CBY	Tyrer Bus
MF03BLU	Bluebird	MX03YCP	TM Travel	MX08UZK	N I B S	MX57CCAY	Tyrer Bus
MFX174W	Shamrock	MX03YDB	Norfolk Green	MX08UZT	Konectbus	MX57OEL	Bluebird
MG53BLU	Bluebird	MX03YDF	TM Travel	MX08UZU	Konectbus	MX58ABF	Safeguard
MH53BLU	Bluebird	MX04AAE	Hulleys	MX08WCJ	JPT	MX58ABV	Konectbus
MHJ347F	Western Greyhound	MX04KUO	W Greyhound	MX08WCK	JPT	MX58SGU	Bluebird
MHS5P	Nu-Venture	MX04VLM	TM Travel	MX08WCT	JPT	MX58SGV	Bluebird
MIB767	Quantock	MX04VLY	Speedwell	MX53FDD	TM Travel	MXX261	Ensignbus
MJI2364	Travel de Courcey	MX04VLZ	Speedwell	MX53FDM	Anglian	N1HAM	Ham's Travel
MJI2365	Travel de Courcey	MX04YLR	Chalkwell	MX53FDN	Anglian	N2FPK	Nu-Venture
MJI2366	Travel de Courcey	MX05EKW	Konectbus	MX53FDO	Anglian	N3ARJ	Andybus
MJI2367	Travel de Courcey	MX05EKY	Konectbus	MX53FDP	Anglian	N3BLU	Pennine
MJI2368	Travel de Courcey	MX05EKZ	Konectbus	MX53SXO	Supertravel	N3FPK	Nu-Venture
MJI2370	Travel de Courcey	MX05ELC	Konectbus	MX53ZWC	Supertravel	N3OCT	Delaine
MJI3529	Travel de Courcey	MX05ELH	Konectbus	MX54BLU	Bluebird	N5BLU	Stansted Transit
MJI3751	Travel de Courcey	MX05ELJ	Konectbus	MX54KXN	Cedar Cs	N9LON	APL Travel
MJI7472	Travel de Courcey	MX05OSW	Supertravel	MX54KXO	Bakerbus	N33SCS	Sanders Cs
MJI7861	Travel de Courcey	MX05OTM	Tyrer Bus	MX54KXY	Bakerbus	N44SCS	Sanders Cs
MJI7862	Travel de Courcey	MX05OTN	Bakerbus	MX54KXZ	Somerbus	N50HAM	Ham's Travel
MJI7863	Travel de Courcey	MX05OTP	Bakerbus	MX54KYH	Speedwell	N53KBW	Pennine
MK04VMK	TM Travel	MX06ABF	Swanbrook	MX54WMJ	Norfolk Green	N60HAM	Ham's Travel
ML53YKU	Sargeants	MX06ABO	Speedwell	MX54WMK	Marshalls	N63MDW	Sanders Cs
MLL528	Western Greyhound	MX06ABU	Speedwell	MX55BXN	Supertravel	N67FWU	Sanders Cs
MLI735	Ensignbus	MX06ACV	Hulleys	MX55BYC	Supertravel	N68MDW	Sanders Cs
MM51XVB	Bodmans	MX06ACY	Compass Bus	MX55BYD	Supertravel	N70SLF	Stephensons
MM51YCB	Bodmans	MX06BOV	Speedwell	MX55BYF	Supertravel	N101CKN	Nu-Venture
MMY991C	South Lancs	MX06BPU	Speedwell	MX55BYO	Supertravel	N101UTT	Plymouth
MNZ1138	Wardle	MX06BSZ	Norfolk Green	MX55BYP	Supertravel	N102CKN	Nu-Venture
MOI4000	DRM	MX06YXU	Konectbus	MX55WCU	Norfolk Green	N102UTT	Plymouth
MOI5055	DRM	MX07BAV	Bakerbus	MX55WCV	Norfolk Green	N103UTT	Plymouth
MOI9565	DRM	MX07BBE	Bakerbus	MX55WCW	Speedwell	N104UTT	Plymouth
MP04BLU	Bluebird	MX07NTJ	Weardale	MX56AAV	Supertravel	N105UTT	Plymouth
MPP747	Cedar Cs	MX07NTK	Ross Travel	MX56AAY	Supertravel	N106LCK	Rossendale
MRD1	Reading	MX07NTL	Sanders Cs	MX56AAZ	Supertravel	N107LCK	Rossendale
MRJ8W	Heddingham	MX07NTM	Sanders Cs	MX56ACF	Hulleys	N107UTT	Plymouth
MRJ233W	Stephensons	MX08DGE	Tyrer Bus	MX56BZY	Uno	N108LCK	Rossendale

Reg	Operator	Reg	Operator	Reg	Operator	Reg	Operator
N108UTT	Plymouth	N419JBV	Aintree Coachlines	N722FKK	Brylaine	OFS702Y	Weavaway
N109LCK	Rossendale	N421ENM	Halton	N723FKK	Brylaine	OFV14X	Shamrock
N109UTT	Plymouth	N421JBV	Aintree Coachlines	N724FKK	Brylaine	OFV23X	Shamrock
N110LCK	Rossendale	N422ENM	Halton	N725FKK	Brylaine	OFV620X	Fishwick
N110UTT	Plymouth	N422JBV	Aintree Coachlines	N726KGF	Country Liner	OGR625T	Weardale
N112UTT	Plymouth	N423ENM	Halton	N731RDD	Emsworth & D	OGR647	Sanders Cs
N126XEG	Bluebird	N423ENM	Uno	N734EOT	Aintree Coachlines	OHV806Y	Speedwell
N128LMW	Thamesdown	N427JBV	Heddingham	N748YVR	Tyrer Bus	OIJ721	Ross Travel
N131XEG	Bodmans	N429FKK	Uno	N777ELK	Premiere	OIL4201	Weavaway
N134XND	Pennine	N429GBV	Preston	N784EUA	D & G	OJD822Y	Speedwell
N139GMJ	Olympus	N430FKK	Uno	N784JBM	Red Rose	OKP980	Thornes
N189EMJ	D & G	N430GBV	Preston	N801SJU	Yeomans	OKZ9847	Wardle
N209ONL	Quantock	N431FKK	Uno	N802GRV	Red Rose	OLZ7276	Wardle
N211HGO	Yeomans	N450DWJ	Midland Rider	N803GRV	Red Rose	OO06BLU	Bluebird
N222LFR	Andybus	N451DWJ	Midland Rider	N805NHS	Renown	OU04FMV	Thames Travel
N241EWC	Heddingham	N453VOD	Travel de Courcey	N808PDS	D & G	OU05AVB	Weavaway
N258DUR	South Lancs	N462VDD	Green Bus	N819RFP	South Lancs	OU05AVD	Weavaway
N260PRJ	Pennine	N464VOD	Travel de Courcey	N890SBB	Bodmans	OU05AVY	Weavaway
N275PDV	Plymouth	N472EHA	D & G	N904HWY	Country Liner	OU05AWJ	Weavaway
N276PDV	Plymouth	N506LUA	Compass Bus	N905HWY	Nu-Venture	OU05KKB	Weavaway
N277PDV	Plymouth	N529LHG	Chambers	N952KBJ	Chambers	OU05KLA	Weavaway
N278PDV	Plymouth	N531LHG	Chambers	N967ENA	Norfolk Green	OU07FKH	Red Rose
N279PDV	Plymouth	N532LHG	Chambers	N988FWT	Sanders Cs	OU07FKJ	Red Rose
N280PDV	Plymouth	N540LHG	Heddingham	NBZ1676	Green Bus	OU08EHO	Red Rose
N281PDV	Plymouth	N548LHG	Ensignbus	NBZ70	Weavaway	OU08EHP	Red Rose
N282PDV	Plymouth	N561UPF	Safeguard	NDZ70	Weavaway	OU08HGM	Thames Travel
N283PDV	Plymouth	N584JND	N Warrington	NEL1F	West End Travel	OU08HGN	Thames Travel
N284PDV	Plymouth	N590GRN	Blackpool	NEZ9506	Wardle	OU08HGO	Thames Travel
N286PDV	Plymouth	N593DWY	Sanders Cs	NFR748T	Green Bus	OU54PGZ	Thames Travel
N287PDV	Plymouth	N593LFV	Blackpool	NHG541	Somerbus	OU57FGV	Thames Travel
N288PDV	Plymouth	N594DWY	Sanders Cs	NHM465X	Ham's Travel	OU57FGX	Thames Travel
N289PDV	Plymouth	N597DWY	Sanders Cs	NIL9517	Premiere	OU57FGZ	Thames Travel
N300EST	Rossendale	N607JNO	Brylaine	NIL9777	Weardale	OU57FHA	Thames Travel
N326HUA	Heddingham	N611WND	Safeguard	NIW6517	Beestons	OU57FKB	Thames Travel
N355OBC	Chalkwell	N618VSS	Green Bus	NK55KBU	Reading	OUC45R	Emsworth & D
N357VRC	Somerbus	N624JNO	Chalkwell	NML616E	Stansted Transit	OUI2376	D & G
N360VRC	Hulleys	N630XBU	Go Whippet	NMW329X	Weavaway	OUI3925	Wardle
N361VRC	Hulleys	N633XBU	Go Whippet	NMY648E	W Greyhound	OUI6274	Regal Busways
N362VRC	Hulleys	N644JNO	Chalkwell	NMY655E	Ensignbus	OUI7148	D & G
N401ARA	Aintree Coachlines	N645HSX	Speedwell	NUW552Y	Nu-Venture	OUI8418	Cedric
N402ARA	Aintree Coachlines	N646FKK	Abus	NUW592Y	Nu-Venture	OUI9120	D & G
N403ARA	Aintree Coachlines	N647JNO	Swanbrook	NUW652Y	Nu-Venture	OUI9143	D & G
N404ARA	Aintree Coachlines	N653THO	Bodmans	NXP775	Ensignbus	OWY197K	Pennine
N411MPN	Country Liner	N664THO	Heddingham	NYH161Y	Emsworth & D	OXK76	Marshalls
N414JBV	Aintree Coachlines	N667THO	Heddingham	OAE954M	Abus	OYD693	Country Liner
N414MPN	Country Liner	N682YAV	Cedar Cs	OEZ2159	Wardle	P10TL	Delaine
N415JBV	Aintree Coachlines	N720FKK	Brylaine	OFS683Y	Stephensons	P2JPT	JPT
N416MPN	Country Liner	N721FKK	Brylaine	OFS684Y	Blackpool	P2OTL	Delaine

P5ACL	Aintree Coachlines	P240EJW	Chambers	P488MBY	Chambers	P685RWU	Olympus
P6JPT	JPT	P255RUM	Swanbrook	P489MBY	Heddingham	P686RWU	Country Liner
P6WMS	Weardale	P255SMW	Thamesdown	P493FRR	Cedar Cs	P688KCC	D & G
P10TLS	Trustybus	P263NRH	Red Rose	P493MBY	Aintree Coachlines	P688RWU	Country Liner
P20VHO	Thornes	P264NRH	Red Rose	P494MBY	Heddingham	P690RUU	Renown
P33TCC	Sanders Cs	P268SWC	N Warrington	P495SWC	Ensignbus	P691RWU	Renown
P34KWA	Sanders Cs	P273NRH	Hulleys	P498GKJ	Olympus	P696HND	Pennine
P35KWA	Sanders Cs	P292MLD	Bodmans	P501OUG	Trustybus	P696RUU	Renown
P71MOV	Yeomans	P302MLD	Bodmans	P501UFR	Blackpool	P697RWU	Renown
P77SCS	Sanders Cs	P302SWC	N Warrington	P502UFR	Blackpool	P699RUU	Renown
P82MOR	Ensignbus	P304MLD	Bodmans	P503MOT	Ensignbus	P71000A	Country Liner
P87SAF	Delaine	P306MLD	Bodmans	P503UFR	Blackpool	P724RYL	Country Liner
P104OLX	Heddingham	P307HDP	Country Liner	P504MOT	Ensignbus	P725RYL	Country Liner
P112SGS	Delaine	P314SWC	N Warrington	P504OUG	Trustybus	P738RYL	Renown
P120GSR	Hulleys	P324SWC	N Warrington	P504UFR	Blackpool	P746HND	Renown
P123HCH	JPT	P337ROO	TM Travel	P505MOT	Ensignbus	P749HND	Stephensons
P128RWR	Brylaine	P337SWC	N Warrington	P505OUG	TM Travel	P766BJU	Leven Valley
P129RWR	Brylaine	P341OEW	Renown	P506OUG	Trustybus	P767PCL	Sanders Cs
P130PPV	Ipswich	P342OEW	Renown	P507NWU	Red Rose	P768PCL	Sanders Cs
P131PPV	Ipswich	P343OEW	Renown	P508OUG	Trustybus	P776BJU	Leven Valley
P131RWR	White Bus	P343ROO	Go Whippet	P509NWU	Ham's Travel	P844KOT	Trustybus
P132PPV	Ipswich	P344OEW	Renown	P509OUG	Premiere	P848YGB	Chalkwell
P134MEH	D & G	P345ROO	Go Whippet	P509PUM	Sargeants	P850YGB	Chalkwell
P151SMW	Thamesdown	P347ROO	South Lancs	P510OUG	Premiere	P861VFG	Travel de Courcey
P152SMW	Thamesdown	P348ROO	Go Whippet	P513UUG	D & G	P862VFG	Travel de Courcey
P153SMW	Thamesdown	P353ROO	Marshalls	P514CVO	Ensignbus	P863VFG	Travel de Courcey
P154SMW	Thamesdown	P368SWC	N Warrington	P514UUG	Richardson	P864VFG	Travel de Courcey
P156SMW	Thamesdown	P384SWC	N Warrington	P515UUG	Chambers	P867PWW	Geldards
P157SMW	Thamesdown	P395AAA	Heddingham	P530CLJ	Heddingham	P881PWW	Hulleys
P158SMW	Thamesdown	P411SWC	N Warrington	P542SCL	Sanders Cs	P894FMO	APL Travel
P159VHR	Thamesdown	P423MLE	Carousel	P543SCL	Sanders Cs	P895PWW	Country Liner
P160VHR	Thamesdown	P425VRG	Country Liner	P549WGT	Chambers	P901RYO	Chambers
P161VHR	Thamesdown	P426SWC	Hopley's	P570APJ	Norfolk Green	P902RYO	Chambers
P175UAD	Wardle	P426UUG	Ensignbus	P647SBH	Uno	P910MOR	Ensignbus
P190SGV	Norfolk Green	P427UUG	Ensignbus	P659BUB	Hulleys	P943EBB	Renown
P191SGV	Ipswich	P429UUG	Ensignbus	P664PNM	Uno	P958YGG	Rossendale
P192SGV	Ipswich	P442SWX	Ipswich	P665PNM	Uno	P959YGG	Rossendale
P195SGV	Norfolk Green	P466GUA	Rossendale	P666WMS	Weardale	P960DNR	Cedric
P196SGV	Ipswich	P475MBY	Chambers	P667PNM	Uno	PA52HAM	Ham's Travel
P201RUM	Brylaine	P476MBY	Geldards	P668PNM	Uno	PAZ3184	Kime's
P202BNR	Carousel	P476SWC	Anglian	P677RWU	Olympus	PAZ9346	Kime's
P205RUM	Brylaine	P478CBV	Supertravel	P678RWU	Olympus	PCZ6034	Wardle
P211DCK	Rossendale	P478MBY	Geldards	P679RWU	JPT	PE51YHF	Preston
P212DCK	Rossendale	P478SWC	Anglian	P680RWU	JPT	PE51YHG	Preston
P213DCK	Rossendale	P479MLE	Carousel	P681RWU	Olympus	PE51YHH	Preston
P214RUM	Bodmans	P480MLE	Carousel	P682RWU	Country Liner	PE51YHJ	Preston
P225EJW	Chambers	P484MBY	Chambers	P684RWU	Olympus	PE51YHK	Preston
P236EJW	Chambers	P487SWC	TM Travel	P685HND	Norfolk Green	PE51YHL	Preston

Reg	Operator	Reg	Operator	Reg	Operator	Reg	Operator
PE51YHM	Preston	PJ03TGE	Blackpool	PN07NTV	Preston	PO51WNG	Reading
PE51YHN	Preston	PJ53OLA	Ipswich	PN08SVK	Preston	PO51WNJ	Reading
PE55WMD	Preston	PJ53OLB	Ipswich	PN08SVL	Preston	PO51WNK	Reading
PE55WMF	Preston	PJ53OLC	Ipswich	PN08SVO	Preston	PO51WNL	Reading
PE55WMG	Preston	PJ53OLE	Ipswich	PN08SVP	Preston	PO51WNM	Reading
PF02XMW	Rossendale	PJ53OLG	Ipswich	PN08SVR	Preston	PO51WNN	Kime's
PF02XMX	Rossendale	PJ53OLH	Ipswich	PN08SVS	Preston	PO53OBM	Rossendale
PF04WML	Ipswich	PJ53OLK	Ipswich	PN08SVT	Preston	PO53OBN	Rossendale
PF06ENL	Thames Travel	PJ53UHV	Rossendale	PN08SVU	Preston	PO53OBP	Rossendale
PF06ENM	Thames Travel	PJ54YZT	Ipswich	PN08SVZ	Uno	PO53OBR	Rossendale
PF06ENN	Thames Travel	PJ54YZU	Ipswich	PN08SWF	Uno	PO53OBT	Rossendale
PF06ENO	Thames Travel	PJ54YZV	Ipswich	PN52WWK	Rossendale	PO53OBU	Rossendale
PF06EZL	Blackpool	PJI9172	DRM	PN52WWL	Rossendale	PO56JDF	Preston
PF06EZM	Blackpool	PL03BPZ	Blackpool	PN52WWM	Rossendale	PO56JDJ	Preston
PF06EZN	Blackpool	PL06RYO	Preston	PN52WWO	Rossendale	PO56JDK	Preston
PF06EZO	Blackpool	PL06RYP	Preston	PN52WWP	Rossendale	PO56JDU	Preston
PF06EZP	Blackpool	PL06TFX	Anglian	PN52WWR	Rossendale	PO56RNY	Preston
PF51KMM	Rossendale	PL08YLZ	Sanders Cs	PN52XBJ	Ipswich	PO56RNZ	Preston
PF51KMO	Rossendale	PL08YMA	Sanders Cs	PN52XBK	Ipswich	PO56ROH	Preston
PF51KMU	Rossendale	PL52MZU	Rossendale	PN52XBM	Ipswich	PO56ROU	Preston
PF51KMV	Rossendale	PL56JDX	Preston	PN52XBO	Ipswich	PO56RPU	Preston
PF51KMX	Rossendale	PM04JAM	Marshalls	PN52XBP	Richardson	PO56RPV	Preston
PFN858	Halifax JC	PM05JAM	Marshalls	PN52XKF	Blackpool	PO56RPX	Preston
PG03YYW	Halton	PM07JAM	Marshalls	PN52XKG	Blackpool	PO56RPY	Preston
PG03YYX	Halton	PN03UGG	Blackpool	PN52XKH	Blackpool	PO56RPZ	Preston
PG03YYZ	Halton	PN03UGH	Blackpool	PN52XKJ	Blackpool	PO56RRU	Preston
PIL4682	Sanders Cs	PN04XDE	Blackpool	PN52XKK	Blackpool	PO56RRV	Preston
PIL7971	West End Travel	PN04XDF	Blackpool	PN52VH	Preston	PO56RRX	Preston
PIL7972	West End Travel	PN04XDG	Blackpool	PN52ZVJ	Preston	PO56RRY	Preston
PJ02PYD	Blackpool	PN04XDH	Blackpool	PN52ZVK	Preston	PO56RRZ	Preston
PJ02PYF	Blackpool	PN04XDJ	Blackpool	PN52ZVL	Preston	PO56RSU	Preston
PJ02PYG	Blackpool	PN04XDK	Blackpool	PN52ZVM	Preston	PO56RSV	Preston
PJ02PYH	Blackpool	PN04XDL	Blackpool	PN52ZVO	Preston	PO56RSX	Preston
PJ02PYL	Blackpool	PN05RNU	Tyrer Bus	PN52ZVP	Preston	PO56RSY	Preston
PJ02PYO	Blackpool	PN05SYF	Richardson	PN52ZVR	Preston	PO56RSZ	Preston
PJ02PYP	Blackpool	PN05SYG	Halton	PN52ZVS	Preston	PO58NPG	Ensignbus
PJ02PYS	Blackpool	PN05SYH	Halton	PN57LGF	Rossendale	PO58NPJ	Ensignbus
PJ02PYT	Blackpool	PN05SYJ	Halton	PN57NFA	Preston	PO58NPK	Ensignbus
PJ02RGZ	Supertravel	PN05SYO	Halton	PN57NFC	Preston	PO58NPN	Ensignbus
PJ02RHE	Heddingham	PN06KTG	Travel de Courcey	PN57NFD	Preston	PO58NPP	Ensignbus
PJ03TFF	Blackpool	PN06TKD	Travel de Courcey	PN57NFE	Preston	PO58NPU	Ensignbus
PJ03TFK	Blackpool	PN07NTJ	Preston	PN57NFF	Preston	PO58NPV	Ensignbus
PJ03TFN	Blackpool	PN07NTK	Preston	PN57NFG	Preston	PO58NPX	Ensignbus
PJ03TFU	Blackpool	PN07NTL	Preston	PO51UMG	Blackpool	PO58NPY	Ensignbus
PJ03TFV	Blackpool	PN07NTM	Preston	PO51UMJ	Blackpool	PO58NRE	Ensignbus
PJ03TFX	Blackpool	PN07NTO	Preston	PO51UMK	Ipswich	POG508Y	Travel de Courcey
PJ03TFY	Blackpool	PN07NTT	Preston	PO51WEC	Rossendale	POG513Y	Travel de Courcey
PJ03TFZ	Blackpool	PN07NTU	Preston	PO51WNF	Reading	POG514Y	Go Whippet

POG536Y	Travel de Courcey	R110VNT	D & G	R315NGM	Thamesdown	R513YWC	Speedwell
POG567Y	Travel de Courcey	R112NTA	Chambers	R317NGM	Thamesdown	R514YWC	Speedwell
POG580Y	Go Whippet	R112RLY	Renown	R319NGM	Thamesdown	R561UOT	Ensignbus
POG583Y	Marshalls	R113NTA	Swanbrook	R358XVX	Marshalls	R576NFX	Bodmans
PP57BLU	Bluebird	R113OFJ	Plymouth	R361GDX	Chambers	R578GDS	Bakerbus
PRN909	Preston	R114OFJ	Plymouth	R362DJN	Reliance	R619VEG	Ensignbus
PVF377	Sanders Cs	R115OFJ	Plymouth	R363DJN	Reliance	R620VEG	Ensignbus
PWY39W	Heddingham	R116OFJ	Plymouth	R374DJN	TM Travel	R621VEG	Ensignbus
Q723GHG	TM Travel	R117OFJ	Plymouth	R379DJN	TM Travel	R622VEG	Ensignbus
R3YRR	Marshalls	R117RLY	Renown	R396XDA	Renown	R623VEG	Ensignbus
R4OCT	Delaine	R118OFJ	Plymouth	R397XDA	Renown	R624VEG	Ensignbus
R6BLU	Rossendale	R119OFJ	Plymouth	R399EOS	Sargeants	R625VEG	Ensignbus
R7BLU	Rossendale	R120OFJ	Plymouth	R410XFL	Rossendale	R626VEG	Ensignbus
R8BLU	Rossendale	R121OFJ	Plymouth	R412XFL	Country Liner	R627VEG	Ensignbus
R9CCC	Cedar Cs	R122OFJ	Plymouth	R415XFL	Chambers	R629VEG	Ensignbus
R15CED	Cedric	R123OFJ	Plymouth	R432NFR	Preston	R630VEG	Ensignbus
R16HAM	Ham's Travel	R124OFJ	Plymouth	R433FWT	Safeguard	R631VEG	Ensignbus
R17HAM	Ham's Travel	R125OFJ	Plymouth	R434NFR	Preston	R632VEG	Ensignbus
R30ARJ	Andybus	R126OFJ	Plymouth	R435NFR	Preston	R638VEG	Ensignbus
R31GNW	Sanders Cs	R133FBJ	Ipswich	R436FWT	Bodmans	R648VMJ	Uno
R40TGM	Country Liner	R134FBJ	Ipswich	R436NFR	Preston	R649VBM	Uno
R50TGM	Bodmans	R140XWF	Hulleys	R437NFR	Preston	R650VBM	Uno
R54AWO	Bodmans	R144PCJ	Yeomans	R438RCW	Preston	R651VBM	Uno
R54OCK	D & G	R161GNW	Geldards	R439RCW	Preston	R652VBM	Uno
R55ACL	Aintree Coachlines	R175VLA	Renown	R449PRH	Compass Bus	R653VBM	Uno
R57JSG	Andybus	R178GNW	Geldards	R453FWT	Heddingham	R654VBM	Uno
R58GNW	Country Liner	R179GNW	Bennetts	R466RRA	Nottingham	R674OEB	White Bus
R59EDW	Cedar Cs	R186DDX	Norfolk Green	R467RRA	W Greyhound	R685MEW	Shamrock
R81EDW	Compass Bus	R189DDX	Ipswich	R472RRA	W Greyhound	R686MEW	Brylaine
R81EMB	Anglian	R189NFE	Thamesdown	R473RRA	W Greyhound	R687MEW	Brylaine
R82EMB	Anglian	R197DDX	Ipswich	R475RRA	W Greyhound	R689MEW	Shamrock
R83GNW	Ensignbus	R200PAR	Country Liner	R476RRA	W Greyhound	R692MEW	Ensignbus
R84EDW	Compass Bus	R200VHO	Thornes	R478RRA	W Greyhound	R694WAW	Midland Rider
R84EMB	Anglian	R207DKG	Premiere	R501SJM	Andybus	R695WAW	Midland Rider
R85EMB	Anglian	R20CED	Cedric	R501YWC	Speedwell	R703YWC	Pennine
R86EMB	Anglian	R211DKG	TM Travel	R502SJM	Andybus	R706MEW	Carousel
R89GNW	White Bus	R211VNT	D & G	R502YWC	Red Rose	R706YWC	Pennine
R93HUA	Compass Bus	R213DKG	TM Travel	R503SJM	Andybus	R708MEW	Carousel
R100PAR	Country Liner	R222AJP	Abus	R504YWC	Speedwell	R708NJH	Bodmans
R101NTA	Country Liner	R23GNW	Brylaine	R505YWC	Speedwell	R711MEW	Trustybus
R104GNW	South Lancs	R259LGH	Heddingham	R506YWC	Speedwell	R712MEW	Rossendale
R105CKN	Chalkwell	R261XDA	JPT	R507SJM	N I B S	R713MEW	D & G
R107GKN	Chalkwell	R269XDA	Speedwell	R508YWC	Speedwell	R714KGK	White Bus
R107GNH	Bakerbus	R273LGH	Heddingham	R510SJM	N I B S	R714MEW	D & G
R107TKO	Wardle	R279LHG	Heddingham	R510YWC	Speedwell	R717YWC	Pennine
R108AKP	Chalkwell	R279RAU	Beestons	R511WDC	Country Liner	R718BNF	Nu-Venture
R109AKP	Chalkwell	R280RAU	Hulleys	R511YWC	Speedwell	R720BNF	Nu-Venture
R110JKP	Chalkwell	R314NGM	Thamesdown	R512YWC	Speedwell	R726EGD	D & G

Reg	Operator	Reg	Operator	Reg	Operator	Reg	Operator
R739TMO	Stansted Transit	RL51CXC	Country Liner	S101LBL	Reading	S171UAL	Norfolk Green
R741BUJ	D & G	RL51CXD	Country Liner	S102LBL	Reading	S172UAL	Norfolk Green
R742BUJ	D & G	RO03JVA	Courtney	S103LBL	Reading	S181BMR	Thamesdown
R743BUJ	D & G	ROX630Y	Travel de Courcey	S104LBL	Reading	S182BMR	Thamesdown
R744BUJ	D & G	ROX656Y	Travel de Courcey	S105LBL	Reading	S183BMR	Thamesdown
R785DUB	JPT	ROX663Y	Travel de Courcey	S106LBL	Reading	S184BMR	Thamesdown
R791YFL	Travel de Courcey	ROX666Y	Travel de Courcey	S107LBL	Reading	S185BMR	Thamesdown
R796GSF	Red Rose	RR03BUS	Red Rose	S111AJP	Abus	S186BMR	Thamesdown
R807HWS	WGreyhound	RR57BLU	Bluebird	S112GUB	N Warrington	S194FFM	D & G
R808HWS	WGreyhound	RUI2486	Wardle	S113GUB	N Warrington	S195FFM	D & G
R809HWS	WGreyhound	RX03XKH	Safeguard	S114GUB	N Warrington	S195HOK	Anglian
R810HWS	WGreyhound	RX06WRU	Courtney	S114KRN	Rossendale	S197FFM	D & G
R835FNG	Konectbus	RX06XFD	Courtney	S115GUB	N Warrington	S200VHO	Thornes
R836FNG	Konectbus	RX06XFE	Courtney	S115KRN	Rossendale	S222JDC	Leven Valley
R837FNG	Konectbus	RX07BNF	Carousel	S116GUB	N Warrington	S233RLH	Chambers
R843FWW	Red Rose	RX07RKV	Reading	S116KRN	Rossendale	S234HGU	Olympus
R845VEC	Fishwick	RX55AOT	Courtney	S117GUB	N Warrington	S257JUG	Ross Travel
R846VEC	Fishwick	RX56DWE	Country Liner	S117KRN	Rossendale	S260JUG	Wardle
R847VEC	Fishwick	RX57GOE	Bodmans	S118GUB	N Warrington	S262JUG	Ross Travel
R848VEC	Fishwick	RX57MDZ	Courtney	S118KRN	Rossendale	S268JUG	Wardle
R849JGD	Tyrer Bus	RX58HVJ	Courtney	S119GUB	N Warrington	S285MGB	Sargeants
R864MCE	Ensignbus	S2CLA	D & G	S119KRN	Rossendale	S285UAL	Cedar Cs
R876MCE	Ensignbus	S2WMS	Wardle	S119RCS	D & G	S289UAL	Cedar Cs
R899AVM	JPT	S3CLA	D & G	S120GUB	N Warrington	S290NRB	Central
R941YOV	Chambers	S3HAM	Ham's Travel	S120KRN	Rossendale	S290TVW	Heddingham
R944AMB	Speedwell	S3JPT	JPT	S121GUB	N Warrington	S291NRB	Central
R948AMB	Bakerbus	S3YRR	Marshalls	S121KRN	Rossendale	S291TVW	Heddingham
R976FNW	Carousel	S4HAM	Ham's Travel	S123KRN	Rossendale	S292UAL	Norfolk Green
R901FNW	Carousel	S4YRR	Marshalls	S123RDC	Leven Valley	S294UAL	Norfolk Green
R984FNW	Carousel	S5OCT	Delaine	S124RLE	Ensignbus	S300XHK	Heddingham
R986FNW	Carousel	S6APH	Olympus	S125RLE	Ensignbus	S309DLG	Norfolk Green
R987EWU	D & G	S7STM	Supertravel	S126RLE	Ensignbus	S313DLG	Norfolk Green
R988FNW	Carousel	S8BLU	Rossendale	S127FTA	Plymouth	S333AJP	Abus
R995YJO	Chambers	S18RED	Western Greyhound	S127RLE	Heddingham	S333HEB	Cedar Cs
RAU624R	Marshalls	S20WGL	Western Greyhound	S129RLE	Aintree Coachlines	S342SET	Olympus
RBO508Y	Abus	S24SLT	South Lancs	S131RLE	Chambers	S343SET	Olympus
RCT3	Delaine	S25SLT	South Lancs	S132RLE	Ensignbus	S361AHC	Sargeants
RDV903	Sanders Cs	S26SLT	South Lancs	S134EJE	Go Whippet	S365VKW	Country Liner
REZ8516	Wardle	S30ARJ	Western Greyhound	S136RLE	Ensignbus	S376MVP	Heddingham
RGV284N	Heddingham	S34BMR	Western Greyhound	S138RLE	Ensignbus	S401JUA	Brylaine
RIB6199	Weardale	S41FWY	Norfolk Green	S154NNH	Norfolk Green	S450ATV	Western Greyhound
RIB7002	Nu-Venture	S42FWY	D & G	S156JUA	Sanders Cs	S452ATV	Western Greyhound
RJI8604	Sanders Cs	S45UBO	Stansted Transit	S156UAL	Norfolk Green	S453ATV	Western Greyhound
RJI8611	Yeomans	S48RGA	Chambers	S157JUA	Sanders Cs	S454ATV	Western Greyhound
RK07KDA	Courtney	S54NCW	Avon Buses	S158JUA	White Bus	S455ATV	Western Greyhound
RK08CYJ	Reading	S73TNM	Compass Bus	S162BMR	Thamesdown	S458ATV	Nottingham
RK08CYL	Reading	S84XCJ	Sargeants	S169UAL	Konectbus	S459ATV	Nottingham
RL51CXB	Country Liner	S100PAF	Western Greyhound	S170UAL	Norfolk Green	S460ATV	Nottingham

S461ATV	Nottingham	S855DGX	Chambers	SJI1622	Sanders Cs	SUO575	Western Greyhound
S462ATV	Nottingham	S856DGX	Chambers	SJI1626	Sanders Cs	T1KET	Jim Stones
S463ATV	Nottingham	S857DGX	Courtney	SJI1632	Sanders Cs	T1WET	West End Travel
S464ATV	Nottingham	S858DGX	South Lancs	SJI8098	Beestons	T6OCT	Delaine
S465ATV	Nottingham	S859DGX	Rossendale	SJI9319	Beestons	T9BHN	Leven Valley
S501SRL	Western Greyhound	S860DGX	Rossendale	SJI9320	Beestons	T9RTG	Ross Travel
S502SRL	Western Greyhound	S861DGX	Rossendale	SJI9321	Beestons	T10BHN	Leven Valley
S503SRL	Western Greyhound	S862DGX	Rossendale	SK02TYS	Aintree Coachlines	T11SLT	South Lancs
S503UAK	Safeguard	S863DGX	Rossendale	SK02XGO	Stephensons	T12SBK	Swanbrook
S505LHG	Blackpool	S864DGX	Rossendale	SK02XGP	Stephensons	T13VCC	Stansted Transit
S506LHG	Blackpool	S865DGX	Rossendale	SK07DYA	Bodmans	T20CCH	Wardle
S507LHG	Blackpool	S866DGX	Rossendale	SK51SBK	Swanbrook	T27BXG	Leven Valley
S508LHG	Blackpool	S876BYJ	Reading	SK52USS	Reading	T28BXG	Leven Valley
S509LHG	Blackpool	S877BYJ	Reading	SLK886	Simonds	T30ARJ	Andybus
S509NFR	Avon Buses	S878BYJ	Reading	SMK665F	Ensignbus	T49JBA	Bakerbus
S510LHG	Blackpool	S879BYJ	Reading	SMK702F	Geldards	T57AUA	Geldards
S511KFL	Avon Buses	S880BYJ	Reading	SMK737F	W Greyhound	T63KLD	Renown
S511LHG	Blackpool	S881BYJ	Reading	SMV24	Thornes	T64KLD	Renown
S512KFL	Avon Buses	S925LBL	Reading	SN03LGA	Hornsby Travel	T65KLD	Renown
S512LHG	Blackpool	S926LBL	Reading	SN04CPE	Abus	T70HAM	Ham's Travel
S513KFL	Carousel	S927LBL	Reading	SN04EFL	Renown	T74JBO	Sargeants
S513LHG	Blackpool	S928LBL	Reading	SN04EFW	Thames Travel	T74WWV	Stansted Transit
S514KFL	Carousel	S929LBL	Reading	SN08AAF	Bodmans	T75WWV	Stansted Transit
S514LHG	Blackpool	S930LBL	Reading	SN53ETJ	Compass Bus	T78JBA	Country Liner
S515KFL	Central	S931LBL	Reading	SN53ETK	Compass Bus	T80HAM	Ham's Travel
S515LHG	Blackpool	SA52MYR	Compass Bus	SN53ETL	Compass Bus	T81JBA	Anglian
S516JJH	Reading	SA52MYS	Compass Bus	SN53ETO	Compass Bus	T83JBA	Anglian
S517JJH	Reading	SAC500	Thornes	SN53ETR	Compass Bus	T84JBA	South Lancs
S517LHG	Blackpool	SBZ5810	Sargeants	SN53LWL	Southdown PSV	T85JBA	South Lancs
S518KFL	Trustybus	SED232	Thornes	SN53LWM	Nu-Venture	T86JBA	South Lancs
S518LHG	Blackpool	SF04HXR	JPT	SN53LWO	Nu-Venture	T90HAM	Ham's Travel
S521KFL	Trustybus	SF04RGY	Wardle	SN53LWP	Nu-Venture	T100CBC	Anglian
S522KFL	Trustybus	SF05FNW	Bodmans	SN55DVA	Courtney	T103KCC	Central
S523KFL	Carousel	SF05XDC	Chalkwell	SN55DVB	White Bus	T111PDC	Leven Valley
S527KFL	Central	SF06VYT	Premiere	SN55DVF	Avon Buses	T112AUA	TM Travel
S528KFL	Trustybus	SF06WEC	Thornes	SN55DVG	Avon Buses	T128EFJ	Plymouth
S582PGB	D & G	SF0DFLO	Bodmans	SN55DVP	Avon Buses	T129EFJ	Plymouth
S737RNE	JPT	SF57JSX	Weardale	SN56AXM	Bodmans	T129XVT	Bakerbus
S762XYA	APL Travel	SGF965	Sanders Cs	SN56AYH	Stephensons	T130EFJ	Plymouth
S779RNE	TM Travel	SGR780V	Heddingham	SN58CDK	Aintree Coachlines	T131EFJ	Plymouth
S794XUG	Renown	SHO800	Thornes	SN58EOX	Weavaway	T132AUA	Southdown PSV
S838VAG	Thamesdown	SIA488	Simonds	SN58EOY	Weavaway	T132EFJ	Plymouth
S846DGX	South Lancs	SIB7516	Sanders Cs	SN58EOZ	Weavaway	T133EFJ	Plymouth
S847DGX	Courtney	SJ04KEU	Chalkwell	SNM71R	Minsterley Ms	T134EFJ	Plymouth
S848DGX	Chambers	SJ53AWX	JPT	SNR1H	West End Travel	T135EFJ	Plymouth
S851DGX	Chambers	SJI1615	Sanders Cs	SS55BLU	Bluebird	T135KPV	Ipswich
S853DGX	Courtney	SJI1617	Sanders Cs	SS57BLU	Bluebird	T136EFJ	Plymouth
S853PKH	APL Travel	SJI1621	Sanders Cs	STW30W	Heddingham	T136KPV	Ipswich

Reg	Operator	Reg	Operator	Reg	Operator	Reg	Operator
T137EFJ	Plymouth	T408BNN	Nottingham	T795TWC	Ham's Travel	UJI1761	Swanbrook
T137KPV	Ipswich	T409BNN	Nottingham	T840CCK	Avon Buses	UJI1762	Swanbrook
T138EFJ	Plymouth	T410BNN	Nottingham	T841CCK	Avon Buses	UK02DRM	DRM
T139EFJ	Plymouth	T411BNN	Nottingham	T880RBR	Blackpool	UK54DRM	DRM
T140EFJ	Plymouth	T412BNN	Nottingham	T881RBR	Blackpool	UK56DRM	DRM
T156AUA	Speedwell	T413BNN	Nottingham	T884RBR	Blackpool	UKZ5466	Wardle
T163RMR	Thamesdown	T414BNN	Nottingham	T888EXG	Leven Valley	UL55UNO	Uno
T164AUA	Renown	T414LGP	Stansted Transit	T927PNV	Renown	UMR199T	Thamesdown
T164RMR	Thamesdown	T415BNN	Nottingham	T932EAN	Blackpool	UN55UNO	Uno
T165RMR	Thamesdown	T416BNN	Nottingham	T933EAN	Blackpool	UO55UNO	Uno
T200CBC	Anglian	T417MNH	Renown	T934EAN	Blackpool	UOD541	Hopley's
T200VHO	Thornes	T417XVO	Nottingham	T935EAN	Blackpool	UOI2679	Emsworth & D
T201AFM	N Warrington	T418XVO	Nottingham	T936EAN	Blackpool	URN172Y	Thornes
T202AFM	N Warrington	T419PDG	Bennetts	T936HTR	Olympus	UWR498	Western Greyhound
T203AFM	N Warrington	T419XVO	Nottingham	T961ACC	Western Greyhound	UWW11X	Blackpool
T204AFM	N Warrington	T420XVO	Nottingham	T973TBA	South Lancs	UWW15X	Blackpool
T205AFM	N Warrington	T421ADN	Compass Bus	T974TBA	South Lancs	UWW5X	Blackpool
T206AFM	N Warrington	T421XVO	Nottingham	TAZ4061	Kime's	V2JPT	JPT
T207AFM	N Warrington	T422ADN	Compass Bus	TAZ4062	Kime's	V3JPT	D & G
T208AFM	N Warrington	T422XVO	Nottingham	TAZ4063	Kime's	V6WMS	Weardale
T208KJV	Hornsby Travel	T439EBD	Renown	TAZ4064	Kime's	V11GMT	Velvet
T210HCW	Blackpool	T440EBD	Renown	TCF496	Simonds	V12GMT	Velvet
T211HCW	Blackpool	T445WWT	Speedwell	TCZ6121	Quantock	V14GMT	Velvet
T212HCW	Blackpool	T446HRV	Nu-Venture	TCZ6122	Quantock	V22SLT	South Lancs
T213HCW	Blackpool	T446WWT	Speedwell	TCZ6123	Quantock	V22WGL	Western Greyhound
T214HCW	Blackpool	T455HNH	TM Travel	TCZ6124	Quantock	V31WGL	Western Greyhound
T215HCW	Blackpool	T458HNH	Renown	THX101S	Ensignbus	V33SLT	South Lancs
T216HCW	Blackpool	T458JRH	Red Rose	THX209S	Premiere	V40WGL	Western Greyhound
T217HCW	Blackpool	T459HNH	Renown	THX646S	Ensignbus	V67SVY	Thames Travel
T218HCW	Blackpool	T461HNH	Renown	TIL8148	Quantock	V78JKG	APL Travel
T223SAS	Stansted Transit	T466HNH	D & G	TM52BUS	TM Travel	V81JKG	APL Travel
T270EWW	Sanders Cs	T467HNH	D & G	TND409X	Emsworth & D	V108DCF	Reading
T289CGU	Olympus	T504RPN	Chalkwell	TRN808V	Premiere	V108LVH	Red Rose
T290ROF	Claribels	T548HNH	Avon Buses	TRN809V	Konectbus	V109DCF	Reading
T310MBU	D & G	T550HNH	Country Liner	TSV302	Hopley's	V110DCF	Reading
T314SMV	Country Liner	T551HNH	D & G	TT57BLU	Bluebird	V112DCF	Reading
T322ELG	N Warrington	T552HNH	D & G	TUB250R	Aintree Coachlines	V113DCF	Reading
T323ELG	N Warrington	T553ADN	Reading	TVG397	Simonds	V114DCF	Reading
T329KDM	Minsterley Ms	T553HNH	Renown	TYR95	Quantock	V115DCF	Reading
T330KDM	Minsterley Ms	T554ADN	Reading	UAR219Y	Stephensons	V116ESL	Bakerbus
T337TVM	Country Liner	T556ADN	Reading	UDZ7580	Tyrer Bus	V117ESL	Bakerbus
T341FWR	Red Rose	T581JTD	Minsterley Ms	UH55UNO	Uno	V187EAM	Thamesdown
T342PNV	Renown	T619VEW	Norfolk Green	UHG149V	Fishwick	V188EAM	Thamesdown
T400CBC	Anglian	T648KPU	Kime's	UHG150V	Fishwick	V189EAM	Thamesdown
T402LGP	Stansted Transit	T731DGD	South Lancs	UHY384	Abus	V190EAM	Thamesdown
T405BNN	Nottingham	T760LFM	Halton	UIB3987	Shamrock	V190EBV	Preston
T406BNN	Nottingham	T789XVO	Konectbus	UIL2724	Quantock	V191EAM	Thamesdown
T407BNN	Nottingham	T793TUX	Sanders Cs	UJ55UNO	Uno	V191EBV	Preston

Reg	Operator	Reg	Operator	Reg	Operator	Reg	Operator
V192EBV	Preston	V267HEC	Blackpool	V444AJP	Abus	VLZ9237	Wardle
V193EBV	Preston	V268BNV	Supertravel	V544JBH	Aintree Coachlines	VO03DZC	Bennetts
V194EBV	Preston	V268DRC	Nottingham	V547JBH	Aintree Coachlines	VO03DZD	APL Travel
V195EBV	Preston	V268HEC	Blackpool	V549JBH	Aintree Coachlines	VO03DZE	APL Travel
V196EBV	Preston	V269DRC	Marshalls	V561JFL	Norfolk Green	VOY182X	TM Travel
V200CHO	Thornes	V269HEC	Blackpool	V563JBH	Uno	VRA124Y	Aintree Coachlines
V203ENU	Konectbus	V270BNV	Nu-Venture	V564JBH	Uno	VRY841	Simonds
V204ENU	Konectbus	V270DRC	Marshalls	V565JBH	Uno	VU02TPZ	Red Rose
V205ENU	Konectbus	V271DRC	Nottingham	V618EVU	Supertravel	VU02TSZ	Brylaine
V208EAL	Cedar Cs	V271HEC	Blackpool	V649EEF	Renown	VU02TTJ	Safeguard
V209JLG	N Warrington	V272HEC	Blackpool	V682FPO	Stansted Transit	VU02TTZ	Speedwell
V210JLG	N Warrington	V273HEC	Blackpool	V708GRY	Felix	VU03ZPS	Andybus
V211JLG	N Warrington	V274HEC	Blackpool	V723GGE	Bakerbus	VU03ZPY	Andybus
V212JLG	N Warrington	V275HEC	Blackpool	V844OOF	Swanbrook	VUB396H	West End Travel
V213JLG	N Warrington	V276DRC	Nottingham	V845OOF	Swanbrook	VVJ808	Ensignbus
V214JLG	N Warrington	V276HEC	Blackpool	V895DNB	Uno	VWK239	Travel de Courcey
V215JLG	N Warrington	V282SBW	Olympus	V896DNB	Uno	VX04ULT	Sargeants
V216JLG	N Warrington	V283SBW	Olympus	V896LOH	Compass Bus	VX05TZO	Bluebird
V217JLG	N Warrington	V293UVY	Go Whippet	V901LOH	Cedar Cs	VX05UHS	Compass Bus
V218JLG	N Warrington	V331CVV	Supertravel	V909SEG	Norfolk Green	VX51RHZ	Konectbus
V227KAH	Anglian	V332CVV	D & G	V917TAV	White Bus	VX51RJZ	Konectbus
V237LWU	Stansted Transit	V361DLH	Country Liner	V928FMS	South Lancs	VX53AVJ	Simonds
V247BNV	Country Liner	V362HKG	Chalkwell	V929FMS	Kime's	VX54CLU	Supertravel
V250BNV	Heddingham	V378SVV	Simonds	V931VUB	Renown	VX55FWE	Swanbrook
V252JRR	Nottingham	V380HGG	Anglian	V932VUB	Renown	VYD333	Quantock
V253JRR	Nottingham	V380SVV	Southdown PSV	V939ECU	Wardle	W1DRM	DRM
V254JRR	Nottingham	V385KVY	Claribels	V944DNB	Central	W3YRR	Marshalls
V255JRR	Nottingham	V386KVY	Claribels	V946DCF	Reading	W5ACL	Aintree Coachlines
V256JRR	Nottingham	V387HGG	JPT	V946DNB	Central	W5HAM	Ham's Travel
V257BNV	Country Liner	V387KVY	Claribels	V967RCX	Ross Travel	W5JPT	Norfolk Green
V257DRB	Nottingham	V389HGG	JPT	V983DNB	Speedwell	W6WMS	Weardale
V258DRB	Nottingham	V389KVY	TM Travel	V993DNB	Norfolk Green	W19SLT	South Lancs
V259DRB	Nottingham	V392SVV	Southdown PSV	V993LLG	Halton	W20SLT	South Lancs
V260DRB	Nottingham	V415KMY	Stephensons	V994DNB	Speedwell	W20VHO	Thornes
V261BNV	D & G	V423DRC	Nottingham	V994LLG	Halton	W29WGL	Western Greyhound
V261DRB	Nottingham	V424DRC	Nottingham	V995LLG	Halton	W31COM	Cedar Cs
V261HEC	Blackpool	V425DRC	Nottingham	VA51SAR	Sargeants	W44SLT	South Lancs
V262DRC	Nottingham	V426DRC	Nottingham	VDV752	Quantock	W44WGL	Western Greyhound
V262HEC	Blackpool	V427DRC	Nottingham	VFN53	Thornes	W50MBH	Western Greyhound
V263BNV	Supertravel	V428DRC	Nottingham	VH02OOH	Thornes	W55SLT	South Lancs
V263DRC	Nottingham	V429DRC	Nottingham	VH0200	Thornes	W80HOD	Sanders Cs
V263HEC	Blackpool	V430DRC	Nottingham	VIB5072	Yeomans	W82NDW	D & G
V264BNV	Supertravel	V431DRC	Nottingham	VIB5237	Chalkwell	W107RNC	Southdown PSV
V264HEC	Blackpool	V432DRC	Nottingham	VIL8577	D & G	W109MTL	Hornsby Travel
V265HEC	Blackpool	V433DRC	Nottingham	VIL8677	D & G	W116SRX	Reading
V266DRC	Nottingham	V434DRC	Nottingham	VJO202X	Heddingham	W117SRX	Reading
V266HEC	Blackpool	V435DRC	Nottingham	VJO206X	Heddingham	W119UCF	Norfolk Green
V267DRC	Nottingham	V436DRC	Nottingham	VLT25	Ensignbus	W176CDN	Renown

Reg	Operator	Reg	Operator	Reg	Operator	Reg	Operator
W178ODN	Kime's	W397PRC	South Lancs	W654SNN	Nottingham	WA08LDN	Plymouth
W187YBN	South Lancs	W398PRC	South Lancs	W656SNN	Nottingham	WA08LDU	Plymouth
W195CDN	Emsworth & D	W408UCJ	Sargeants	W657SNN	Nottingham	WA08LDV	Plymouth
W200VHO	Thornes	W426CWX	Compass Bus	W658SNN	Nottingham	WA08LDX	Plymouth
W203YAP	Safeguard	W443CWX	Yeomans	W659SNN	Nottingham	WA08LDZ	Plymouth
W209CDN	Hornsby Travel	W471VMA	Halton	W666WMS	Weardale	WA08LEF	Plymouth
W216JND	Wardle	W475RKS	Reliance	W671TNV	Renown	WA08LEJ	Plymouth
W216PRB	Konectbus	W478EUB	Wardle	W709PTO	Felix	WA51ACO	Plymouth
W217PRB	Konectbus	W482EUB	Wardle	W772URP	Regal Busways	WA51ACU	Plymouth
W218PRB	Konectbus	W483EUB	Wardle	W773URP	Country Liner	WA51ACV	Plymouth
W219PRB	Konectbus	W483OUF	Stansted Transit	W821RJT	Compass Bus	WA51ACX	Plymouth
W242PAU	Nottingham	W554JVV	Southdown PSV	W829UMJ	Brylaine	WA51ACY	Plymouth
W243PAU	Nottingham	W556JVV	Tyrer Bus	W84NDW	Renown	WA54JVV	Plymouth
W244PAU	Nottingham	W558WCD	Renown	W892WDT	Country Liner	WA54JVW	Plymouth
W246PAU	Nottingham	W563JVV	Tyrer Bus	W895AGA	D & G	WA54JVX	Plymouth
W247PAU	Nottingham	W566JVV	Renown	W898RFA	Hornsby Travel	WA54JVY	Plymouth
W248PAU	Nottingham	W567XRO	Uno	W921JNF	Country Liner	WA54JVZ	Plymouth
W249PAU	Nottingham	W568XRO	Uno	W923JNF	TM Travel	WA54JWC	Plymouth
W251PAU	Nottingham	W569JVV	Southdown PSV	W936JNF	Renown	WA54JWD	Plymouth
W261CDN	Sanders Cs	W569XRO	Uno	W937JNF	Country Liner	WA54JWE	Plymouth
W261EWU	TM Travel	W571XRO	Uno	W941SNR	Nottingham	WA56ENV	Weavaway
W262CDN	Sanders Cs	W572XRO	Uno	W942SNR	Nottingham	WA56HHN	Plymouth
W263CDN	Sanders Cs	W573XRO	Uno	W943SNR	Nottingham	WA56HHO	Plymouth
W265CDN	Sanders Cs	W574XRO	Uno	W958PAU	Nottingham	WA56HHP	Plymouth
W266CDN	Sanders Cs	W575XRO	Uno	W959PAU	Nottingham	WA56OZM	Plymouth
W282EYG	D & G	W576XRO	Uno	W966JNF	Hornsby Travel	WA56OZO	Plymouth
W283EYG	Renown	W577XRO	Uno	W967JNF	Brylaine	WA56OZP	Plymouth
W284EYG	TM Travel	W578XRO	Uno	W985XFM	Halton	WA56OZR	Plymouth
W285EYG	TM Travel	W586YDM	D & G	W986XFM	Halton	WA56OZS	Plymouth
W286EYG	TM Travel	W599PTO	Nottingham	W987XFM	Halton	WA56OZT	Plymouth
W287EYG	D & G	W601PTO	Nottingham	W992BDP	Go Whippet	WA56OZU	Plymouth
W288EYG	TM Travel	W602PTO	Nottingham	WA03BHW	Plymouth	WA57CYY	Safeguard
W289EYG	TM Travel	W603PTO	Nottingham	WA03BHX	Plymouth	WA57JZW	Bodmans
W291EYG	D & G	W604PTO	Nottingham	WA03BHY	Plymouth	WA58EOO	Safeguard
W292EYG	TM Travel	W605PTO	Nottingham	WA03BHZ	Plymouth	WAL782	Quantock
W295EYG	TM Travel	W606PTO	Nottingham	WA03BJE	Plymouth	WAZ8278	Kime's
W298EYG	D & G	W607PTO	Nottingham	WA03BJF	Plymouth	WBZ8737	Wardle
W299EYG	D & G	W608KFE	APL Travel	WA03MGE	Plymouth	WDL693Y	Country Liner
W304EYG	D & G	W608PTO	Nottingham	WA03MGJ	Plymouth	WDL695Y	Country Liner
W311CJN	Heddingham	W609PTO	Nottingham	WA05DFG	Weavaway	WDL696Y	Country Liner
W312CJN	Heddingham	W634MKY	Yeomans	WA05DFJ	Weavaway	WET1K	West End Travel
W348WCS	W Greyhound	W647SNN	Nottingham	WA06GSU	Weavaway	WJ52GNY	Plymouth
W349WCS	W Greyhound	W648SNN	Nottingham	WA06GSV	Weavaway	WJ52GNZ	Plymouth
W364ABD	Country Liner	W649FUM	Heddingham	WA07KXX	Safeguard	WJ52GOA	Plymouth
W365ABD	Supertravel	W649SNN	Nottingham	WA08LDF	Plymouth	WJ52GOC	Plymouth
W378SVV	Renown	W651SNN	Nottingham	WA08LDJ	Plymouth	WJ52GOE	Plymouth
W385WGE	W Greyhound	W652SNN	Nottingham	WA08LDK	Plymouth	WJ52GOH	Plymouth
W396PRC	South Lancs	W653SNN	Nottingham	WA08LDL	Plymouth	WJ52GOK	Plymouth

Reg	Operator	Reg	Operator	Reg	Operator	Reg	Operator
WJ53HLG	Plymouth	WK53EUU	W Greyhound	WX04CZK	Thamesdown	X97LBJ	Ipswich
WJ53HLH	Plymouth	WK54BHL	W Greyhound	WX04CZL	Thamesdown	X98LBJ	Ipswich
WJ53HLK	Plymouth	WK54BHN	W Greyhound	WX05NFY	Trustybus	X131JCW	Rossendale
WJ53HLM	Plymouth	WK54BHO	W Greyhound	WX06JXR	Thamesdown	X132JCW	Rossendale
WJ53HLN	Plymouth	WK54BHP	W Greyhound	WX06JXS	Thamesdown	X133JCW	Rossendale
WJ53HLO	Plymouth	WK55BUS	W Greyhound	WX06JXT	Thamesdown	X134JCW	Rossendale
WJ53HLP	Plymouth	WK55PEN	W Greyhound	WX06JXZ	Thamesdown	X141CDV	Plymouth
WJ53HLR	Plymouth	WK56SET	W Greyhound	WX06JYA	Thamesdown	X142CDV	Plymouth
WJ55TRV	Simonds	WK56SUN	W Greyhound	WX06JYB	Thamesdown	X143CFJ	Plymouth
WJY759	Ensignbus	WK57DNX	Hopley's	WX06JYC	Thamesdown	X151LBJ	Ipswich
WK02BUS	W Greyhound	WK57TAF	W Greyhound	WX06JYD	Thamesdown	X152LBJ	Ipswich
WK02SAT	W Greyhound	WL03AUL	D & G	WX06JYE	Thamesdown	X153LBJ	Ipswich
WK02SUN	W Greyhound	WLT307	Ensignbus	WX08RZJ	Bodmans	X181BNH	Simonds
WK02TUE	W Greyhound	WLT428	Ensignbus	WX08SXD	Thamesdown	X182BNH	Simonds
WK02XLN	W Greyhound	WLT896	Carousel	WX08SXE	Thamesdown	X182RRN	Preston
WK02XLO	W Greyhound	WOA521	Sanders Cs	WX08SXF	Thamesdown	X183RRN	Preston
WK02XLR	W Greyhound	WP52YZA	Andybus	WX08SXG	Thamesdown	X184RRN	Preston
WK03BTE	Hopley's	WPF926	Safeguard	WX08SXH	Thamesdown	X185BNH	Nu-Venture
WK03BTF	Hopley's	WPH135Y	Heddingham	WX53OXZ	Renown	X185RRN	Preston
WK03BUS	W Greyhound	WR02XXO	Andybus	WX55ZZR	Thamesdown	X186RRN	Preston
WK04BUS	W Greyhound	WR53VYW	Ham's Travel	WX57TLN	Thamesdown	X187RRN	Preston
WK04CCC	W Greyhound	WSV503	Sanders Cs	WX57TLO	Thamesdown	X188BNH	Southdown PSV
WK04CUA	W Greyhound	WSV537	Western Greyhound	WX57TLU	Thamesdown	X188RRN	Preston
WK04CUC	W Greyhound	WT08BUS	Wardle	WX57TLV	Thamesdown	X189RRN	Preston
WK04DAD	W Greyhound	WT58BUS	Wardle	WX57TLY	Thamesdown	X195FOR	South Lancs
WK04HSD	W Greyhound	WT58SOT	Wardle	WX57TLZ	Thamesdown	X195LBJ	Ipswich
WK04OTA	W Greyhound	WU52YWE	Thamesdown	WYV64T	Ham's Travel	X196FOR	South Lancs
WK05AOX	W Greyhound	WU52YWF	Thamesdown	X2JPT	Rossendale	X197RRN	Preston
WK05CFD	W Greyhound	WU52YWG	Thamesdown	X2PCC	Wardle	X198FOR	South Lancs
WK05CFE	W Greyhound	WU52YWH	Thamesdown	X7HAM	Ham's Travel	X198RRN	Preston
WK05CFF	W Greyhound	WU52YWJ	Thamesdown	X7OCT	Delaine	X199LBJ	Ipswich
WK05FJE	W Greyhound	WU52YWK	Thamesdown	X8HAM	Ham's Travel	X199RRN	Preston
WK06SET	W Greyhound	WU52YWL	Thamesdown	X9HAM	Ham's Travel	X201CDV	Plymouth
WK06SUN	W Greyhound	WU52YWM	Thamesdown	X33WGL	Western Greyhound	X202CDV	Plymouth
WK07AOJ	Hopley's	WUK155	Hornsby Travel	X46CNY	Bodmans	X203CDV	Plymouth
WK08CZV	W Greyhound	WV02NNA	Thamesdown	X49VVY	Norfolk Green	X204CDV	Plymouth
WK08ESU	W Greyhound	WV02NNB	Thamesdown	X56LRY	Marshalls	X204NKR	Abus
WK08ESV	W Greyhound	WV02NNC	Thamesdown	X70OXF	Weavaway	X209ONH	D & G
WK08ESY	W Greyhound	WVE284	Simonds	X80SLT	South Lancs	X211ONH	Regal Busways
WK08ETA	W Greyhound	WW58BLU	Bluebird	X89LBJ	Ipswich	X212ONH	Renown
WK51AVP	W Greyhound	WWL212X	Heddingham	X91LBJ	Ipswich	X214ONH	Country Liner
WK51BUS	W Greyhound	WWW33	Thornes	X92FOR	Claribels	X215ONH	Renown
WK51HNF	W Greyhound	WX03ENM	Chalkwell	X92LBJ	Ipswich	X216HCD	Trustybus
WK53BNA	W Greyhound	WX03YFD	Thamesdown	X93LBJ	Ipswich	X228AWB	Compass Bus
WK53BNB	W Greyhound	WX03YFE	Thamesdown	X94HTL	Hornsby Travel	X228WRA	Anglian
WK53BND	W Greyhound	WX03ZNS	Thamesdown	X94LBJ	Ipswich	X229WRA	Konectbus
WK53BUS	W Greyhound	WX04CZH	Thamesdown	X94USC	Nottingham	X231MBJ	Ipswich
WK53EUT	W Greyhound	WX04CZJ	Thamesdown	X96LBJ	Ipswich	X232MBJ	Ipswich

Reg	Operator	Reg	Operator	Reg	Operator	Reg	Operator
X233MBJ	Ipswich	X822NWX	Fishwick	XSL596A	Aintree Coachlines	Y302KNB	D & G
X234MBJ	Ipswich	X822XCK	Avon Buses	XTF98D	Quantock	Y313NYD	Plymouth
X235MBJ	Ipswich	X823NWX	Fishwick	XUD367	Bodmans	Y314NYD	Plymouth
X236MBJ	Ipswich	X901LBJ	Ipswich	Y1WET	West End Travel	Y358LCK	Red Rose
X239HBC	Nottingham	X906RHG	Country Liner	Y4CCC	Cedar Cs	Y359LCK	Red Rose
X241HBC	Nottingham	X939NUB	D & G	Y6WMS	Weardale	Y371TGE	Western Greyhound
X242ABU	Shamrock	X941NUB	TM Travel	Y8BLU	Bluebird	Y381HKE	D & G
X249VWR	Norfolk Green	X943NUB	D & G	Y8OCT	Delaine	Y382HKE	D & G
X282MSP	Bakerbus	X947HBC	Nottingham	Y10HAM	Ham's Travel	Y383HKE	D & G
X291ABU	Bodmans	X961BPA	Reading	Y23OXF	Travel de Courcey	Y384HKE	D & G
X292ABU	Bodmans	X962BPA	Reading	Y24OXF	Travel de Courcey	Y385HKE	D & G
X307CBT	Safeguard	X963BPA	Reading	Y28WGL	Western Greyhound	Y386HKE	D & G
X308CBT	Safeguard	X964BPA	Reading	Y39WVL	Hornsby Travel	Y387HKE	D & G
X313FVV	Supertravel	X965ULG	Halton	Y54HBT	Norfolk Green	Y388HKE	D & G
X332ABU	D & G	X966ULG	Halton	Y56HBT	Norfolk Green	Y389HKE	D & G
X349AUX	D & G	X967ULG	Halton	Y57HBT	Norfolk Green	Y391HKE	D & G
X381XON	Reliance	X968ULG	Halton	Y58HBT	Norfolk Green	Y392HKE	D & G
X383VVY	Nu-Venture	X984SKH	Supertravel	Y59HBT	Yeomans	Y393HKE	D & G
X384VVY	Norfolk Green	XAU701Y	Blackpool	Y66SLT	South Lancs	Y400BCC	Bennetts
X385VVY	TM Travel	XAU702Y	Blackpool	Y77SLT	South Lancs	Y402GKJ	Chalkwell
X463UKS	Rossendale	XAU703Y	Blackpool	Y129TBF	Uno	Y403GKJ	Chalkwell
X464UKS	Rossendale	XAU704Y	Blackpool	Y184KNB	Supertravel	Y466HUA	Hornsby Travel
X465UKS	Rossendale	XAU705Y	Blackpool	Y185KNB	Supertravel	Y486TSU	Avon Buses
X466UKS	Rossendale	XAU706Y	Blackpool	Y192YMR	Thamesdown	Y546DTO	Nottingham
X469XUT	Swanbrook	XAZ1310	Brylaine	Y193YMR	Thamesdown	Y547LRB	Nottingham
X533NWT	Yeomans	XAZ1312	Brylaine	Y194YMR	Thamesdown	Y548LRB	Nottingham
X534NWT	Yeomans	XAZ1314	Brylaine	Y195YMR	Thamesdown	Y568LRN	Rossendale
X601AHE	Country Liner	XAZ1316	Brylaine	Y196YMR	Thamesdown	Y594HPK	Reading
X602AHE	Country Liner	XAZ1408	Brylaine	Y197YMR	Thamesdown	Y595HPK	Reading
X613HNV	Wardle	XBZ7729	Thamesdown	Y198KNB	TM Travel	Y619GFM	N Warrington
X618VWH	Reliance	XBZ7730	Thamesdown	Y200BCC	Bennetts	Y621GFM	N Warrington
X645RDA	South Lancs	XBZ7731	Thamesdown	Y202PFM	Halton	Y622GFM	N Warrington
X661WCH	Nottingham	XBZ7732	Thamesdown	Y203PFM	Halton	Y623GFM	N Warrington
X662WCH	Nottingham	XHK221X	Abus	Y204PFM	Halton	Y624GFM	N Warrington
X663WCH	Nottingham	XHY378	Safeguard	Y207PFM	Halton	Y626GFM	N Warrington
X664WCH	Nottingham	XIL1273	Renown	Y213BGB	Avon Buses	Y627GFM	N Warrington
X665WCH	Nottingham	XIL1274	Renown	Y236LRR	Nottingham	Y628GFM	N Warrington
X674OBA	Hornsby Travel	XIL4697	Chalkwell	Y237LRR	Nottingham	Y629GFM	N Warrington
X685REC	TM Travel	XIL4698	Chalkwell	Y238LRR	Nottingham	Y631GFM	N Warrington
X712CCA	Supertravel	XIL7250	Chalkwell	Y251KNB	Bluebird	Y644NYD	Plymouth
X713CCA	Central	XIL8438	Quantock	Y252KNB	Bluebird	Y645NYD	Plymouth
X731FPO	D & G	XIL8793	Wardle	Y253KNB	Bluebird	Y646NYD	Plymouth
X732FPO	TM Travel	XIL9100	Chalkwell	Y254KNB	Bluebird	Y647NYD	Plymouth
X773EVS	Uno	XIL9300	Chalkwell	Y284HUA	Kime's	Y648NYD	Plymouth
X774EVS	Uno	XIL9400	Chalkwell	Y293HUA	Hulleys	Y667DRC	Nottingham
X776EVS	Uno	XJO46	Simonds	Y297HUA	Emsworth & D	Y668DRC	Nottingham
X821NWX	Fishwick	XMW120	Thamesdown	Y300BCC	Bennetts	Y687HPG	Yeomans
X821XCK	Avon Buses	XS2210	Coastal	Y301KNB	D & G	Y701LRB	Nottingham

Reg	Operator	Reg	Operator	Reg	Operator	Reg	Operator
Y747HWT	Heddingham	YGO2FVZ	Ipswich	YJ05JWO	Rossendale	YJ07EJK	Blackpool
Y748HWT	Safeguard	YGO2FWA	Reading	YJ05JXD	Brylaine	YJ07EJL	Blackpool
Y751HVY	Ross Travel	YGO2FWB	Reading	YJ05JXE	Brylaine	YJ07EJN	Blackpool
Y753HVY	Ross Travel	YGO2FWC	Reading	YJ05JXF	Brylaine	YJ07JDZ	Fishwick
Y758HWT	Safeguard	YGO2FWD	Reading	YJ05JXG	Brylaine	YJ07JND	Sanders Cs
Y773OEE	Supertravel	YGO2FWE	Reading	YJ05JXH	Brylaine	YJ07JNN	TM Travel
Y783WHH	Bakerbus	YGO2FWL	Marshalls	YJ05JXK	Brylaine	YJ07JPV	Sanders Cs
Y784WHH	Anglian	YG52CFY	Fishwick	YJ05JXL	Brylaine	YJ07JVC	TM Travel
Y867PWT	TM Travel	YG52CGE	Geldards	YJ05JXP	Wardle	YJ07JVE	TM Travel
Y876PWT	Sargeants	YG52CGY	Hornsby Travel	YJ05JXR	D & G	YJ07JWD	Fishwick
Y877PWT	Sargeants	YG52CGZ	Hornsby Travel	YJ05JXS	D & G	YJ07JWE	Fishwick
Y942GEU	Supertravel	YG52DGE	Ipswich	YJ05PWN	Fishwick	YJ07JWF	Fishwick
Y957DRR	Nottingham	YG52DGF	Ipswich	YJ05PXA	Konectbus	YJ07JWG	Fishwick
Y966DRC	Nottingham	YG52DGO	Ipswich	YJ05PXB	Konectbus	YJ07JWU	TM Travel
YAF65	Hopley's	YG52DHD	Renown	YJ05PXC	Konectbus	YJ07JWW	TM Travel
YAZ4142	Kime's	YG52DHF	Ipswich	YJ05PXD	Konectbus	YJ07XND	Halifax JC
YAZ4143	Kime's	YG52DHL	Norfolk Green	YJ05PXE	Konectbus	YJ08DUA	Geldards
YAZ8773	Kime's	YG52EVY	Fishwick	YJ05PZA	Sanders Cs	YJ08DUU	Geldards
YAZ8774	Kime's	YIL1720	Sargeants	YJ05PZE	Sanders Cs	YJ08EBZ	Beestons
YC53NZZ	Stansted Transit	YIL4540	Chalkwell	YJ05UKR	Reliance	YJ08EFL	Felix
YCT187	Yeomans	YIL7758	Cedric	YJ05WCA	Renown	YJ08EFS	Fishwick
YCT502	Yeomans	YIL8824	Chalkwell	YJ05WCC	Andybus	YJ08EFW	Fishwick
YCV834	Yeomans	YIL8825	Chalkwell	YJ05WCD	Renown	YJ08PFA	Blackpool
YDO2RCO	Cedric	YIL8826	Chalkwell	YJ05WCE	Renown	YJ08PFD	Blackpool
YDO2RCX	Claribels	YIL9717	Chalkwell	YJ05WCM	Tyrer Bus	YJ08PFE	Blackpool
YDO2RCY	Claribels	YJ03PFF	Fishwick	YJ05WCN	Tyrer Bus	YJ08PFF	Blackpool
YDO2RCZ	Claribels	YJ03PFG	Fishwick	YJ05WCO	Tyrer Bus	YJ08PFG	Blackpool
YDO2RGY	D & G	YJ03PKA	Claribels	YJ05WCP	Ross Travel	YJ08PFK	Blackpool
YDZ2082	Uno	YJ03PKX	Geldards	YJ05WCR	Ross Travel	YJ08PFN	TM Travel
YE06FVH	Geldards	YJ03PKX	Sanders Cs	YJ05WCW	Sargeants	YJ08PGU	Courtney
YE52FGX	Courtney	YJ03PKY	Sanders Cs	YJ05XMT	Courtney	YJ08PGV	Courtney
YE52FHA	Sargeants	YJ03UMK	Reading	YJ05XNA	Courtney	YJ08PHV	Brylaine
YE52FHF	Reading	YJ03UML	Reading	YJ05XNB	Uno	YJ08PHX	Brylaine
YE52FHG	Reading	YJ03UMM	Safeguard	YJ05XNY	Ipswich	YJ08PHY	Brylaine
YGO2DGU	Yeomans	YJ04BJF	Fishwick	YJ05XWV	Simonds	YJ08VPR	Ross Travel
YGO2DGV	Yeomans	YJ04BKG	Geldards	YJ06FXM	Safeguard	YJ51NXK	Reliance
YGO2DGX	Yeomans	YJ04BKK	Geldards	YJ06FZK	Anglian	YJ51XSR	Nottingham
YGO2DGY	Yeomans	YJ04BMV	Kime's	YJ06LGA	Fishwick	YJ51XST	Nottingham
YGO2DGZ	Yeomans	YJ05CCY	Uno	YJ06VSP	Marshalls	YJ51XSU	Nottingham
YGO2FVP	Blackpool	YJ05JWC	Rossendale	YJ06YSK	Regal Busways	YJ51ZVE	Geldards
YGO2FVR	Blackpool	YJ05JWD	Rossendale	YJ07DWM	Ross Travel	YJ51ZVE	Reading
YGO2FVS	Blackpool	YJ05JWE	Rossendale	YJ07EGC	Weardale	YJ51ZVF	Reading
YGO2FVT	Blackpool	YJ05JWF	Rossendale	YJ07EGD	White Bus	YJ51ZVG	Reading
YGO2FVU	Blackpool	YJ05JWG	Rossendale	YJ07EJA	Blackpool	YJ51ZVH	Geldards
YGO2FVV	Reading	YJ05JWK	Rossendale	YJ07EJC	Blackpool	YJ51ZVH	Reading
YGO2FVW	Reading	YJ05JWL	Rossendale	YJ07EJD	Blackpool	YJ51ZVK	Reading
YGO2FVX	Reading	YJ05JWM	Rossendale	YJ07EJE	Blackpool	YJ51ZVK	Renown
YGO2FVY	Reading	YJ05JWN	Rossendale	YJ07EJG	Blackpool	YJ51ZVL	Reading

Reg	Operator	Reg	Operator	Reg	Operator	Reg	Operator
YJ51ZVM	Reading	YJ54UXH	Cedar Cs	YJ57BRX	N Warrington	YK08EPA	Nottingham
YJ51ZVM	Renown	YJ54UXT	Rossendale	YJ57BRZ	N Warrington	YK08EPC	Nottingham
YJ51ZVN	Reading	YJ54UXU	Rossendale	YJ57EGX	Konectbus	YK08EPD	Nottingham
YJ51ZVO	Ensignbus	YJ54UXV	Rossendale	YJ57EGY	Konectbus	YK08EPE	Nottingham
YJ51ZVO	Reading	YJ54UXW	Rossendale	YJ57EHB	TM Travel	YK08EPF	Nottingham
YJ51ZVV	Nottingham	YJ54XGE	Ross Travel	YJ57EHC	TM Travel	YK08EPJ	Nottingham
YJ51ZVW	Nottingham	YJ54ZYA	D & G	YJ57EHD	Marshalls	YK08EPL	Nottingham
YJ51ZVX	Nottingham	YJ54ZYB	D & G	YJ57EHE	Yeomans	YK08EPN	Nottingham
YJ51ZVY	Nottingham	YJ54ZYC	D & G	YJ57EHU	TM Travel	YK08EPO	Nottingham
YJ51ZVZ	Nottingham	YJ54ZYD	D & G	YJ57EKH	Courtney	YK08EPP	Nottingham
YJ51ZWA	Cedar Cs	YJ54ZYE	D & G	YJ57EKK	Courtney	YK08EPU	Nottingham
YJ52HEA	Reliance	YJ54ZYF	D & G	YJ57FHM	Norfolk Green	YK08EPV	Nottingham
YJ52RXL	Renown	YJ55BLV	Courtney	YJ57KFE	TM Travel	YK08EPX	Nottingham
YJ53CFF	Ross Travel	YJ55BLX	Anglian	YJ57NFF	Halifax JC	YK08EPY	Nottingham
YJ53VBN	Uno	YJ55KZP	Fishwick	YJ57WKB	Halifax JC	YK08EPZ	Nottingham
YJ53VBO	Uno	YJ55KZR	Fishwick	YJ57WKC	Halifax JC	YK08ERJ	Nottingham
YJ53VDL	Claribels	YJ55KZT	Bakerbus	YJ57XWM	Courtney	YK08ESF	Nottingham
YJ53VDT	Fishwick	YJ55KZX	Claribels	YJ57XWN	Courtney	YK08ESG	Nottingham
YJ53VDV	Fishwick	YJ55KZZ	Kime's	YJ57XWO	Courtney	YK08ESN	Nottingham
YJ54BSX	D & G	YJ55WPA	Sanders Cs	YJ57XWP	Courtney	YK08ESO	Nottingham
YJ54BSY	D & G	YJ56AOT	Regal Busways	YJ57XWR	Courtney	YK08EUC	Norfolk Green
YJ54BSZ	D & G	YJ56AOU	Regal Busways	YJ57XWS	Courtney	YK53HLV	Bodmans
YJ54BUA	Rossendale	YJ56AOW	Tyrer Bus	YJ57XWU	Courtney	YK53JPU	Stansted Transit
YJ54BUE	Rossendale	YJ56AOX	Tyrer Bus	YJ57XWY	Courtney	YK54AWH	Wardle
YJ54BUF	Rossendale	YJ56AOZ	Tyrer Bus	YJ57XXE	Courtney	YK54AWJ	Wardle
YJ54BUH	Rossendale	YJ56JXX	Claribels	YJ57YCC	Norfolk Green	YL02FKU	TM Travel
YJ54BUO	Rossendale	YJ56JYA	TM Travel	YJ57YCD	Norfolk Green	YL02FKV	TM Travel
YJ54BUP	Rossendale	YJ56KAO	TM Travel	YJ58CCX	Yeomans	YL02FKW	TM Travel
YJ54BUU	Rossendale	YJ56KBF	Felix	YJ58CCY	Yeomans	YM03EOY	Chalkwell
YJ54BUV	Rossendale	YJ56KCK	Courtney	YJ58CCZ	Yeomans	YM52TPU	Aintree Coachlines
YJ54BUW	Rossendale	YJ56KCV	Claribels	YJ58CDE	Regal Busways	YM52TPV	D & G
YJ54CEA	Claribels	YJ56WUG	Norfolk Green	YJ58CDF	Regal Busways	YM52TPX	D & G
YJ54CEF	Claribels	YJ56WUH	Norfolk Green	YJ58FXH	Norfolk Green	YM52TPY	D & G
YJ54CEK	Claribels	YJ56WVA	Konectbus	YJ58PGK	Ross Travel	YN03DDA	Beestons
YJ54CEN	Claribels	YJ56WVB	Konectbus	YJ58PHN	Uno	YN03NCV	Norfolk Green
YJ54CEO	Claribels	YJ56WVC	Regal Busways	YJ58PHO	Uno	YN03NCX	Norfolk Green
YJ54CEY	Bakerbus	YJ56WVW	Yeomans	YK04ENN	Ross Travel	YN03NCY	Norfolk Green
YJ54CFA	Bakerbus	YJ56WVV	Yeomans	YK04GWL	Yeomans	YN03NDK	Ipswich
YJ54CFM	Fishwick	YJ56WVX	Regal Busways	YK04KVU	Country Liner	YN03NDL	Ipswich
YJ54CFN	Fishwick	YJ56WVY	Regal Busways	YK04KVW	Country Liner	YN03NDU	Ipswich
YJ54CKL	Claribels	YJ57BBE	Brylaine	YK04KWA	Bennetts	YN03WRA	Felix
YJ54UBD	D & G	YJ57BLF	Fishwick	YK04KWB	Bennetts	YN03ZXE	TM Travel
YJ54UBG	Yeomans	YJ57BNB	Brylaine	YK04KWC	Bennetts	YN04AGY	Felix
YJ54UBL	Tyrer Bus	YJ57BOF	Claribels	YK04KWD	Bennetts	YN04AHC	Reading
YJ54UCA	APL Travel	YJ57BOH	Claribels	YK05CDU	Country Liner	YN04AHD	Reading
YJ54UWN	South Lancs	YJ57BPK	N Warrington	YK05CDX	Country Liner	YN04AMK	Nottingham
YJ54UWO	South Lancs	YJ57BPO	N Warrington	YK05ENU	Yeomans	YN04AMU	Nottingham
YJ54UXA	Courtney	YJ57BPU	N Warrington	YK07BFU	APL Travel	YN04AMV	Nottingham

Reg	Operator	Reg	Operator	Reg	Operator	Reg	Operator
YN04AMX	Nottingham	YN05GXM	Reading	YN06CYP	TM Travel	YN07EGE	Norfolk Green
YN04ANF	Nottingham	YN05GXO	Reading	YN06CYX	Weavaway	YN07EGF	Norfolk Green
YN04ANP	Nottingham	YN05GXP	Reading	YN06FVR	Weardale	YN07EHO	Norfolk Green
YN04ANR	Nottingham	YN05GXR	Reading	YN06JWC	Reading	YN07EXP	Beestons
YN04ANX	Beestons	YN05GXS	Reading	YN06JWD	Reading	YN07EYA	Nottingham
YN04GMU	Nottingham	YN05GXT	Reading	YN06JWE	Reading	YN07EYB	Nottingham
YN04GMV	Nottingham	YN05GXU	Reading	YN06JWF	Reading	YN07EYC	Nottingham
YN04GMX	Nottingham	YN05GXV	Reading	YN06JWG	Reading	YN07EYD	Nottingham
YN04GMY	Nottingham	YN05GXW	Reading	YN06JWL	Reading	YN07EYH	Nottingham
YN04GMZ	Nottingham	YN05GXX	Reading	YN06JWM	Reading	YN07EYJ	Nottingham
YN04HJV	Marshalls	YN05GZB	Felix	YN06JWO	Reading	YN07EYK	Nottingham
YN04LWJ	Ipswich	YN05GZV	Minsterley Ms	YN06JWP	Reading	YN07EYL	Nottingham
YN04LWK	Ipswich	YN05HFV	Yeomans	YN06JWU	Reading	YN07EYM	Nottingham
YN04OBX	Beestons	YN05HUY	Safeguard	YN06JWV	Reading	YN07EYO	Nottingham
YN04OCJ	Beestons	YN05NGE	Nottingham	YN06JWW	Reading	YN07EYP	Nottingham
YN04PZY	Regal Busways	YN05NGF	Nottingham	YN06JWX	Reading	YN07EYR	Nottingham
YN04PZZ	Regal Busways	YN05NGG	Nottingham	YN06JWY	Reading	YN07EYT	Nottingham
YN04UJF	Nottingham	YN05NGJ	Nottingham	YN06JWZ	Reading	YN07EYU	Nottingham
YN04UJG	Nottingham	YN05NGU	Nottingham	YN06MXK	Richardson	YN07EYW	Nottingham
YN04UJH	Nottingham	YN05NGV	Nottingham	YN06NXP	Reading	YN07EYX	Nottingham
YN04UJJ	Nottingham	YN05NGX	Nottingham	YN06NXR	Reading	YN07EYY	Nottingham
YN04UJK	Nottingham	YN05NGY	Nottingham	YN06NXS	Reading	YN07EYZ	Nottingham
YN04UJL	Nottingham	YN05NGZ	Nottingham	YN06NXT	Reading	YN07EZB	Anglian
YN04UJP	Nottingham	YN05UGO	Marshalls	YN06NXU	Reading	YN07JWA	TM Travel
YN04UJR	Nottingham	YN05UVB	Minsterley Ms	YN06NXV	Reading	YN07KHC	Weardale
YN04UJS	Nottingham	YN05VRR	Premiere	YN06NXW	Reading	YN07LDX	Nottingham
YN04UJT	Nottingham	YN05WFB	Nottingham	YN06NXX	Reading	YN07LDY	Nottingham
YN04UJU	Nottingham	YN05WFC	Nottingham	YN06NXY	Reading	YN07LDZ	Nottingham
YN04UJV	Nottingham	YN05WFD	Nottingham	YN06NXZ	Reading	YN07LFA	Reading
YN04UJW	Nottingham	YN05WFE	Nottingham	YN06NZH	Minsterley Ms	YN07LFB	Reading
YN04UJY	Nottingham	YN05WFF	Nottingham	YN06NZR	Minsterley Ms	YN07LFD	Reading
YN04UJZ	Nottingham	YN05WFG	Nottingham	YN06NZS	Minsterley Ms	YN07LFE	Reading
YN04WTL	Safeguard	YN05WFK	Nottingham	YN06NZV	Minsterley Ms	YN07LFF	Reading
YN04WTM	Safeguard	YN05WFO	Nottingham	YN06NZW	Minsterley Ms	YN07LFG	Reading
YN04XYZ	Blackpool	YN05WFP	Nottingham	YN06RVT	Thornes	YN07LFT	Trustybus
YN04XZA	Ipswich	YN05WFR	Nottingham	YN06TGF	Nottingham	YN07LFU	Anglian
YN05GWX	Nottingham	YN05WFS	Nottingham	YN06TGJ	Nottingham	YN07LGG	Yeomans
YN05GWY	Nottingham	YN05WFT	Nottingham	YN06TGK	Nottingham	YN07LHD	Reading
YN05GXA	Reading	YN05WFU	Nottingham	YN06TGO	Nottingham	YN07LJU	Chalkwell
YN05GXB	Reading	YN05WFV	Nottingham	YN06TGU	Nottingham	YN07NUF	Richardson
YN05GXC	Reading	YN05WFW	Nottingham	YN06TGV	Nottingham	YN07SYS	TM Travel
YN05GXD	Reading	YN05WFX	Nottingham	YN06TGX	Nottingham	YN08DME	Weardale
YN05GXE	Reading	YN05WGC	Nottingham	YN06TGY	Nottingham	YN08DNU	TM Travel
YN05GXF	Reading	YN05WGD	Nottingham	YN06TGZ	Nottingham	YN08HXZ	Beestons
YN05GXG	Reading	YN05WGG	Nottingham	YN07DTO	Sanders Cs	YN08HYM	Reading
YN05GXH	Reading	YN06CJE	Beestons	YN07DVF	Sanders Cs	YN08HYO	Reading
YN05GXJ	Reading	YN06CKY	Trustybus	YN07DVO	TM Travel	YN08HYP	Reading
YN05GXL	Reading	YN06CYO	TM Travel	YN07ECC	Weardale	YN08HYR	Reading

YN08HYS	Reading	YN53CEA	Nottingham	YN53ZWP	Blackpool	YN55NDY	Beestons
YN08HYT	Reading	YN53CEF	Nottingham	YN53ZWR	Blackpool	YN55NJZ	Reading
YN08HYU	Reading	YN53CEJ	Nottingham	YN53ZWT	Blackpool	YN55RDV	Quantock
YN08HYW	Reading	YN53CEK	Nottingham	YN53ZWU	Blackpool	YN55YSC	TM Travel
YN08HYX	Reading	YN53CEO	Nottingham	YN53ZWV	Blackpool	YN55YSE	Felix
YN08HYY	Reading	YN53CEU	Nottingham	YN53ZWW	Blackpool	YN55YSF	Felix
YN08JWC	TM Travel	YN53CEV	Nottingham	YN53ZWX	Blackpool	YN56AHY	TM Travel
YN08JWD	TM Travel	YN53CEX	Trustybus	YN53ZWY	Blackpool	YN56BGV	Weardale
YN08JWE	TM Travel	YN53CFA	Nottingham	YN53ZWZ	Blackpool	YN56EZV	DRM
YN08JWY	Hulleys	YN53CFD	Nottingham	YN54AEP	Reading	YN56FAA	Reading
YN08MKM	Reading	YN53CFE	Nottingham	YN54AET	Reading	YN56FAF	Reading
YN08MKO	Reading	YN53CFF	Nottingham	YN54AEU	Reading	YN56FAJ	Reading
YN08MKP	Reading	YN53CFG	Nottingham	YN54AEV	Reading	YN56FAK	Reading
YN08MKU	Reading	YN53CFJ	Nottingham	YN54AEW	Reading	YN56FAM	Reading
YN08MKV	Reading	YN53CFK	Nottingham	YN54AEX	Reading	YN56FAO	Reading
YN08MKX	Reading	YN53CFL	Nottingham	YN54AEY	Reading	YN56FAU	Reading
YN08MKZ	Reading	YN53CFO	Nottingham	YN54AEZ	Reading	YN56FBF	Reading
YN08MLE	Nottingham	YN53CFP	Nottingham	YN54AFA	Reading	YN56FBG	Reading
YN08MLF	Nottingham	YN53CFU	Nottingham	YN54AFE	Reading	YN56FBJ	Reading
YN08MLJ	Nottingham	YN53CFV	Nottingham	YN54AFF	Reading	YN56FBK	Reading
YN08MLK	Nottingham	YN53CFX	Nottingham	YN54AFJ	Reading	YN56FBL	Reading
YN08MLL	Nottingham	YN53CFY	Nottingham	YN54AFK	Reading	YN56NHA	Avon Buses
YN08MLO	Nottingham	YN53CFZ	Nottingham	YN54AFO	Reading	YN56NRY	Uno
YN08MLU	Nottingham	YN53CHF	Nottingham	YN54AFU	Reading	YN56NRZ	Uno
YN08MLV	Nottingham	YN53EJG	Chalkwell	YN54AFV	Reading	YN56NVB	Ipswich
YN08MLX	Nottingham	YN53EJL	Minsterley Ms	YN54AFX	Reading	YN56NVC	Ipswich
YN08MLY	Nottingham	YN53EJX	TM Travel	YN54AHA	Nottingham	YN56NVD	Ipswich
YN08MMA	Reading	YN53EKX	Yeomans	YN54AHC	Nottingham	YN56NVE	Ipswich
YN08MME	Reading	YN53EKY	Yeomans	YN54AHD	Nottingham	YN56NVF	Ipswich
YN08MMF	Reading	YN53EKZ	Yeomans	YN54AHE	Nottingham	YN56NVG	Ipswich
YN08MMJ	Reading	YN53ELC	Sargeants	YN54AHF	Nottingham	YN56OWP	TM Travel
YN08MMK	Reading	YN53ELV	D & G	YN54AHG	Nottingham	YN57AAF	Weardale
YN08MMO	Reading	YN53ELW	Ipswich	YN54AHJ	Nottingham	YN57AAY	Weardale
YN08MMU	Reading	YN53EMF	D & G	YN54AHK	Nottingham	YN57BWE	Richardson
YN08MSO	Nottingham	YN53EMK	D & G	YN54AHL	Nottingham	YN57BWW	Safeguard
YN08MSU	Nottingham	YN53EMV	D & G	YN54AHO	Nottingham	YN57BXG	TM Travel
YN08MSV	Nottingham	YN53EMX	D & G	YN54AHP	Nottingham	YN57EHV	White Bus
YN08MSX	Nottingham	YN53GHH	Trustybus	YN54AHU	Nottingham	YN57FWG	Reading
YN08MSY	Nottingham	YN53SSZ	TM Travel	YN54AHV	Nottingham	YN57FWH	Reading
YN08MTE	Nottingham	YN53SVT	D & G	YN54NXK	Nottingham	YN57FWJ	Reading
YN08NKH	Minsterley Ms	YN53SVW	Cedar Cs	YN54NXL	Nottingham	YN57FWK	Reading
YN08NKW	TM Travel	YN53VBO	Western Greyhound	YN54OBB	Beestons	YN57FWL	Reading
YN08NLG	Minsterley Ms	YN53VCD	Richardson	YN54SYG	TM Travel	YN57FWM	Reading
YN08NLJ	Minsterley Ms	YN53ZWH	Supertravel	YN54WWR	Richardson	YN57FXA	Reading
YN08OCR	N I B S	YN53ZWJ	Supertravel	YN54WWU	TM Travel	YN57FXB	Reading
YN51MGU	South Lancs	YN53ZWK	Blackpool	YN54WWV	TM Travel	YN57FXC	Reading
YN51MHO	Cedric	YN53ZWL	Blackpool	YN55KLX	Thornes	YN57FXD	Reading
YN51MJE	Kime's	YN53ZWM	Blackpool	YN55KMV	TM Travel	YN57FXE	Reading

YN57FXF	Reading	YN58NCC	Safeguard	YS02YYD	Richardson	YT55TMT	TM Travel
YN57FXH	Reading	YN58NDD	Safeguard	YS03ZHL	Nottingham	YU06KFA	Wardle
YN57FXJ	Reading	YN58NDY	TM Travel	YS03ZHM	Nottingham	YUP6	Weardale
YN57FXK	Reading	YO53OUG	Chalkwell	YS03ZHN	Nottingham	YV03UTX	Renown
YN57FXL	Reading	YO53OVC	Chalkwell	YS03ZHP	Nottingham	YV03UTY	Renown
YN57FXM	Reading	YOI2747	Yeomans	YS03ZLN	Nottingham	YVF158	Simonds
YN57FYV	Nottingham	YOI298	Yeomans	YS03ZLU	Nottingham	YVJ677	Sanders Cs
YN57FYW	Nottingham	YP02AAN	Nottingham	YS03ZLV	Nottingham	YX04AWV	Supertravel
YN57FYX	Nottingham	YP02BRF	APL Travel	YS03ZLX	Nottingham	YX05DWA	Swanbrook
YN57FZA	Uno	YP52CTV	Yeomans	YS03ZLY	Nottingham	YX05FFS	Sargeants
YN57FZB	Uno	YP52KSK	Yeomans	YT51DZY	Country Liner	YX53CYZ	Wardle
YN57HPU	Anglian	YP52VEO	Yeomans	YT51EAC	Country Liner	YX55BGY	Wardle
YN57HPV	Anglian	YR02UMU	Bennetts	YT51EAE	Country Liner	YX56DJD	Wardle
YN57HPX	Anglian	YR52MBU	Bodmans	YT51EAF	Country Liner	YX56DJE	Wardle
YN57HPZ	Anglian	YR52MDY	Richardson	YT51EAG	Stansted Transit	YX58FPP	Weavaway
YN57HRA	Anglian	YR52UNB	Sargeants	YT51EAK	Stansted Transit	XI5860	Compass Bus
YN57PYJ	Richardson	YS02XDO	Beestons	YT51EAM	Stansted Transit		

ISBN 9781904875635   © Published by *British Bus Publishing Ltd* , February 2009

British Bus Publishing Ltd, 16 St Margaret's Drive, Telford, TF1 3PH
Telephone: 01952 255669

www.britishbuspublishing.co.uk - e-mail: sales@britishbuspublishing.co.uk